Means Residential Repair & Remodeling Costs

Contractor's Pricing Guide 2006

Senior Editor
Robert W. Mewis, CCC

Contributing Editors
Christopher Babbitt
Ted Baker
Barbara Balboni
Robert A. Bastoni
John H. Chiang, PE
John Kane
Robert J. Kuchta
Robert C. McNichols
Melville J. Mossman, PE
John J. Moylan
Jeannene D. Murphy
Stephen C. Plotner
Eugene R. Spencer
Marshall J. Stetson
Phillip R. Waier, PE

*Senior Engineering
Operations Manager*
John H. Ferguson, PE

*Senior Vice President
& General Manager*
John Ware

Vice President of Sales
John M. Shea

Production Manager
Michael Kokernak

Technical Support
Wayne D. Anderson
Thomas J. Dion
Jonathan Forgit
Mary Lou Geary
Gary L. Hoitt
Genevieve Medeiros
Paula Reale-Camelio
Kathryn S. Rodriguez
Sheryl A. Rose
Laurie Thom

Book & Cover Design
Norman R. Forgit

RSMeans

Means Residential Repair & Remodeling Costs

Contractor's Pricing Guide 2006

- Quick Costs to Help You Estimate
- New, Easy-to-Use Format . . . Organized the Way You Build
- Hundreds of Work Items with Material, Labor & Equipment

$39.95 per copy (in United States).
Price subject to change without prior notice.

Reed Construction Data®

Copyright © 2005

*RS*Means
Construction Publishers & Consultants
63 Smiths Lane
Kingston, MA 02364-0800
(781) 422-5000

Printed in the United States of America.

10 9 8 7 6 5 4 3 2 1

ISSN 1542-9156

ISBN 0-87629-807-2

Foreword

RSMeans is a product line of Reed Construction Data, Inc., a leading provider of construction information, products, and services in North America and globally. Reed Construction Data's project information products include more than 100 regional editions, national construction data, sales leads, and local plan rooms in major business centers. Reed Construction Data's PlansDirect provides surveys, plans, and specifications. The First Source suite of products consists of *First Source for Products*, SPEC-DATA™, MANU-SPEC™, CADBlocks, Manufacturer Catalogs, and First Source Exchange (www.firstsourceexchange.com) for the selection of nationally available building products. Reed Construction Data also publishes ProFile, a database of more than 20,000 U.S. architectural firms. RSMeans provides construction cost data, training, and consulting services in print, CD-ROM, and online. Reed Construction Data, headquartered in Atlanta, is owned by Reed Business Information (www.reedconstructiondata.com), a leading provider of critical information and marketing solutions to business professionals in the media, manufacturing, electronics, construction, and retail industries. Its market-leading properties include more than 135 business-to-business publications and over 125 Webzines and Web portals, as well as online services, custom publishing, directories, research, and direct-marketing lists. Reed Business Information is a member of the Reed Elsevier plc group (NYSE: RUK and ENL)—a world-leading publisher and information provider operating in the science and medical, legal, education, and business-to-business industry sectors.

Our Mission

Since 1942, RSMeans has been actively engaged in construction cost publishing and consulting throughout North America.

Today, over 60 years after RSMeans began, our primary objective remains the same: to provide you, the construction and facilities professional, with the most current and comprehensive construction cost data possible.

Whether you are a contractor, an owner, an architect, an engineer, a facilities manager, or anyone else who needs a fast and reliable construction cost estimate, you'll find this publication to be a highly useful and necessary tool.

Today, with the constant flow of new construction methods and materials, it's difficult to find the time to look at and evaluate all the different construction cost possibilities. In addition, because labor and material costs keep changing, last year's cost information is not a reliable basis for today's estimate or budget.

That's why so many construction professionals turn to RSMeans. We keep track of the costs for you, along with a wide range of other key information, from city cost indexes . . . to productivity rates . . . to crew composition . . . to contractor's overhead and profit rates.

RSMeans performs these functions by collecting data from all facets of the industry and organizing it in a format that is instantly accessible to you. From the preliminary budget to the detailed unit price estimate, you'll find the data in this book useful for all phases of construction cost determination.

The Staff, the Organization, and Our Services

When you purchase one of RSMeans' publications, you are, in effect, hiring the services of a full-time staff of construction and engineering professionals.

Our thoroughly experienced and highly qualified staff works daily at collecting, analyzing, and disseminating comprehensive cost information for your needs. These staff members have years of practical construction experience and engineering training prior to joining the firm. As a result, you can count on them not only for the cost figures, but also for additional background reference information that will help you create a realistic estimate.

The RSMeans organization is always prepared to help you solve construction problems through its five major divisions: Construction and Cost Data Publishing, Electronic Products and Services, Consulting Services, Insurance Services, and Educational Services.

Besides a full array of construction cost estimating books, RSMeans also publishes a number of other reference works for the construction industry. Subjects include construction estimating and project and business management; special topics such as HVAC, roofing, plumbing, and hazardous waste remediation; and a library of facility management references.

In addition, you can access all of our construction cost data through your computer with *Means CostWorks 2006* CD-ROM, an electronic tool that offers over 50,000 lines of RSMeans detailed construction cost data, along with assembly and whole building cost data. You can also access RSMeans cost information from our Web site at www.rsmeans.com.

What's more, you can increase your knowledge and improve your construction estimating and management performance with an RSMeans Construction Seminar or In-House Training Program. These two-day seminar programs offer unparalleled opportunities for everyone in your organization to get updated on a wide variety of construction-related issues.

RSMeans also is a worldwide provider of construction cost management and analysis services for commercial and government owners and of claims and valuation services for insurers.

In short, RSMeans can provide you with the tools and expertise for constructing accurate and dependable construction estimates and budgets in a variety of ways.

Robert Snow Means Established a Tradition of Quality That Continues Today

Robert Snow Means spent years building RSMeans, making certain he always delivered a quality product.

Today, at RSMeans, we do more than talk about the quality of our data and the usefulness of our books. We stand behind all of our data, from historical cost indexes to construction materials and techniques to current costs.

If you have any questions about our products or services, please call us toll-free at 1-800-334-3509. Our customer service representatives will be happy to assist you. You can also visit our Web site at www.rsmeans.com.

Table of Contents

How the Book is Built: An Overview

A Powerful Construction Tool

You have in your hands one of the most powerful construction tools available today. A successful project is built on the foundation of an accurate and dependable estimate. This book will enable you to construct just such an estimate.

For the casual user the book is designed to be:

- quickly and easily understood so you can get right to your estimate.
- filled with valuable information so you can understand the necessary factors that go into the cost estimate.

For the regular user, the book is designed to be:

- a handy desk reference that can be quickly referred to for key costs.
- a comprehensive, fully reliable source of current construction costs, so you'll be prepared to estimate any project.
- a source book for project cost, product selections, and alternate materials and methods.

To meet all of these requirements we have organized the book into the following clearly defined sections.

Unit Price Section

The cost data has been divided into 23 sections representing the order of construction. Within each section is a listing of components and variations to those components applicable to that section. Costs are shown for the various actions that must be done to that component.

Reference Section

This section includes information on Location Factors, and a listing of Abbreviations.

Location Factors: You can adjust total project costs to over 900 locations throughout the U.S. and Canada by using the data in this section.

Abbreviations: A listing of the abbreviations used throughout this book, along with the terms they represent, is included.

Index

A comprehensive listing of all terms and subjects in this book will help you quickly find what you need when you are not sure where it falls in the order of construction.

The Scope of This Book

This book is designed to be as comprehensive and as easy to use as possible. To that end we have made certain assumptions and limited its scope in three key ways:

1. We have established material prices based on a national average.
2. We have computed labor costs based on a seven major-region average of open shop wage rates.
3. We have targeted the data for projects of a certain size range.

Project Size

This book is intended for use by those involved primarily in Residential Repair and Remodeling construction costing between $10,000-$100,000.

With reasonable exercise of judgment the figures can be used for any building work. For other types of projects, such as new home construction or commercial buildings, consult the appropriate RSMeans publication for more information.

How to Use the Book: The Details

What's Behind the Numbers? The Development of Cost Data

The staff at RSMeans continuously monitors developments in the construction industry in order to ensure reliable, thorough, and up-to-date cost information.

While *overall* construction costs may vary relative to general economic conditions, price fluctuations within the industry are dependent upon many factors. Individual price variations may, in fact, be opposite to overall economic trends. Therefore, costs are continually monitored and complete updates are published yearly. Also, new items are frequently added in response to changes in materials and methods.

Costs—$ (U.S.)

All costs represent U.S. national averages and are given in U.S. dollars. The RSMeans Location Factors can be used to adjust costs to a particular location. The Location Factors for Canada can be used to adjust U.S. national averages to local costs in Canadian dollars. No exchange rate conversion is necessary.

Material Costs

The RSMeans staff contacts manufacturers, dealers, distributors, and contractors all across the U.S. and Canada to determine national average material costs. If you have access to current material costs for your specific location, you may wish to make adjustments to reflect differences from the national average.

Included within material costs are fasteners for a normal installation. RSMeans engineers use manufacturers' recommendations, written specifications, and/or standard construction practice for size and spacing of fasteners. Adjustments to material costs may be required for your specific application or location. Material costs do not include sales tax.

Labor Costs

Labor costs are based on the average of residential wages from across the U.S. for the current year. Rates, along with overhead and profit markups, are listed on the inside back cover of this book.

- If wage rates in your area vary from those used in this book, or if rate increases are expected within a given year, labor costs should be adjusted accordingly.

Labor costs reflect productivity based on actual working conditions. These figures include time spent during a normal workday on tasks other than actual installation, such as material receiving and handling, mobilization at site, site movement, breaks, and cleanup.

Productivity data is developed over an extended period so as not to be influenced by abnormal variations and reflects a typical average.

Equipment Costs

Equipment costs include not only rental costs, but also operating costs for equipment under normal use. Equipment and rental rates are obtained from industry sources throughout North America—contractors, suppliers, dealers, manufacturers, and distributors.

Factors Affecting Costs

Costs can vary depending upon a number of variables. Here's how we have handled the main factors affecting costs.

Quality—The prices for materials and the workmanship upon which productivity is based represent sound construction work. They are also in line with U.S. government specifications.

Overtime—We have made no allowance for overtime. If you anticipate premium time or work beyond normal working hours, be sure to make an appropriate adjustment to your labor costs.

Productivity—The labor costs for each line item are based on working an eight-hour day in daylight hours in moderate temperatures. For work that extends beyond normal work hours or is performed under adverse conditions, productivity may decrease.

Size of Project—The size, scope of work, and type of construction project will have a significant impact on cost. Economies of scale can reduce costs for large projects. Unit costs can often run higher for small projects. Costs in this book are intended for the size and type of project as previously described in "How the Book Is Built: An Overview." Costs for projects of a significantly different size or type should be adjusted accordingly.

Location—Material prices in this book are for metropolitan areas. However, in dense urban areas, traffic and site storage limitations may increase costs. Beyond a 20-mile radius of large cities, extra trucking or transportation charges may also increase the material costs slightly. On the other hand, lower wage rates may be in effect. Be sure to consider both of these factors when preparing an estimate, particularly if the job site is located in a central city or remote rural location.

In addition, highly specialized subcontract items may require travel and per diem expenses for mechanics.

Other Factors—

- season of year
- contractor management
- weather conditions
- building code requirements
- availability of:
  - adequate energy
  - skilled labor
  - building materials
- owner's special requirements/ restrictions
- safety requirements
- environmental considerations

General Conditions—The extreme right-hand column of each chart gives the "Totals." These figures contain the installing contractor's O&P. Therefore, it is necessary for a general contractor to add a percentage of all subcontracted items. For a detailed breakdown of O&P see the inside back cover of this book.

Unpredictable Factors—General business conditions influence "in-place" costs of all items. Substitute materials and construction methods may have to be employed. These may affect the installed cost and/or life cycle costs. Such factors may be difficult to evaluate and cannot necessarily be predicted on the basis of the job's location in a particular section of the country. Thus, where these factors apply, you may find significant but unavoidable cost variations for which you will have to apply a measure of judgment to your estimate.

Final Checklist

Estimating can be a straightforward process provided you remember the basics. Here's a checklist of some of the steps you should remember to complete before finalizing your estimate.

Did you remember to . . .

• factor in the Location Factor for your locale?
• take into consideration which items have been marked up and by how much?
• mark up the entire estimate sufficiently for your purposes?
• include all components of your project in the final estimate?
• double check your figures to be sure of your accuracy?
• call RSMeans if you have any questions about your estimate or the data you've found in our publications?

Remember, RSMeans stands behind its publications. If you have any questions about your estimate . . . about the costs you've used from our books . . . or even about the technical aspects of the job that may affect your estimate, feel free to call the RSMeans editors at 1-800-334-3509.

Unit Price Section

Table of Contents

Unit Price Section

Table of Contents

How to Use the Unit Price Pages

The prices in this publication include all overhead and profit mark-ups for the installing contractor. If a general contractor is involved, an additional 10% should be added to the cost shown in the "Total" column.

The costs for material, labor and any equipment are shown here. Material costs include delivery, and a 10% mark-up. No sales taxes are included. Labor costs include all mark-ups as shown on the inside back cover of this book. Equipment costs also include a 10% mark-up.

The cost in the total column should be multiplied by the quantity of each item to determine the total cost for that item. The sum of all item's total costs becomes the project's total cost.

The book is arranged into 23 categories that follow the order of construction for residential projects. The category is shown as white letters on a black background at the top of each page.

A brief specification of the cost item is shown here.

Graphics representing some of the cost items are found at the beginning of each category.

The cost data is arranged in a tree structure format. Major cost sections are left justified and are shown as bold headings within a grey background. Cost items and sub-cost items are found under the bold headings. The complete description should read as "bold heading, cost item, sub-cost item."
For example, the identified item's description is "Beam/Girder, Solid Wood Beam, 2" x 8"."

Each action performed to the cost item is conveniently listed to the right of the description. See the explanation below for the scope of work included as part of the action.

The unit of measure upon which all the costs are based is shown here. An abbreviations list is included within the reference section found at the back of this book.

Action	Scope
Demolish	Includes removal of the item and hauling the debris to a truck, dumpster, or storage area.
Install	Includes the installation of the item with normal placement and spacing of fasteners.
Demolish and Install	Includes the demolish action and the install action.
Reinstall	Includes the reinstallation of a removed item.
Clean	Includes cleaning of the item.
Paint	Includes up to two coats of finish on the item (unless otherwise indicated).
Minimum Charge	Includes the minimum labor or the minimum labor and equipment cost. Use this charge for small quantities of material.

Scaffolding	Temporary Electrical Service	Air Compressor

Construction Fees		Unit	Material	Labor	Equip.	Total	Specification
Asbestos Test	Minimum Charge	Job		197		197	Minimum charge for testing.
Architectural	Minimum Charge	Day	480			480	Minimum labor charge for architect.
Engineering	Minimum Charge	Day	256			256	Minimum labor charge for field engineer.

Construction Permits		Unit	Material	Labor	Equip.	Total	Specification
Job (less than 5K)	Minimum Charge	Job	110			110	Permit fees, job less than 5K.
Job (5K to 10K)	Minimum Charge	Job	220			220	Permit fees, job more than 5K, but less than 10K.
Job (10K to 25K)	Minimum Charge	Job	550			550	Permit fees, job more than 10K, but less than 25K.
Job (25K to 50K)	Minimum Charge	Job	1100			1100	Permit fees, job more than 25K, but less than 50K.
Job (50K to 100K)	Minimum Charge	Job	2200			2200	Permit fees, job more than 50K, but less than 100K.
Job (100K to 250K)	Minimum Charge	Job	5500			5500	Permit fees, job more than 100K, but less than 250K.
Job (250K to 500K)	Minimum Charge	Job	11000			11000	Permit fees, job more than 250K, but less than 500K.
Job (500K to 1M)	Minimum Charge	Job	22000			22000	Permit fees, job more than 500K, but less than 1 million.

Job Costs

Trade Labor		Unit	Material	Labor	Equip.	Total	Specification
Carpenter							
Daily (8 hours)							
	Install	Day		340		340	Daily labor rate for one carpenter.
	Minimum Charge	Job		169		169	
Weekly (40 hours)							
	Install	Week		1700		1700	Weekly labor rate for one carpenter.
	Minimum Charge	Job		169		169	
Drywaller							
Daily (8 hours)							
	Install	Day		340		340	Daily labor rate for one carpenter.
	Minimum Charge	Job		169		169	
Weekly (40 hours)							
	Install	Week		1700		1700	Weekly labor rate for one carpenter.
	Minimum Charge	Job		169		169	
Roofer							
Daily (8 hours)							
	Install	Day		310		310	Daily labor rate for one skilled roofer.
	Minimum Charge	Job		155		155	
Weekly (40 hours)							
	Install	Week		1550		1550	Weekly labor rate for one skilled
	Minimum Charge	Job		155		155	roofer.
Painter							
Daily (8 hours)							
	Install	Day		296		296	Daily labor rate for one painter.
	Minimum Charge	Job		148		148	
Weekly (40 hours)							
	Install	Week		1475		1475	Weekly labor rate for one painter.
	Minimum Charge	Job		148		148	
Mason							
Daily (8 hours)							
	Install	Day		340		340	Daily labor rate for one skilled mason.
	Minimum Charge	Job		170		170	
Weekly (40 hours)							
	Install	Week		1700		1700	Weekly labor rate for one skilled
	Minimum Charge	Job		170		170	mason.
Electrician							
Daily (8 hours)							
	Install	Day		365		365	Daily labor rate for an electrician.
	Minimum Charge	Job		184		184	
Weekly (40 hours)							
	Install	Week		1825		1825	Weekly labor rate for an electrician.
	Minimum Charge	Job		184		184	
Plumber							
Daily (8 hours)							
	Install	Day		370		370	Daily labor rate for a plumber.
	Minimum Charge	Job		186		186	
Weekly (40 hours)							
	Install	Week		1850		1850	Weekly labor rate for a plumber.
	Minimum Charge	Job		186		186	
Common Laborer							
Daily (8 hours)							
	Install	Day		242		242	Daily labor rate for a common
	Minimum Charge	Job		121		121	laborer.

Job Costs

Trade Labor		Unit	Material	Labor	Equip.	Total	Specification
Weekly (40 hours)		Week		1200		1200	Weekly labor rate for a common
	Install						laborer.
	Minimum Charge	Job		121		121	

Temporary Utilities		Unit	Material	Labor	Equip.	Total	Specification
Power Pole							
Job Site Electricity							
	Install	Month	39.50			39.50	Includes average cost of electricity per
	Minimum Charge	Job		184		184	month.
Water Service							
Job Site Water							
	Install	Month	66			66	Includes average costs for water for
	Minimum Charge	Month	66			66	one month.
Telephone							
	Install	Month	231			231	Includes average cost for local and
	Minimum Charge	Month	231			231	long distance telephone service for
							one month.
Chemical Toilet							
	Install	Month			256	256	Monthly rental of portable toilet.
	Minimum Charge	Day			12.80	12.80	
Office Trailer							
Place and Remove							
	Install	Job	263			263	Includes cost to a contractor to deliver,
	Minimum Charge	Job		134		134	place and level an office trailer and
							remove at end of job.
Storage Container							
	Install	Job	160			160	Includes cost to a contractor to deliver
	Minimum Charge	Job		134		134	and place a storage container and
							remove at end of job.
Fencing							
	Install	L.F.	2.41			2.41	Includes material costs for temporary
	Minimum Charge	Job	275			275	fencing.
Railing							
	Install	L.F.	11.10	9.60	.56	21.26	Includes material and labor to install
	Minimum Charge	Job	275	7		275	primed steel pipe.

Equipment Rental		Unit	Material	Labor	Equip.	Total	Specification
Flatbed Truck							
Daily Rental							
	Install	Day			142	142	Daily rental of flatbed truck.
	Minimum Charge	Job			142	142	
Weekly Rental							
	Install	Week			710	710	Weekly rental of flatbed truck.
	Minimum Charge	Job			142	142	
Monthly Rental							
	Install	Month			2825	2825	Monthly rental of flatbed truck.
	Minimum Charge	Job			142	142	

Job Costs

Equipment Rental		Unit	Material	Labor	Equip.	Total	Specification
Dump Truck							
Daily Rental							
	Install	Day			360	360	Daily rental of dump truck.
	Minimum Charge	Day			360	360	
Weekly Rental							
	Install	Week			1800	1800	Weekly rental of dump truck.
	Minimum Charge	Day			360	360	
Monthly Rental							
	Install	Month			7175	7175	Monthly rental of dump truck.
	Minimum Charge	Day			360	360	
Forklift							
Daily Rental							
	Install	Day			254	254	Daily rental of forklift.
	Minimum Charge	Day			254	254	
Weekly Rental							
	Install	Week			1275	1275	Weekly rental of forklift.
	Minimum Charge	Day			254	254	
Monthly Rental							
	Install	Month			5075	5075	Monthly rental of forklift.
	Minimum Charge	Day			254	254	
Bobcat							
Daily Rental							
	Install	Day			169	169	Daily rental of wheeled, skid steer loader.
	Minimum Charge	Day			169	169	
Weekly Rental							
	Install	Week			845	845	Weekly rental of wheeled, skid steer loader.
	Minimum Charge	Day			169	169	
Monthly Rental							
	Install	Month			3400	3400	Monthly rental of wheeled, skid steer loader.
	Minimum Charge	Day			169	169	
Air Compressor							
Daily Rental							
	Install	Day			65	65	Daily rental of air compressor.
	Minimum Charge	Day			65	65	
Weekly Rental							
	Install	Week			325	325	Weekly rental of air compressor.
	Minimum Charge	Day			65	65	
Monthly Rental							
	Install	Month			1300	1300	Monthly rental of air compressor.
	Minimum Charge	Day			65	65	
Jackhammer							
Daily Rental							
	Install	Day			9.90	9.90	Daily rental of air tool, jackhammer.
	Minimum Charge	Day			9.90	9.90	
Weekly Rental							
	Install	Week			49.50	49.50	Weekly rental of air tool, jackhammer.
	Minimum Charge	Day			9.90	9.90	
Monthly Rental							
	Install	Month			198	198	Monthly rental of air tool, jackhammer.
	Minimum Charge	Day			9.90	9.90	
Concrete Mixer							
Daily Rental							
	Install	Day			116	116	Daily rental of concrete mixer.
	Minimum Charge	Day			116	116	

Job Costs

Equipment Rental		Unit	Material	Labor	Equip.	Total	Specification
Weekly Rental							
	Install	Week			580	580	Weekly rental of concrete mixer.
	Minimum Charge	Day			116	116	
Monthly Rental							
	Install	Month			2325	2325	Monthly rental of concrete mixer.
	Minimum Charge	Day			116	116	
Concrete Bucket							
Daily Rental							
	Install	Day			15.45	15.45	Daily rental of concrete bucket.
	Minimum Charge	Day			15.45	15.45	
Weekly Rental							
	Install	Week			77	77	Weekly rental of concrete bucket.
	Minimum Charge	Day			15.45	15.45	
Monthly Rental							
	Install	Month			310	310	Monthly rental of concrete bucket.
	Minimum Charge	Day			15.45	15.45	
Concrete Pump							
Daily Rental							
	Install	Day			770	770	Daily rental of concrete pump.
	Minimum Charge	Day			770	770	
Weekly Rental							
	Install	Week			3850	3850	Weekly rental of concrete pump.
	Minimum Charge	Day			770	770	
Monthly Rental							
	Install	Month			15400	15400	Monthly rental of concrete pump.
	Minimum Charge	Day			770	770	
Generator							
Daily Rental							
	Install	Day			24.50	24.50	Daily rental of generator.
	Minimum Charge	Day			24.50	24.50	
Weekly Rental							
	Install	Week			123	123	Weekly rental of generator.
	Minimum Charge	Day			24.50	24.50	
Monthly Rental							
	Install	Month			495	495	Monthly rental of generator.
	Minimum Charge	Day			24.50	24.50	
Welder							
Daily Rental							
	Install	Day			59	59	Daily rental of welder.
	Minimum Charge	Day			59	59	
Weekly Rental							
	Install	Week			294	294	Weekly rental of welder.
	Minimum Charge	Day			59	59	
Monthly Rental							
	Install	Month			1175	1175	Monthly rental of welder.
	Minimum Charge	Day			59	59	
Sandblaster							
Daily Rental							
	Install	Day			17.20	17.20	Daily rental of sandblaster.
	Minimum Charge	Day			17.20	17.20	
Weekly Rental							
	Install	Week			86	86	Weekly rental of sandblaster.
	Minimum Charge	Day			17.20	17.20	

Job Costs

Equipment Rental

		Unit	Material	Labor	Equip.	Total	Specification
Monthly Rental							
	Install	Month			345	345	Monthly rental of sandblaster.
	Minimum Charge	Day			17.20	17.20	
Space Heater							
Daily Rental							
	Install	Day			15.90	15.90	Daily rental of space heater.
	Minimum Charge	Day			19.50	19.50	
Weekly Rental							
	Install	Day			79.50	79.50	Weekly rental of space heater.
	Minimum Charge	Day			19.50	19.50	
Monthly Rental							
	Install	Day			320	320	Monthly rental of space heater.
	Minimum Charge	Day			19.50	19.50	

Move / Reset Contents

		Unit	Material	Labor	Equip.	Total	Specification
Small Room							
	Install	Room		29.50		29.50	Includes cost to remove and reset the
	Minimum Charge	Job		121		121	contents of a small room.
Average Room							
	Install	Room		35.50		35.50	Includes cost to remove and reset the
	Minimum Charge	Job		121		121	contents of an average room.
Large Room							
	Install	Room		49		49	Includes cost to remove and reset the
	Minimum Charge	Job		121		121	contents of a large room.
Extra Large Room							
	Install	Room		77.50		77.50	Includes cost to remove and reset the
	Minimum Charge	Job		121		121	contents of a extra large room.
Store Contents							
Weekly							
	Install	Week	165			165	Storage fee, per week.
Monthly							
	Install	Month	495			495	Storage fee, per month.

Cover and Protect

		Unit	Material	Labor	Equip.	Total	Specification
Cover / Protect Walls							
	Install	SF Wall		.09		.09	Includes labor costs to cover and
	Minimum Charge	Job		121		121	protect walls.
Cover / Protect Floors							
	Install	SF Flr.		.09		.09	Includes labor costs to cover and
	Minimum Charge	Job		121		121	protect floors.

Temporary Bracing

		Unit	Material	Labor	Equip.	Total	Specification
Shim Support Piers							
	Install	Ea.	3.59	97.50		101.09	Includes labor and material to install
	Minimum Charge	Job		292		292	shims support for piers.

Job Costs

Jack and Relevel

Jack and Relevel		Unit	Material	Labor	Equip.	Total	Specification
Single Story							
	Install	S.F.	1.32	1.17		2.49	Includes labor and equipment to jack
	Minimum Charge	Job		292		292	and relevel single-story building.
Two-story							
	Install	S.F.	1.32	1.95		3.27	Includes labor and equipment to jack
	Minimum Charge	Job		292		292	and relevel two-story building.

Dumpster Rental

Dumpster Rental		Unit	Material	Labor	Equip.	Total	Specification
Small Load							
	Install	Week			465	465	Dumpster, weekly rental, 2 dumps, 5
	Minimum Charge	Week			465	465	yard container.
Medium Load							
	Install	Week			560	560	Dumpster, weekly rental, 2 dumps, 10
	Minimum Charge	Week			560	560	yard container.
Large Load							
	Install	Week			985	985	Dumpster, weekly rental, 2 dumps, 30
	Minimum Charge	Week			985	985	yard container.

Debris Hauling

Debris Hauling		Unit	Material	Labor	Equip.	Total	Specification
Per Ton							
	Install	Ton		75.50	39	114.50	Includes hauling of debris by trailer or
	Minimum Charge	Ton		565	291	856	dump truck.
Per Cubic Yard							
	Install	C.Y.		25	12.95	37.95	Includes hauling of debris by trailer or
	Minimum Charge	Job		565	291	856	dump truck.
Per Pick-up Truck Load							
	Install	Ea.		134		134	Includes hauling of debris by pick-up
	Minimum Charge	Ea.		134		134	truck.

Dump Fees

Dump Fees		Unit	Material	Labor	Equip.	Total	Specification
Per Cubic Yard							
	Install	C.Y.	7.45			7.45	Dump charges, building materials.
	Minimum Charge	C.Y.	7.45			7.45	
Per Ton							
	Install	Ton	66			66	Dump charges, building materials.
	Minimum Charge	Ton	66			66	

Scaffolding

Scaffolding		Unit	Material	Labor	Equip.	Total	Specification
4' to 6' High							
	Install	Month	54.50			54.50	Includes monthly rental of scaffolding,
	Minimum Charge	Month	54.50			54.50	steel tubular, 30" wide, 7' long, 5' high.
7' to 11' High							
	Install	Month	64.50			64.50	Includes monthly rental of scaffolding,
	Minimum Charge	Month	64.50			64.50	steel tubular, 30" wide, 7' long, 10' high.

Job Costs

Scaffolding

Scaffolding		Unit	Material	Labor	Equip.	Total	Specification
12' to 16' High							
	Install	Month	75			75	Includes monthly rental of scaffolding, steel tubular, 30" wide, 7' long, 15' high.
	Minimum Charge	Month	75			75	
17' to 21' High							
	Install	Month	100			100	Includes monthly rental of scaffolding, steel tubular, 30" wide, 7' long, 20' high.
	Minimum Charge	Month	100			100	
22' to 26' High							
	Install	Month	110			110	Includes monthly rental of scaffolding, steel tubular, 30" wide, 7' long, 25' high.
	Minimum Charge	Month	110			110	
27' to 30' High							
	Install	Month	120			120	Includes monthly rental of scaffolding, steel tubular, 30" wide, 7' long, 30' high.
	Minimum Charge	Month	120			120	

Construction Clean-up

Construction Clean-up		Unit	Material	Labor	Equip.	Total	Specification
Initial							
	Clean	S.F.		.30		.30	Includes labor for general clean-up.
	Minimum Charge	Job		121		121	
Progressive							
	Clean	S.F.		.15		.15	Includes labor for progressive site/job clean-up on a weekly basis.
	Minimum Charge	Job		121		121	

Building Demolition		Unit	Material	Labor	Equip.	Total	Specification
Wood-framed Building							
1 Story Demolition							
	Demolish	S.F.		1.14	1.35	2.49	Includes demolition of 1 floor only
	Minimum Charge	Job		1800	2125	3925	light wood-framed building, including hauling of debris. Dump fees not included.
2nd Story Demolition							
	Demolish	S.F.		1.16	1.37	2.53	Includes demolition of 2nd floor only
	Minimum Charge	Job		1800	2125	3925	light wood-framed building, including hauling of debris. Dump fees not included.
3rd Story Demolition							
	Demolish	S.F.		1.51	1.79	3.30	Includes demolition of 3rd floor only
	Minimum Charge	Job		1800	2125	3925	light wood-framed building, including hauling of debris. Dump fees not included.
Masonry Building							
	Demolish	S.F.		1.62	1.92	3.54	Includes demolition of one story
	Minimum Charge	Job		1800	2125	3925	masonry building, including hauling of debris. Dump fees not included.
Concrete Building							
Unreinforced							
	Demolish	S.F.		3.33	.97	4.30	Includes demolition of a one story
	Minimum Charge	Job		1800	2125	3925	non-reinforced concrete building, including hauling of debris. Dump fees not included.
Reinforced							
	Demolish	S.F.		3.82	1.11	4.93	Includes demolition of a one story
	Minimum Charge	Job		1800	2125	3925	reinforced concrete building, including hauling of debris. Dump fees not included.
Steel Building							
	Demolish	S.F.		2.16	2.56	4.72	Includes demolition of a metal
	Minimum Charge	Job		1800	2125	3925	building, including hauling of debris. Dump fees not included. No salvage value assumed for scrap metal.

Job Preparation

General Clean-up

General Clean-up		Unit	Material	Labor	Equip.	Total	Specification
Remove Debris							
	Clean	S.F.		.74	.14	.88	Includes removal of debris caused by the loss including contents items and loose debris.
	Minimum Charge	Job		370	71	441	
Muck-out							
	Clean	S.F.		.47		.47	Includes labor to remove 2" of muck and mud after a flood loss.
	Minimum Charge	Job		121		121	
Deodorize / Disinfect							
	Clean	SF Flr.		.16		.16	Includes deodorizing building with application of mildicide and disinfectant to floors and walls - will vary depending on the extent of deodorization required.
	Minimum Charge	Job		121		121	
Clean Walls							
Heavy							
	Clean	S.F.		.12		.12	Includes labor and material for heavy cleaning (multiple applications) with detergent and solvent.
	Minimum Charge	Job		121		121	
Clean Ceiling							
Light							
	Clean	S.F.		.20		.20	Includes labor and material to clean light to moderate smoke and soot.
	Minimum Charge	Job		121		121	
Heavy							
	Clean	S.F.		.30		.30	Includes labor and material to clean heavy smoke and soot.
	Minimum Charge	Job		121		121	

Flood Clean-up

Flood Clean-up		Unit	Material	Labor	Equip.	Total	Specification
Emergency Service Call							
After Hours / Weekend							
	Minimum Charge	Job		161		161	Includes minimum charges for deflooding clean-up after hours or on weekends.
Grey Water / Sewage							
	Minimum Charge	Job		121		121	Includes minimum charges for deflooding clean-up.
Water Extraction							
Clean (non-grey)							
	Clean	SF Flr.	.03	.48		.51	Includes labor to remove clean uncontaminated water from the loss.
	Minimum Charge	Job		121		121	
Grey (non-solids)							
	Clean	SF Flr.	.03	.97		1	Includes labor to remove non-solid grey water from the loss.
	Minimum Charge	Job		121		121	
Sewage / Solids							
	Clean	SF Flr.	.04	1.21		1.25	Includes labor to remove water from the loss that may contain sewage or solids.
	Minimum Charge	Job		121		121	

Job Preparation

Flood Clean-up

Flood Clean-up		Unit	Material	Labor	Equip.	Total	Specification
Mildicide Walls							
Topographical							
	Clean	SF Wall		.22		.22	Clean-up of deflooding including
	Minimum Charge	Job		121		121	topographical mildicide application.
Injection Treatments							
	Clean	SF Flr.	26.50	3.72		30.22	Clean-up of deflooding including
	Minimum Charge	Job		121		121	mildicide injection.
Carpet Cleaning							
Uncontaminated							
	Clean	SF Flr.		.15		.15	Includes labor to clean carpet.
	Minimum Charge	Job		121		121	
Contaminated							
	Clean	SF Flr.		.24		.24	Includes labor to clean contaminated
	Minimum Charge	Job		121		121	carpet.
Stairway							
	Clean	Ea.	.10	2.84		2.94	Includes labor to clean contaminated
	Minimum Charge	Job		121		121	carpet on a stairway.
Carpet Treatment							
Disinfect / Deodorize							
	Clean	SF Flr.		.15		.15	Includes disinfecting and deodorizing
	Minimum Charge	Job		121		121	of carpet.
Disinfect For Sewage							
	Clean	SF Flr.		.39		.39	Includes labor to disinfect and
	Minimum Charge	Job		121		121	deodorize carpet exposed to sewage.
Mildicide							
	Clean	SF Flr.	.01	.15		.16	Includes labor and materials to apply
	Minimum Charge	Job		121		121	mildicide to carpet.
Thermal Fog Area							
	Clean	SF Flr.	.03	.37		.40	Includes labor and materials to
	Minimum Charge	Job		121		121	perform thermal fogging during
							deflooding clean-up.
Wet Fog / ULV Area							
	Clean	SF Flr.	.02	.21		.23	Includes labor and materials to
	Minimum Charge	Job		121		121	perform wet fogging / ULV during
							deflooding clean-up.
Steam Clean Fixtures							
	Clean	Ea.	.01	24		24.01	Includes labor and materials to steam
	Minimum Charge	Job		121		121	clean fixtures during deflooding
							clean-up.
Pressure Wash Sewage							
	Clean	SF Flr.		2.20		2.20	Includes labor to pressure wash
	Minimum Charge	Job		121		121	sewage contamination.
Airmover							
	Minimum Charge	Day			19.50	19.50	Daily rental of air mover.

Job Preparation

Flood Clean-up

	Unit	Material	Labor	Equip.	Total	Specification
Dehumidifier						
Medium Size						
Minimum Charge	Day			36.50	36.50	Daily rental of dehumidifier, medium.
Large Size						
Minimum Charge	Day			73	73	Daily rental of dehumidifier, large.
Air Purifier						
HEPA Filter-Bacteria						
Minimum Charge	Ea.	198			198	HEPA filter for air purifier.

Asbestos Removal

	Unit	Material	Labor	Equip.	Total	Specification
Asbestos Analysis						
Minimum Charge	Job		197		197	Minimum charge for testing.
Asbestos Pipe Insulation						
1/2" to 3/4" Diameter						
Demolish	L.F.	.32	5.10		5.42	Minimum charge to remove asbestos. Includes full Tyvek suits for workers changed four times per eight-hour shift with respirators, air monitoring and supervision by qualified professionals.
Minimum Charge	Job		3150		3150	
1" to 3" Diameter						
Demolish	L.F.	.98	15.80		16.78	Minimum charge to remove asbestos. Includes full Tyvek suits for workers changed four times per eight-hour shift with respirators, air monitoring and supervision by qualified professionals.
Minimum Charge	Job		3150		3150	
Asbestos-based Siding						
Demolish	S.F.	2.18	35		37.18	Minimum charge to remove asbestos. Includes full Tyvek suits for workers changed four times per eight-hour shift with respirators, air monitoring and supervision by qualified professionals.
Minimum Charge	Job		3150		3150	
Asbestos-based Plaster						
Demolish	S.F.	1.67	27		28.67	Minimum charge to remove asbestos. Includes full Tyvek suits for workers changed four times per eight-hour shift with respirators, air monitoring and supervision by qualified professionals.
Minimum Charge	Job		3150		3150	
Encapsulate Asbestos Ceiling						
Install	S.F.	.30	.15		.45	Includes encapsulation with penetrating sealant sprayed onto ceiling and walls with airless sprayer.
Minimum Charge	Job		3150		3150	
Scrape Acoustical Ceiling						
Demolish	S.F.	.07	.89		.96	Minimum charge to remove asbestos. Includes full Tyvek suits for workers changed four times per eight-hour shift with respirators, air monitoring and supervision by qualified professionals.
Minimum Charge	Job		3150		3150	

Job Preparation

Asbestos Removal

Asbestos Removal		Unit	Material	Labor	Equip.	Total	Specification
Seal Ceiling Asbestos							
	Install	S.F.	1.12	2.97		4.09	Includes attaching plastic cover to ceiling area and sealing where asbestos exists, taping all seams, caulk or foam insulation in cracks and negative air system.
	Minimum Charge	Job		3150		3150	
Seal Wall Asbestos							
	Install	S.F.	1.12	2.64		3.76	Includes attaching plastic cover to wall area and sealing where asbestos exists, taping all seams, caulk or foam insulation in cracks and negative air system.
	Minimum Charge	Job		3150		3150	
Seal Floor Asbestos							
	Install	S.F.	1.79	5.30		7.09	Includes attaching plastic cover to floor area and sealing where asbestos exists, taping all seams, caulk or foam insulation in cracks and negative air system.
	Minimum Charge	Job		3150		3150	
Negative Air Vent System							
	Install	Day			43.50	43.50	Daily rental of negative air vent system.
	Minimum Charge	Week			218	218	
HEPA Vacuum Cleaner							
	Install	Day			18.25	18.25	Daily rental of HEPA vacuum, 16 gallon.
	Minimum Charge	Week			91	91	
Airless Sprayer							
	Install	Day			49	49	Daily rental of airless sprayer.
	Minimum Charge	Week			246	246	
Decontamination Unit							
	Install	Day			133	133	Daily rental of decontamination unit.
	Minimum Charge	Week			665	665	
Light Stand							
	Install	Day			10.55	10.55	Daily rental of floodlight w/tripod.
	Minimum Charge	Week			53	53	

Room Demolition

Room Demolition		Unit	Material	Labor	Equip.	Total	Specification
Strip Typical Room							
	Demolish	S.F.		2.42		2.42	Includes 4 labor hours for general demolition.
	Minimum Charge	Job		134		134	
Strip Typical Bathroom							
	Demolish	S.F.		3.22		3.22	Includes 4 labor hours for general demolition.
	Minimum Charge	Job		134		134	
Strip Typical Kitchen							
	Demolish	S.F.		3.10		3.10	Includes 4 labor hours for general demolition.
	Minimum Charge	Job		134		134	
Strip Typical Laundry							
	Demolish	S.F.		2.76		2.76	Includes 4 labor hours for general demolition.
	Minimum Charge	Job		134		134	

Job Preparation

Selective Demolition		Unit	Material	Labor	Equip.	Total	Specification
Concrete Slab							
4″ Unreinforced							
	Demolish	S.F.		1.61	.21	1.82	Includes labor and equipment to remove unreinforced concrete slab and haul debris to truck or dumpster.
	Minimum Charge	Job		685	90	775	
6″ Unreinforced							
	Demolish	S.F.		2.29	.30	2.59	Includes labor and equipment to remove unreinforced concrete slab and haul debris to truck or dumpster.
	Minimum Charge	Job		685	90	775	
8″ Unreinforced							
	Demolish	S.F.		3.76	.49	4.25	Includes labor and equipment to remove unreinforced concrete slab and haul debris to truck or dumpster.
	Minimum Charge	Job		685	90	775	
4″ Reinforced							
	Demolish	S.F.		2.03	.27	2.30	Includes labor and equipment to remove reinforced concrete slab and haul debris to truck or dumpster.
	Minimum Charge	Job		685	90	775	
6″ Reinforced							
	Demolish	S.F.		3.05	.40	3.45	Includes labor and equipment to remove reinforced concrete slab and haul debris to truck or dumpster.
	Minimum Charge	Job		685	90	775	
8″ Reinforced							
	Demolish	S.F.		4.99	.65	5.64	Includes labor and equipment to remove reinforced concrete slab and haul debris to truck or dumpster.
	Minimum Charge	Job		685	90	775	
Brick Wall							
4″ Thick							
	Demolish	S.F.		2.25	.29	2.54	Includes labor and equipment to remove brick wall and haul debris to a central location for disposal.
	Minimum Charge	Job		685	90	775	
8″ Thick							
	Demolish	S.F.		3.09	.41	3.50	Includes labor and equipment to remove brick wall and haul debris to a central location for disposal.
	Minimum Charge	Job		685	90	775	
12″ Thick							
	Demolish	S.F.		4.12	.54	4.66	Includes labor and equipment to remove brick wall and haul debris to a central location for disposal.
	Minimum Charge	Job		685	90	775	
Concrete Masonry Wall							
4″ Thick							
	Demolish	S.F.		2.50	.33	2.83	Includes minimum labor and equipment to remove 4″ block wall and haul debris to a central location for disposal.
	Minimum Charge	Job		685	90	775	
6″ Thick							
	Demolish	S.F.		3.09	.41	3.50	Includes minimum labor and equipment to remove 6″ block wall and haul debris to a central location for disposal.
	Minimum Charge	Job		685	90	775	

Job Preparation

Selective Demolition		Unit	Material	Labor	Equip.	Total	Specification
8" Thick							
	Demolish	S.F.		3.80	.50	4.30	Includes minimum labor and
	Minimum Charge	Job		685	90	775	equipment to remove 8" block wall and haul debris to a central location for disposal.
12" Thick							
	Demolish	S.F.		5.50	.72	6.22	Includes minimum labor and
	Minimum Charge	Job		685	90	775	equipment to remove 12" block wall and haul debris to a central location for disposal.
Tree Removal							
Small, 4" - 8"							
	Demolish	Ea.		87	63	150	Minimum charge, tree removal.
	Minimum Charge	Job		121		121	
Medium, 9" - 12"							
	Demolish	Ea.		131	95	226	Minimum charge, tree removal.
	Minimum Charge	Job		121		121	
Large, 13" - 18"							
	Demolish	Ea.		157	114	271	Minimum charge, tree removal.
	Minimum Charge	Job		400	129	529	
Very Large, 19" - 24"							
	Demolish	Ea.		196	142	338	Minimum charge, tree removal.
	Minimum Charge	Job		400	129	529	
Stump Removal							
	Demolish	Ea.		54.50	115	169.50	Includes minimum charges for a site
	Minimum Charge	Job		223	69	292	demolition crew and backhoe / loader.

Excavation

Backhoe/Loader - Wheel Type

Tractor Loader - Wheel Type

Backhoe - Crawler Type

Excavation		Unit	Material	Labor	Equip.	Total	Specification
Footings							
Hand Trenching							
	Install	L.F.		1.51		1.51	Includes hand excavation for footing
	Minimum Charge	Job		60.50		60.50	or trench up to 10″ deep and 12″ wide.
Machine Trenching							
	Install	L.F.		.58	.25	.83	Includes machine excavation for
	Minimum Charge	Job		288	124	412	footing or trench up to 10″ deep and 12″ wide.
Trenching w / Backfill							
Hand							
	Install	L.F.		2.42		2.42	Includes hand excavation, backfill and
	Minimum Charge	Job		60.50		60.50	compaction.
Machine							
	Install	L.F.		2.88	8.05	10.93	Includes machine excavation, backfill
	Minimum Charge	Job		121	26	147	and compaction.

Backfill		Unit	Material	Labor	Equip.	Total	Specification
Hand (no compaction)							
	Install	L.C.Y.		20		20	Includes backfill by hand in average
	Minimum Charge	Job		121		121	soil without compaction.
Hand (w / compaction)							
	Install	E.C.Y.		28.50		28.50	Includes backfill by hand in average
	Minimum Charge	Job		121		121	soil with compaction.
Machine (no compaction)							
	Install	L.C.Y.		2.88	1.24	4.12	Includes backfill, by 55 HP wheel
	Minimum Charge	Job		288	124	412	loader, of trenches from loose material piled adjacent without compaction.
Machine (w / compaction)							
	Install	L.C.Y.		7.65	3.32	10.97	Includes backfill, by 55 HP wheel
	Minimum Charge	Job		288	124	412	loader, of trenches from loose material piled adjacent with compaction by vibrating tampers.

Concrete Footing and Formwork

Wood Pile

Piles	Unit	Material	Labor	Equip.	Total	Specification
Treated Wood, 12″ Butt						
To 30′						
Demolish	L.F.		3.88	1.20	5.08	Includes material, labor and
Install	V.L.F.	10.65	3.76	2.77	17.18	equipment to install up to 30′ long
Demolish and Install	V.L.F.	10.65	7.64	3.97	22.26	treated wood piling.
Paint	L.F.	.39	1.69		2.08	
Minimum Charge	Job		2350	1725	4075	
30′ to 40′						
Demolish	L.F.		3.88	1.20	5.08	Includes material, labor and
Install	V.L.F.	10.40	3.36	2.47	16.23	equipment to install 30′ to 40′ long
Demolish and Install	V.L.F.	10.40	7.24	3.67	21.31	treated wood piling.
Paint	L.F.	.39	1.69		2.08	
Minimum Charge	Job		2350	1725	4075	
40′ to 50′						
Demolish	L.F.		3.88	1.20	5.08	Includes material, labor and
Install	V.L.F.	11.15	3.27	2.40	16.82	equipment to install 40′ to 50′ long
Demolish and Install	V.L.F.	11.15	7.15	3.60	21.90	treated wood piling.
Paint	L.F.	.39	1.69		2.08	
Minimum Charge	Job		2350	1725	4075	
50′ to 60′						
Demolish	L.F.		3.88	1.20	5.08	Includes material, labor and
Install	V.L.F.	12.30	2.94	2.16	17.40	equipment to install 50′ to 60′ long
Demolish and Install	V.L.F.	12.30	6.82	3.36	22.48	treated wood piling.
Paint	L.F.	.39	1.69		2.08	
Minimum Charge	Job		2350	1725	4075	
60′ to 80′						
Demolish	L.F.		3.88	1.20	5.08	Includes material, labor and
Install	V.L.F.	21	3.28	2.63	26.91	equipment to install 60′ to 80′ long
Demolish and Install	V.L.F.	21	7.16	3.83	31.99	treated wood piling.
Paint	L.F.	.39	1.69		2.08	
Minimum Charge	Job		2350	1725	4075	
Untreated Wood, 12″ Butt						
To 30′						
Demolish	L.F.		3.88	1.20	5.08	Includes material, labor and
Install	V.L.F.	7.20	3.76	2.77	13.73	equipment to install up to 30′ long
Demolish and Install	V.L.F.	7.20	7.64	3.97	18.81	untreated wood piling.
Paint	L.F.	.39	1.69		2.08	
Minimum Charge	Job		2350	1725	4075	

Foundation

Piles

	Unit	Material	Labor	Equip.	Total	Specification
30' to 40'						
Demolish	L.F.		3.88	1.20	5.08	Includes material, labor and
Install	V.L.F.	7.20	3.36	2.47	13.03	equipment to install 30' to 40' long
Demolish and Install	V.L.F.	7.20	7.24	3.67	18.11	untreated wood piling.
Paint	L.F.	.39	1.69		2.08	
Minimum Charge	Job		2350	1725	4075	
40' to 50'						
Demolish	L.F.		3.88	1.20	5.08	Includes material, labor and
Install	V.L.F.	7.20	3.27	2.40	12.87	equipment to install 40' to 50' long
Demolish and Install	V.L.F.	7.20	7.15	3.60	17.95	untreated wood piling.
Paint	L.F.	.39	1.69		2.08	
Minimum Charge	Job		2350	1725	4075	
50' to 60'						
Demolish	L.F.		3.88	1.20	5.08	Includes material, labor and
Install	V.L.F.	7.30	2.94	2.16	12.40	equipment to install 50' to 60' long
Demolish and Install	V.L.F.	7.30	6.82	3.36	17.48	untreated wood piling.
Paint	L.F.	.39	1.69		2.08	
Minimum Charge	Job		2350	1725	4075	
60' to 80'						
Demolish	L.F.		3.88	1.20	5.08	Includes material, labor and
Install	V.L.F.	9.15	2.80	2.06	14.01	equipment to install 60' to 80' long
Demolish and Install	V.L.F.	9.15	6.68	3.26	19.09	untreated wood piling.
Paint	L.F.	.39	1.69		2.08	
Minimum Charge	Job		2350	1725	4075	

Precast Concrete

	Unit	Material	Labor	Equip.	Total	Specification
10" Square						
Demolish	L.F.		3.97	1.23	5.20	Includes material, labor and
Install	V.L.F.	10	3.36	2.47	15.83	equipment to install a precast,
Demolish and Install	V.L.F.	10	7.33	3.70	21.03	prestressed, 40' long, 10" square
Paint	L.F.	.39	1.69		2.08	concrete pile.
Minimum Charge	Job		2350	1725	4075	
12" Square						
Demolish	L.F.		4.25	1.32	5.57	Includes material, labor and
Install	V.L.F.	12.60	3.46	2.54	18.60	equipment to install a precast,
Demolish and Install	V.L.F.	12.60	7.71	3.86	24.17	prestressed, 40' long, 12" square
Paint	L.F.	.39	1.69		2.08	concrete pile.
Minimum Charge	Job		2350	1725	4075	
14" Square						
Demolish	L.F.		4.36	1.35	5.71	Includes material, labor and
Install	V.L.F.	14.90	3.92	2.88	21.70	equipment to install a precast,
Demolish and Install	V.L.F.	14.90	8.28	4.23	27.41	prestressed, 40' long, 14" square
Paint	L.F.	.39	1.69		2.08	concrete pile.
Minimum Charge	Job		2350	1725	4075	
16" Square						
Demolish	L.F.		4.58	1.42	6	Includes material, labor and
Install	V.L.F.	23.50	4.20	3.09	30.79	equipment to install a precast,
Demolish and Install	V.L.F.	23.50	8.78	4.51	36.79	prestressed, 40' long, 16" square
Paint	L.F.	.39	1.69		2.08	concrete pile.
Minimum Charge	Job		2350	1725	4075	
18" Square						
Demolish	L.F.		4.96	1.54	6.50	Includes material, labor and
Install	V.L.F.	28	5.30	4.25	37.55	equipment to install a precast,
Demolish and Install	V.L.F.	28	10.26	5.79	44.05	prestressed, 40' long, 18" square
Paint	L.F.	.39	1.69		2.08	concrete pile.
Minimum Charge	Job		2350	1725	4075	

Foundation

Piles

	Unit	Material	Labor	Equip.	Total	Specification
Cross Bracing						
Treated 2″ x 6″						
Demolish	L.F.		.79		.79	Cost includes material and labor to install 2″ x 6″ lumber for cross-bracing of foundation piling or posts.
Install	L.F.	.83	4.71		5.54	
Demolish and Install	L.F.	.83	5.50		6.33	
Reinstall	L.F.		3.77		3.77	
Clean	S.F.		.30		.30	
Paint	L.F.	.17	1.14		1.31	
Minimum Charge	Job		585		585	
Treated 2″ x 8″						
Demolish	L.F.		.79		.79	Cost includes material and labor to install 2″ x 8″ lumber for cross-bracing of foundation piling or posts.
Install	L.F.	1.08	4.87		5.95	
Demolish and Install	L.F.	1.08	5.66		6.74	
Reinstall	L.F.		3.90		3.90	
Clean	S.F.		.30		.30	
Paint	L.F.	.22	1.19		1.41	
Minimum Charge	Job		585		585	
Treated 2″ x 10″						
Demolish	L.F.		.81		.81	Cost includes material and labor to install 2″ x 10″ lumber for cross-bracing of foundation piling or posts.
Install	L.F.	1.64	5.05		6.69	
Demolish and Install	L.F.	1.64	5.86		7.50	
Reinstall	L.F.		4.03		4.03	
Clean	S.F.		.30		.30	
Paint	L.F.	.26	1.21		1.47	
Minimum Charge	Job		585		585	
Treated 2″ x 12″						
Demolish	L.F.		.98		.98	Cost includes material and labor to install 2″ x 12″ lumber for cross-bracing of foundation piling or posts.
Install	L.F.	2.15	5.20		7.35	
Demolish and Install	L.F.	2.15	6.18		8.33	
Reinstall	L.F.		4.17		4.17	
Clean	S.F.		.30		.30	
Paint	L.F.	.31	1.23		1.54	
Minimum Charge	Job		585		585	
Metal Pipe						
6″ Concrete Filled						
Demolish	L.F.		5.50	1.71	7.21	Includes material, labor and equipment to install concrete filled steel pipe pile.
Install	V.L.F.	17.75	4.90	3.60	26.25	
Demolish and Install	V.L.F.	17.75	10.40	5.31	33.46	
Paint	L.F.	.30	1.69		1.99	
Minimum Charge	Job		2350	1725	4075	
6″ Unfilled						
Demolish	L.F.		5.50	1.71	7.21	Includes material, labor and equipment to install unfilled steel pipe pile.
Install	V.L.F.	15.15	4.48	3.29	22.92	
Demolish and Install	V.L.F.	15.15	9.98	5	30.13	
Paint	L.F.	.30	1.69		1.99	
Minimum Charge	Job		2350	1725	4075	
8″ Concrete Filled						
Demolish	L.F.		5.50	1.71	7.21	Includes material, labor and equipment to install concrete filled steel pipe pile.
Install	V.L.F.	16.15	5.10	3.76	25.01	
Demolish and Install	V.L.F.	16.15	10.60	5.47	32.22	
Paint	L.F.	.30	1.69		1.99	
Minimum Charge	Job		2350	1725	4075	
8″ Unfilled						
Demolish	L.F.		5.50	1.71	7.21	Includes material, labor and equipment to install unfilled steel pipe pile.
Install	V.L.F.	15.25	4.70	3.46	23.41	
Demolish and Install	V.L.F.	15.25	10.20	5.17	30.62	
Paint	L.F.	.30	1.69		1.99	
Minimum Charge	Job		2350	1725	4075	

Piles		Unit	Material	Labor	Equip.	Total	Specification
10″ Concrete Filled							
	Demolish	L.F.		5.70	1.76	7.46	Includes material, labor and
	Install	V.L.F.	21	5.20	3.84	30.04	equipment to install concrete filled
	Demolish and Install	V.L.F.	21	10.90	5.60	37.50	steel pipe pile.
	Paint	L.F.	.30	1.69		1.99	
	Minimum Charge	Job		2350	1725	4075	
10″ Unfilled							
	Demolish	L.F.		5.70	1.76	7.46	Includes material, labor and
	Install	V.L.F.	19	4.70	3.46	27.16	equipment to install unfilled steel pipe
	Demolish and Install	V.L.F.	19	10.40	5.22	34.62	pile.
	Paint	L.F.	.30	1.69		1.99	
	Minimum Charge	Job		2350	1725	4075	
12″ Concrete Filled							
	Demolish	L.F.		5.70	1.76	7.46	Includes material, labor and
	Install	V.L.F.	24	5.65	4.16	33.81	equipment to install concrete filled
	Demolish and Install	V.L.F.	24	11.35	5.92	41.27	steel pipe pile.
	Paint	L.F.	.30	1.69		1.99	
	Minimum Charge	Job		2350	1725	4075	
12″ Unfilled							
	Demolish	L.F.		5.70	1.76	7.46	Includes material, labor and
	Install	V.L.F.	23.50	4.95	3.64	32.09	equipment to install unfilled steel pipe
	Demolish and Install	V.L.F.	23.50	10.65	5.40	39.55	pile.
	Paint	L.F.	.30	1.69		1.99	
	Minimum Charge	Job		2350	1725	4075	
Steel H Section							
H Section 8 x 8 x 36#							
	Demolish	L.F.		3.85	1.19	5.04	Includes material, labor and
	Install	V.L.F.	14.10	3.67	2.70	20.47	equipment to install steel H section
	Demolish and Install	V.L.F.	14.10	7.52	3.89	25.51	pile.
	Paint	L.F.	.30	1.69		1.99	
	Minimum Charge	Job		2350	1725	4075	
H Section 10 x 10 x 57#							
	Demolish	L.F.		4.21	1.30	5.51	Includes material, labor and
	Install	V.L.F.	22.50	3.85	2.83	29.18	equipment to install steel H section
	Demolish and Install	V.L.F.	22.50	8.06	4.13	34.69	pile.
	Paint	L.F.	.30	1.69		1.99	
	Minimum Charge	Job		2350	1725	4075	
H Section 12 x 12 x 74#							
	Demolish	L.F.		4.36	1.35	5.71	Includes material, labor and
	Install	V.L.F.	29	4.66	3.75	37.41	equipment to install steel H section
	Demolish and Install	V.L.F.	29	9.02	5.10	43.12	pile.
	Paint	L.F.	.30	1.69		1.99	
	Minimum Charge	Job		2350	1725	4075	
H Section 14 x 14 x 89#							
	Demolish	L.F.		4.58	1.42	6	Includes material, labor and
	Install	V.L.F.	35	5.10	4.09	44.19	equipment to install steel H section
	Demolish and Install	V.L.F.	35	9.68	5.51	50.19	pile.
	Paint	L.F.	.30	1.69		1.99	
	Minimum Charge	Job		2350	1725	4075	

Foundation

Foundation Post

		Unit	Material	Labor	Equip.	Total	Specification
Wood Foundation Post							
4" x 4"							
	Demolish	L.F.		.62	.19	.81	Cost includes material and labor to
	Install	L.F.	.99	4.03		5.02	install 4" x 4" treated wood post by
	Demolish and Install	L.F.	.99	4.65	.19	5.83	hand in normal soil conditions.
	Reinstall	L.F.		3.22		3.22	
	Minimum Charge	Job		585		585	
4" x 6"							
	Demolish	L.F.		.62	.19	.81	Cost includes material and labor to
	Install	L.F.	1.49	4.03		5.52	install 4" x 6" treated wood post by
	Demolish and Install	L.F.	1.49	4.65	.19	6.33	hand in normal soil conditions.
	Reinstall	L.F.		3.22		3.22	
	Minimum Charge	Job		585		585	
6" x 6"							
	Demolish	L.F.		.62	.19	.81	Cost includes material and labor to
	Install	L.F.	2.21	4.03		6.24	install 6" x 6" treated wood post by
	Demolish and Install	L.F.	2.21	4.65	.19	7.05	hand in normal soil conditions.
	Reinstall	L.F.		3.22		3.22	
	Minimum Charge	Job		585		585	
Water Jet							
	Demolish	Ea.		71.50		71.50	Includes material, labor and
	Install	Ea.	58.50	550		608.50	equipment to water jet a 6" x 6" wood
	Demolish and Install	Ea.	58.50	621.50		680	post under an existing building. Cost
	Reinstall	Ea.		552.23		552.23	may vary depending upon location
	Minimum Charge	Job		485		485	and access.
Shim							
	Install	Ea.	3.25	58.50		61.75	Includes labor and material to trim or
	Minimum Charge	Job		585		585	readjust existing foundation posts
							under an existing building. Access of
							2 - 4 foot assumed.
Cross-bracing							
Treated 2" x 6"							
	Demolish	L.F.		.79		.79	Cost includes material and labor to
	Install	L.F.	.83	4.71		5.54	install 2" x 6" lumber for cross-bracing
	Demolish and Install	L.F.	.83	5.50		6.33	of foundation piling or posts.
	Reinstall	L.F.		3.77		3.77	
	Clean	S.F.		.30		.30	
	Paint	L.F.	.17	1.14		1.31	
	Minimum Charge	Job		585		585	
Treated 2" x 8"							
	Demolish	L.F.		.79		.79	Cost includes material and labor to
	Install	L.F.	1.08	4.87		5.95	install 2" x 8" lumber for cross-bracing
	Demolish and Install	L.F.	1.08	5.66		6.74	of foundation piling or posts.
	Reinstall	L.F.		3.90		3.90	
	Clean	S.F.		.30		.30	
	Paint	L.F.	.22	1.19		1.41	
	Minimum Charge	Job		585		585	
Treated 2" x 10"							
	Demolish	L.F.		.81		.81	Cost includes material and labor to
	Install	L.F.	1.64	5.05		6.69	install 2" x 10" lumber for
	Demolish and Install	L.F.	1.64	5.86		7.50	cross-bracing of foundation piling or
	Reinstall	L.F.		4.03		4.03	posts.
	Clean	S.F.		.30		.30	
	Paint	L.F.	.26	1.21		1.47	
	Minimum Charge	Job		585		585	

Foundation

Foundation Post	Unit	Material	Labor	Equip.	Total	Specification
Treated 2″ x 12″						
Demolish	L.F.		.98		.98	Cost includes material and labor to
Install	L.F.	2.15	5.20		7.35	install 2″ x 12″ lumber for
Demolish and Install	L.F.	2.15	6.18		8.33	cross-bracing of foundation piling or
Reinstall	L.F.		4.17		4.17	posts.
Clean	S.F.		.30		.30	
Paint	L.F.	.31	1.23		1.54	
Minimum Charge	Job		585		585	

Concrete Footing	Unit	Material	Labor	Equip.	Total	Specification
Continuous						
12″ w x 6″ d						
Demolish	L.F.		1.18	.92	2.10	Includes material and labor to install
Install	L.F.	3.87	2.77	.01	6.65	continuous reinforced concrete footing
Demolish and Install	L.F.	3.87	3.95	.93	8.75	poured by chute cast against earth
Minimum Charge	Job		675		675	including reinforcing, forming and
						finishing.
12″ w x 12″ d						
Demolish	L.F.		2.27	1.77	4.04	Includes material and labor to install
Install	L.F.	8.70	5.55	.03	14.28	continuous reinforced concrete footing
Demolish and Install	L.F.	8.70	7.82	1.80	18.32	poured by chute cast against earth
Minimum Charge	Job		675		675	including reinforcing, forming and
						finishing.
18″ w x 10″ d						
Demolish	L.F.		2.95	2.30	5.25	Includes material and labor to install
Install	L.F.	8.05	6.80	.04	14.89	continuous reinforced concrete footing
Demolish and Install	L.F.	8.05	9.75	2.34	20.14	poured by chute cast against earth
Minimum Charge	Job		675		675	including reinforcing, forming and
						finishing.
24″ w x 24″ d						
Demolish	L.F.		4.92	3.84	8.76	Includes material and labor to install
Install	L.F.	23.50	22	.12	45.62	continuous reinforced concrete footing
Demolish and Install	L.F.	23.50	26.92	3.96	54.38	poured by chute cast against earth
Minimum Charge	Job		675		675	including reinforcing, forming and
						finishing.
Stem Wall						
Single Story						
Demolish	L.F.		10.90	8.50	19.40	Includes material and labor to install
Install	L.F.	19.85	57.50	45	122.35	stem wall 24″ above and 18″ below
Demolish and Install	L.F.	19.85	68.40	53.50	141.75	grade for a single story structure.
Minimum Charge	Job		675		675	Includes 12″ wide and 8″ deep
						footing.
Two Story						
Demolish	L.F.		10.90	8.50	19.40	Includes material and labor to install
Install	L.F.	24	83	64.50	171.50	stem wall 24″ above and 18″ below
Demolish and Install	L.F.	24	93.90	73	190.90	grade typical for a two story structure,
Minimum Charge	Job		675		675	includes 18″ wide and 10″ deep
						footing.

Foundation

Concrete Slab

Concrete Slab	Unit	Material	Labor	Equip.	Total	Specification
Reinforced						
4″						
Demolish	S.F.		2.03	.27	2.30	Cost includes material and labor to
Install	S.F.	1.96	.43		2.39	install 4″ reinforced concrete
Demolish and Install	S.F.	1.96	2.46	.27	4.69	slab-on-grade poured by chute
Clean	S.F.	.04	.27		.31	including forms, vapor barrier, wire
Paint	S.F.	.12	.64		.76	mesh, 3000 PSI concrete, float finish,
Minimum Charge	Job		156		156	and curing.
6″						
Demolish	S.F.		3.05	.40	3.45	Cost includes material and labor to
Install	S.F.	2.67	.65		3.32	install 6″ reinforced concrete
Demolish and Install	S.F.	2.67	3.70	.40	6.77	slab-on-grade poured by chute
Clean	S.F.	.04	.27		.31	including forms, vapor barrier, wire
Paint	S.F.	.12	.64		.76	mesh, 3000 PSI concrete, float finish,
Minimum Charge	Job		156		156	and curing.
8″						
Demolish	S.F.		4.99	.65	5.64	Cost includes material and labor to
Install	S.F.	3.48	.85	.01	4.34	install 8″ reinforced concrete
Demolish and Install	S.F.	3.48	5.84	.66	9.98	slab-on-grade poured by chute
Clean	S.F.	.04	.27		.31	including forms, vapor barrier, wire
Paint	S.F.	.12	.64		.76	mesh, 3000 PSI concrete, float finish,
Minimum Charge	Job		156		156	and curing.
Unreinforced						
4″						
Demolish	S.F.		1.61	.21	1.82	Cost includes material and labor to
Install	S.F.	1.28	.77	.01	2.06	install 4″ concrete slab-on-grade
Demolish and Install	S.F.	1.28	2.38	.22	3.88	poured by chute, including forms,
Clean	S.F.	.04	.27		.31	vapor barrier, 3000 PSI concrete, float
Paint	S.F.	.12	.64		.76	finish, and curing.
Minimum Charge	Job		156		156	
6″						
Demolish	S.F.		2.29	.30	2.59	Cost includes material and labor to
Install	S.F.	1.86	.78	.01	2.65	install 6″ concrete slab-on-grade
Demolish and Install	S.F.	1.86	3.07	.31	5.24	poured by chute, including forms,
Clean	S.F.	.04	.27		.31	vapor barrier, 3000 PSI concrete, float
Paint	S.F.	.12	.64		.76	finish, and curing.
Minimum Charge	Job		156		156	
8″						
Demolish	S.F.		3.76	.49	4.25	Cost includes material and labor to
Install	S.F.	2.54	.82	.01	3.37	install 8″ concrete slab-on-grade
Demolish and Install	S.F.	2.54	4.58	.50	7.62	poured by chute, including forms,
Clean	S.F.	.04	.27		.31	vapor barrier, 3000 PSI concrete, float
Paint	S.F.	.12	.64		.76	finish, and curing.
Minimum Charge	Job		156		156	
Lightweight						
Demolish	S.F.		.72	.16	.88	Includes material and labor to install
Install	S.F.	1.63	2.31		3.94	lightweight concrete, 3″ - 4″ thick.
Demolish and Install	S.F.	1.63	3.03	.16	4.82	
Clean	S.F.	.04	.27		.31	
Paint	S.F.	.12	.64		.76	
Minimum Charge	Job		156		156	

Foundation

Concrete Slab

	Unit	Material	Labor	Equip.	Total	Specification
Gunite						
Demolish	S.F.		.89	.28	1.17	Includes material and labor to install
Install	S.F.	.30	.98	.39	1.67	gunite on flat plane per inch of
Demolish and Install	S.F.	.30	1.87	.67	2.84	thickness using 1" as base price. No
Clean	S.F.	.04	.27		.31	forms or reinforcing are included.
Paint	S.F.	.12	.64		.76	
Minimum Charge	Job		865		865	
Epoxy Inject						
Install	L.F.	4.80	3.12		7.92	Includes labor and material to inject
Minimum Charge	Job		156		156	epoxy into cracks.
Pressure Grout						
Install	S.F.	1.93	2.31		4.24	Includes labor, material and
Minimum Charge	Job		156		156	equipment to install pressure grout
						under an existing building.
Scrape and Paint						
Paint	S.F.	.19	.74		.93	Includes labor and materials to scrape
Minimum Charge	Job		148		148	and paint a concrete floor.
Vapor Barrier						
Install	S.F.	.03	.09		.12	Cost includes material and labor to
Minimum Charge	Job		169		169	install 4 mil polyethylene vapor barrier
						10' wide sheets with 6" overlaps.

Rough Framing

Beam / Girder	Unit	Material	Labor	Equip.	Total	Specification
Solid Wood Beam						
2″ x 8″						
Demolish	L.F.		.65		.65	Cost includes material and labor to
Install	L.F.	.94	1.04		1.98	install 2″ x 8″ beam.
Demolish and Install	L.F.	.94	1.69		2.63	
Clean	L.F.	.11	.22		.33	
Paint	L.F.	.12	1.60		1.72	
Minimum Charge	Job		169		169	
2″ x 10″						
Demolish	L.F.		.81		.81	Cost includes material and labor to
Install	L.F.	1.42	1.13		2.55	install 2″ x 10″ beam.
Demolish and Install	L.F.	1.42	1.94		3.36	
Clean	L.F.	.13	.26		.39	
Paint	L.F.	.17	.99		1.16	
Minimum Charge	Job		169		169	
2″ x 12″						
Demolish	L.F.		.98		.98	Cost includes material and labor to
Install	L.F.	1.76	1.23		2.99	install 2″ x 12″ beam.
Demolish and Install	L.F.	1.76	2.21		3.97	
Clean	L.F.	.15	.30		.45	
Paint	L.F.	.19	2.37		2.56	
Minimum Charge	Job		169		169	
4″ x 8″						
Demolish	L.F.		2.02		2.02	Cost includes material and labor to
Install	L.F.	2.85	1.52	.70	5.07	install 4″ x 8″ beam.
Demolish and Install	L.F.	2.85	3.54	.70	7.09	
Clean	L.F.	.13	.26		.39	
Paint	L.F.	.17	.99		1.16	
Minimum Charge	Job		169		169	
4″ x 10″						
Demolish	L.F.		2.53		2.53	Cost includes material and labor to
Install	L.F.	3.59	1.60	.74	5.93	install 4″ x 10″ beam.
Demolish and Install	L.F.	3.59	4.13	.74	8.46	
Clean	L.F.	.15	.30		.45	
Paint	L.F.	.19	1.15		1.34	
Minimum Charge	Job		169		169	
4″ x 12″						
Demolish	L.F.		3.04		3.04	Cost includes material and labor to
Install	L.F.	4.44	1.68	.78	6.90	install 4″ x 12″ beam.
Demolish and Install	L.F.	4.44	4.72	.78	9.94	
Clean	L.F.	.18	.35		.53	
Paint	L.F.	.19	2.37		2.56	
Minimum Charge	Job		169		169	

Rough Frame / Structure

Beam / Girder		Unit	Material	Labor	Equip.	Total	Specification
6″ x 8″							
	Demolish	L.F.		3.04		3.04	Cost includes material and labor to
	Install	L.F.	6.65	2.89	1.33	10.87	install 6″ x 8″ beam.
	Demolish and Install	L.F.	6.65	5.93	1.33	13.91	
	Clean	L.F.	.15	.30		.45	
	Paint	L.F.	.19	1.15		1.34	
	Minimum Charge	Job		169		169	
6″ x 10″							
	Demolish	L.F.		3.80		3.80	Cost includes material and labor to
	Install	L.F.	4.26	1.35		5.61	install 6″ x 10″ triple beam.
	Demolish and Install	L.F.	4.26	5.15		9.41	
	Clean	L.F.	.18	.35		.53	
	Paint	L.F.	.22	1.32		1.54	
	Minimum Charge	Job		169		169	
6″ x 12″							
	Demolish	L.F.		4.52		4.52	Cost includes material and labor to
	Install	L.F.	5.30	1.42		6.72	install 6″ x 12″ triple beam.
	Demolish and Install	L.F.	5.30	5.94		11.24	
	Clean	L.F.	.20	.39		.59	
	Paint	L.F.	.24	1.48		1.72	
	Minimum Charge	Job		169		169	
Steel I-type							
W8 x 31							
	Demolish	L.F.		7.05	3.18	10.23	Includes material and labor to install
	Install	L.F.	36	2.27	1.45	39.72	W shaped steel beam / girder.
	Demolish and Install	L.F.	36	9.32	4.63	49.95	
	Minimum Charge	Job		415		415	
W8 x 48							
	Demolish	L.F.		7.05	3.18	10.23	Includes material and labor to install
	Install	L.F.	55	2.38	1.52	58.90	W shaped steel beam / girder.
	Demolish and Install	L.F.	55	9.43	4.70	69.13	
	Minimum Charge	Job		415		415	
W8 x 67							
	Demolish	L.F.		7.05	3.18	10.23	Includes material and labor to install
	Install	L.F.	77	2.49	1.60	81.09	W shaped steel beam / girder.
	Demolish and Install	L.F.	77	9.54	4.78	91.32	
	Minimum Charge	Job		415		415	
W10 x 45							
	Demolish	L.F.		7.05	3.18	10.23	Includes material and labor to install
	Install	L.F.	51.50	2.38	1.52	55.40	W shaped steel beam / girder.
	Demolish and Install	L.F.	51.50	9.43	4.70	65.63	
	Minimum Charge	Job		415		415	
W10 x 68							
	Demolish	L.F.		7.05	3.18	10.23	Includes material and labor to install
	Install	L.F.	78	2.49	1.60	82.09	W shaped steel beam / girder.
	Demolish and Install	L.F.	78	9.54	4.78	92.32	
	Minimum Charge	Job		415		415	
W12 x 50							
	Demolish	L.F.		7.05	3.18	10.23	Includes material and labor to install
	Install	L.F.	58	2.38	1.52	61.90	W shaped steel beam / girder.
	Demolish and Install	L.F.	58	9.43	4.70	72.13	
	Minimum Charge	Job		415		415	
W12 x 87							
	Demolish	L.F.		10.35	4.68	15.03	Includes material and labor to install
	Install	L.F.	100	2.49	1.60	104.09	W shaped steel beam / girder.
	Demolish and Install	L.F.	100	12.84	6.28	119.12	
	Minimum Charge	Job		415		415	

Rough Frame / Structure

Beam / Girder

Beam / Girder	Unit	Material	Labor	Equip.	Total	Specification
W12 x 120						
Demolish	L.F.		10.35	4.68	15.03	Includes material and labor to install
Install	L.F.	138	2.55	1.64	142.19	W shaped steel beam / girder.
Demolish and Install	L.F.	138	12.90	6.32	157.22	
Minimum Charge	Job		415		415	
W14 x 74						
Demolish	L.F.		10.35	4.68	15.03	Includes material and labor to install
Install	L.F.	85.50	2.49	1.60	89.59	W shaped steel beam / girder.
Demolish and Install	L.F.	85.50	12.84	6.28	104.62	
Minimum Charge	Job		415		415	
W14 x 120						
Demolish	L.F.		10.35	4.68	15.03	Includes material and labor to install
Install	L.F.	138	2.55	1.64	142.19	W shaped steel beam / girder.
Demolish and Install	L.F.	138	12.90	6.32	157.22	
Minimum Charge	Job		415		415	

Column

Column	Unit	Material	Labor	Equip.	Total	Specification
Concrete						
Small Diameter						
Demolish	L.F.		11.45	1.50	12.95	Includes material and labor to install
Install	L.F.	60.50	10.45	2.59	73.54	small diameter concrete column.
Demolish and Install	L.F.	60.50	21.90	4.09	86.49	
Minimum Charge	Job		525	129	654	
Large Diameter						
Demolish	L.F.		34.50	4.50	39	Includes material and labor to install
Install	L.F.	79	12.30	3.05	94.35	large diameter concrete column.
Demolish and Install	L.F.	79	46.80	7.55	133.35	
Minimum Charge	Job		525	129	654	

Floor Framing System

Floor Framing System	Unit	Material	Labor	Equip.	Total	Specification
12" O.C.						
2" x 6" Joists						
Demolish	S.F.		.55		.55	Cost includes material and labor to
Install	S.F.	.76	.27		1.03	install 2" x 6" joists, 12" O.C.
Demolish and Install	S.F.	.76	.82		1.58	including box or band joist. Blocking
Reinstall	S.F.		.22		.22	or bridging not included.
Minimum Charge	Job		169		169	
2" x 8" Joists						
Demolish	S.F.		.57		.57	Cost includes material and labor to
Install	S.F.	1.10	.31		1.41	install 2" x 8" joists, 12" O.C.
Demolish and Install	S.F.	1.10	.88		1.98	including box or band joist. Blocking
Reinstall	S.F.		.25		.25	or bridging not included.
Minimum Charge	Job		169		169	
2" x 10" Joists						
Demolish	S.F.		.59		.59	Cost includes material and labor to
Install	S.F.	1.66	.38		2.04	install 2" x 10" joists, 12" O.C.
Demolish and Install	S.F.	1.66	.97		2.63	including box or band joist. Blocking
Reinstall	S.F.		.30		.30	or bridging not included.
Minimum Charge	Job		169		169	

Rough Frame / Structure

Floor Framing System		Unit	Material	Labor	Equip.	Total	Specification
2″ x 12″ Joists							
	Demolish	S.F.		.61		.61	Cost includes material and labor to
	Install	S.F.	2.06	.39		2.45	install 2″ x 12″ joists, 12″ O.C.
	Demolish and Install	S.F.	2.06	1		3.06	including box or band joist. Blocking
	Reinstall	S.F.		.31		.31	or bridging not included.
	Minimum Charge	Job		169		169	
Block / Bridge							
	Demolish	Ea.		.84		.84	Includes material and labor to install
	Install	Ea.	1.90	1.90		3.80	set of cross bridging or per block of
	Demolish and Install	Ea.	1.90	2.74		4.64	solid bridging for 2″ x 10″ joists cut to
	Minimum Charge	Job		169		169	size on site.
16″ O.C.							
2″ x 6″ Joists							
	Demolish	S.F.		.41		.41	Cost includes material and labor to
	Install	S.F.	.58	.40		.98	install 2″ x 6″ joists including box or
	Demolish and Install	S.F.	.58	.81		1.39	band joist installed 16″ O.C. Does not
	Reinstall	S.F.		.32		.32	include beams, blocking or bridging.
	Minimum Charge	Job		169		169	
2″ x 8″ Joists							
	Demolish	S.F.		.43		.43	Cost includes material and labor to
	Install	S.F.	.85	.46		1.31	install 2″ x 8″ joists including box or
	Demolish and Install	S.F.	.85	.89		1.74	band joist installed 16″ O.C. Does not
	Reinstall	S.F.		.37		.37	include beams, blocking or bridging.
	Minimum Charge	Job		169		169	
2″ x 10″ Joists							
	Demolish	S.F.		.44		.44	Cost includes material and labor to
	Install	S.F.	1.28	.56		1.84	install 2″ x 10″ joists including box or
	Demolish and Install	S.F.	1.28	1		2.28	band joist 16″ O.C. Does not include
	Reinstall	S.F.		.45		.45	beams, blocking or bridging.
	Minimum Charge	Job		169		169	
2″ x 12″ Joists							
	Demolish	S.F.		.46		.46	Cost includes material and labor to
	Install	S.F.	1.58	.58		2.16	install 2″ x 12″ joists including box or
	Demolish and Install	S.F.	1.58	1.04		2.62	band joist installed 16″ O.C. Does not
	Reinstall	S.F.		.46		.46	include beams, blocking or bridging.
	Minimum Charge	Job		169		169	
Block / Bridge							
	Demolish	Ea.		.84		.84	Includes material and labor to install
	Install	Ea.	1.90	1.90		3.80	set of cross bridging or per block of
	Demolish and Install	Ea.	1.90	2.74		4.64	solid bridging for 2″ x 10″ joists cut to
	Minimum Charge	Job		169		169	size on site.
24″ O.C.							
2″ x 6″ Joists							
	Demolish	S.F.		.28		.28	Cost includes material and labor to
	Install	S.F.	.42	.13		.55	install 2″ x 6″ joists including box or
	Demolish and Install	S.F.	.42	.41		.83	band joist installed 24″ O.C. Does not
	Reinstall	S.F.		.11		.11	include beams, blocking or bridging.
	Minimum Charge	Job		169		169	
2″ x 8″ Joists							
	Demolish	S.F.		.29		.29	Cost includes material and labor to
	Install	S.F.	.61	.15		.76	install 2″ x 8″ joists including box or
	Demolish and Install	S.F.	.61	.44		1.05	band joist installed 24″ O.C. Does not
	Reinstall	S.F.		.12		.12	include beams, blocking or bridging.
	Minimum Charge	Job		169		169	
2″ x 10″ Joists							
	Demolish	S.F.		.29		.29	Cost includes material and labor to
	Install	S.F.	.91	.19		1.10	install 2″ x 10″ joists including box or
	Demolish and Install	S.F.	.91	.48		1.39	band joist installed 24″ O.C. Does not
	Reinstall	S.F.		.15		.15	include beams, blocking or bridging.
	Minimum Charge	Job		169		169	

Floor Framing System	Unit	Material	Labor	Equip.	Total	Specification
2" x 12" Joists						
Demolish	S.F.		.30		.30	Cost includes material and labor to
Install	S.F.	1.12	.19		1.31	install 2" x 12" joists including box or
Demolish and Install	S.F.	1.12	.49		1.61	band joist installed 24" O.C. Does not
Reinstall	S.F.		.15		.15	include beams, blocking or bridging.
Minimum Charge	Job		169		169	
Block / Bridge						
Demolish	Ea.		.84		.84	Includes material and labor to install
Install	Ea.	1.90	1.90		3.80	set of cross bridging or per block of
Demolish and Install	Ea.	1.90	2.74		4.64	solid bridging for 2" x 10" joists cut to
Minimum Charge	Job		169		169	size on site.
Engineered Lumber, Joist						
9-1/2"						
Demolish	S.F.		.44		.44	Includes material and labor to install
Install	S.F.	1.79	.34		2.13	engineered lumber truss / joists per
Demolish and Install	S.F.	1.79	.78		2.57	S.F. of floor area based on joists 16"
Reinstall	S.F.		.27		.27	O.C. Beams, supports and bridging
Minimum Charge	Job		169		169	are not included.
11-7/8"						
Demolish	S.F.		.46		.46	Includes material and labor to install
Install	S.F.	1.94	.35		2.29	engineered lumber truss / joists per
Demolish and Install	S.F.	1.94	.81		2.75	S.F. of floor area based on joists 16"
Reinstall	S.F.		.28		.28	O.C. Beams, supports and bridging
Minimum Charge	Job		169		169	are not included.
14"						
Demolish	S.F.		.47		.47	Includes material and labor to install
Install	S.F.	2.23	.37		2.60	engineered lumber truss / joists per
Demolish and Install	S.F.	2.23	.84		3.07	S.F. of floor area based on joists 16"
Reinstall	S.F.		.30		.30	O.C. Beams, supports and bridging
Minimum Charge	Job		169		169	are not included.
16"						
Demolish	S.F.		.49		.49	Includes material and labor to install
Install	S.F.	2.52	.39		2.91	engineered lumber truss / joists per
Demolish and Install	S.F.	2.52	.88		3.40	S.F. of floor area based on joists 16"
Reinstall	S.F.		.31		.31	O.C. Beams, supports and bridging
Minimum Charge	Job		169		169	are not included.
Block / Bridge						
Demolish	Pr.		.84		.84	Includes material and labor to install
Install	Pr.	1.30	2.50		3.80	set of steel, one-nail type cross
Demolish and Install	Pr.	1.30	3.34		4.64	bridging for trusses placed 16" O.C.
Minimum Charge	Job		169		169	
Block / Bridge						
2" x 6"						
Demolish	Ea.		.84		.84	Includes material and labor to install
Install	Ea.	.87	1.52		2.39	set of cross bridging or per block of
Demolish and Install	Ea.	.87	2.36		3.23	solid bridging for 2" x 6" joists cut to
Minimum Charge	Job		169		169	size on site.
2" x 8"						
Demolish	Ea.		.84		.84	Includes material and labor to install
Install	Ea.	1.25	1.69		2.94	set of cross bridging or per block of
Demolish and Install	Ea.	1.25	2.53		3.78	solid bridging for 2" x 8" joists cut to
Minimum Charge	Job		169		169	size on site.

Rough Frame / Structure

Floor Framing System		Unit	Material	Labor	Equip.	Total	Specification
2″ x 10″							
	Demolish	Ea.		.84		.84	Includes material and labor to install
	Install	Ea.	1.90	1.90		3.80	set of cross bridging or per block of
	Demolish and Install	Ea.	1.90	2.74		4.64	solid bridging for 2″ x 10″ joists cut to
	Minimum Charge	Job		169		169	size on site.
2″ x 12″							
	Demolish	Ea.		.84		.84	Includes material and labor to install
	Install	Ea.	2.35	2.24		4.59	set of cross bridging or per block of
	Demolish and Install	Ea.	2.35	3.08		5.43	solid bridging for 2″ x 12″ joists cut to
	Minimum Charge	Job		169		169	size on site.
Ledger Strips							
1″ x 2″							
	Demolish	L.F.		.22		.22	Cost includes material and labor to
	Install	L.F.	.21	.98		1.19	install 1″ x 2″ ledger strip nailed to the
	Demolish and Install	L.F.	.21	1.20		1.41	face of studs, beams or joist.
	Minimum Charge	Job		169		169	
1″ x 3″							
	Demolish	L.F.		.22		.22	Includes material and labor to install
	Install	L.F.	.51	1.13		1.64	up to 1″ x 4″ ledger strip nailed to the
	Demolish and Install	L.F.	.51	1.35		1.86	face of studs, beams or joist.
	Minimum Charge	Job		169		169	
1″ x 4″							
	Demolish	L.F.		.22		.22	Includes material and labor to install
	Install	L.F.	.51	1.13		1.64	up to 1″ x 4″ ledger strip nailed to the
	Demolish and Install	L.F.	.51	1.35		1.86	face of studs, beams or joist.
	Minimum Charge	Job		169		169	
2″ x 2″							
	Demolish	L.F.		.24		.24	Cost includes material and labor to
	Install	L.F.	.42	1.02		1.44	install 2″ x 2″ ledger strip nailed to the
	Demolish and Install	L.F.	.42	1.26		1.68	face of studs, beams or joist.
	Minimum Charge	Job		169		169	
2″ x 4″							
	Demolish	L.F.		.27		.27	Cost includes material and labor to
	Install	L.F.	.41	1.35		1.76	install 2″ x 4″ ledger strip nailed to the
	Demolish and Install	L.F.	.41	1.62		2.03	face of studs, beams or joist.
	Reinstall	L.F.		1.08		1.08	
	Minimum Charge	Job		169		169	
Ledger Boards							
2″ x 6″							
	Demolish	L.F.		.27		.27	Cost includes material and labor to
	Install	L.F.	1.85	2.11		3.96	install 2″ x 6″ ledger board fastened to
	Demolish and Install	L.F.	1.85	2.38		4.23	a wall, joists or studs.
	Reinstall	L.F.		1.69		1.69	
	Minimum Charge	Job		169		169	
4″ x 6″							
	Demolish	L.F.		.45		.45	Cost includes material and labor to
	Install	L.F.	3.55	2.11		5.66	install 4″ x 6″ ledger board fastened to
	Demolish and Install	L.F.	3.55	2.56		6.11	a wall, joists or studs.
	Reinstall	L.F.		1.69		1.69	
	Minimum Charge	Job		169		169	
4″ x 8″							
	Demolish	L.F.		.60		.60	Cost includes material and labor to
	Install	L.F.	4.35	2.11		6.46	install 4″ x 8″ ledger board fastened to
	Demolish and Install	L.F.	4.35	2.71		7.06	a wall, joists or studs.
	Reinstall	L.F.		1.69		1.69	
	Minimum Charge	Job		169		169	

Rough Frame / Structure

Floor Framing System

		Unit	Material	Labor	Equip.	Total	Specification
Sill Plate Per L.F.							
2″ x 4″							
	Demolish	L.F.		.23		.23	Cost includes material and labor to
	Install	L.F.	.54	1.23		1.77	install 2″ x 4″ pressure treated lumber,
	Demolish and Install	L.F.	.54	1.46		2	drilled and installed on foundation
	Reinstall	L.F.		.98		.98	bolts 48″ O.C. Bolts, nuts and washers
	Minimum Charge	Job		169		169	are not included.
2″ x 6″							
	Demolish	L.F.		.34		.34	Cost includes material and labor to
	Install	L.F.	.83	1.35		2.18	install 2″ x 6″ pressure treated lumber,
	Demolish and Install	L.F.	.83	1.69		2.52	drilled and installed on foundation
	Reinstall	L.F.		1.08		1.08	bolts 48″ O.C. Bolts, nuts and washers
	Minimum Charge	Job		169		169	not included.
2″ x 8″							
	Demolish	L.F.		.46		.46	Cost includes material and labor to
	Install	L.F.	1.08	1.50		2.58	install 2″ x 8″ pressure treated lumber,
	Demolish and Install	L.F.	1.08	1.96		3.04	drilled and installed on foundation
	Reinstall	L.F.		1.20		1.20	bolts 48″ O.C. Bolts, nuts and washers
	Minimum Charge	Job		169		169	not included.
Earthquake Strapping							
	Install	Ea.	2.17	2.11		4.28	Includes labor and material to replace
	Minimum Charge	Job		169		169	an earthquake strap.

Subflooring

		Unit	Material	Labor	Equip.	Total	Specification
Plywood							
1/2″							
	Demolish	S.F.		.70		.70	Cost includes material and labor to
	Install	SF Flr.	.51	.36		.87	install 1/2″ plywood subfloor, CD
	Demolish and Install	SF Flr.	.51	1.06		1.57	standard interior grade, plugged and
	Minimum Charge	Job		169		169	touch sanded.
5/8″							
	Demolish	S.F.		.71		.71	Cost includes material and labor to
	Install	SF Flr.	.55	.40		.95	install 5/8″ plywood subfloor, CD
	Demolish and Install	SF Flr.	.55	1.11		1.66	standard interior grade, plugged and
	Reinstall	SF Flr.		.32		.32	touch sanded.
	Minimum Charge	Job		169		169	
3/4″							
	Demolish	S.F.		.72		.72	Cost includes material and labor to
	Install	SF Flr.	.69	.44		1.13	install 3/4″ plywood subfloor, CD
	Demolish and Install	SF Flr.	.69	1.16		1.85	standard interior grade, plugged and
	Reinstall	SF Flr.		.35		.35	touch sanded.
	Minimum Charge	Job		169		169	
Particle Board							
1/2″							
	Demolish	S.F.		.35		.35	Cost includes material and labor to
	Install	SF Flr.	.41	.45		.86	install 1/2″ particle board subfloor.
	Demolish and Install	SF Flr.	.41	.80		1.21	
	Minimum Charge	Job		169		169	
5/8″							
	Demolish	S.F.		.35		.35	Cost includes material and labor to
	Install	S.F.	.45	.49		.94	install 5/8″ particle board subfloor.
	Demolish and Install	S.F.	.45	.84		1.29	
	Minimum Charge	Job		169		169	

Rough Frame / Structure

Subflooring		Unit	Material	Labor	Equip.	Total	Specification
3/4"							
	Demolish	S.F.		.36		.36	Cost includes material and labor to
	Install	SF Flr.	.62	.54		1.16	install 3/4" particle board subfloor.
	Demolish and Install	SF Flr.	.62	.90		1.52	
	Minimum Charge	Job		169		169	
Plank Board							
	Demolish	S.F.		1.16		1.16	Cost includes material and labor to
	Install	S.F.	1.32	.38		1.70	install 1" x 6" standard grade plank
	Demolish and Install	S.F.	1.32	1.54		2.86	flooring.
	Reinstall	S.F.		.38		.38	
	Minimum Charge	Job		169		169	
Felt Underlay							
	Demolish	S.F.		.06		.06	Includes material and labor to install
	Install	S.F.	.03	.09		.12	#15 felt building paper.
	Demolish and Install	S.F.	.03	.15		.18	
	Minimum Charge	Job		169		169	
Prep (for flooring)							
	Install	S.F.		.29		.29	Preparation of subflooring for
	Minimum Charge	Job		154		154	installation of new finished flooring. Cost reflects average time to prep area.

Wall Framing System		Unit	Material	Labor	Equip.	Total	Specification
2" x 4"							
8' High							
	Demolish	L.F.		3.58		3.58	Cost includes material and labor to
	Install	L.F.	4.21	3.98		8.19	install 2" x 4" x 8' high wall system,
	Demolish and Install	L.F.	4.21	7.56		11.77	including studs, treated bottom plate,
	Reinstall	L.F.		3.18		3.18	double top plate and one row of fire
	Minimum Charge	Job		169		169	blocking.
9' High							
	Demolish	L.F.		4.01		4.01	Cost includes material and labor to
	Install	L.F.	4.52	3.98		8.50	install 2" x 4" x 9' wall system
	Demolish and Install	L.F.	4.52	7.99		12.51	including studs, treated bottom plate,
	Reinstall	L.F.		3.18		3.18	double top plate and one row of fire
	Minimum Charge	Job		169		169	blocking.
10' High							
	Demolish	L.F.		4.47		4.47	Cost includes material and labor to
	Install	L.F.	4.83	3.98		8.81	install 2" x 4" x 10' wall system
	Demolish and Install	L.F.	4.83	8.45		13.28	including studs, treated bottom plate,
	Reinstall	L.F.		3.18		3.18	double top plate and one row of fire
	Minimum Charge	Job		169		169	blocking.
12' High							
	Demolish	L.F.		5.35		5.35	Cost includes material and labor to
	Install	L.F.	5.45	4.83		10.28	install 2" x 4" x 12' wall system
	Demolish and Install	L.F.	5.45	10.18		15.63	including studs, treated bottom plate,
	Reinstall	L.F.		3.86		3.86	double top plate and one row of fire
	Minimum Charge	Job		169		169	blocking.

Rough Frame / Structure

Wall Framing System

Wall Framing System	Unit	Material	Labor	Equip.	Total	Specification
2" x 6"						
8' High						
Demolish	L.F.		3.36		3.36	Cost includes material and labor to
Install	L.F.	6.70	7.30		14	install 2" x 6" x 8' wall system
Demolish and Install	L.F.	6.70	10.66		17.36	including studs, treated bottom plate,
Reinstall	L.F.		5.84		5.84	double top plate and one row of fire
Minimum Charge	Job		169		169	blocking.
9' High						
Demolish	L.F.		3.78		3.78	Cost includes material and labor to
Install	L.F.	7.15	7.30		14.45	install 2" x 6" x 9' wall system
Demolish and Install	L.F.	7.15	11.08		18.23	including studs, treated bottom plate,
Reinstall	L.F.		5.84		5.84	double top plate and one row of fire
Minimum Charge	Job		169		169	blocking.
10' High						
Demolish	L.F.		4.13		4.13	Cost includes material and labor to
Install	L.F.	7.65	7.30		14.95	install 2" x 6" x 10' wall system
Demolish and Install	L.F.	7.65	11.43		19.08	including studs, treated bottom plate,
Reinstall	L.F.		5.84		5.84	double top plate and one row of fire
Minimum Charge	Job		169		169	blocking.
12' High						
Demolish	L.F.		4.88		4.88	Cost includes material and labor to
Install	L.F.	8.65	9		17.65	install 2" x 6" x 12' wall system
Demolish and Install	L.F.	8.65	13.88		22.53	including studs, treated bottom plate,
Reinstall	L.F.		7.19		7.19	double top plate and one row of fire
Minimum Charge	Job		169		169	blocking.
Fireblock						
2" x 4", 16" O.C. System						
Demolish	L.F.		.45		.45	Cost includes material and labor to
Install	L.F.	.41	1.13		1.54	install 2" x 4" fireblocks in wood frame
Demolish and Install	L.F.	.41	1.58		1.99	walls per L.F. of wall to be blocked.
Reinstall	L.F.		.90		.90	
Minimum Charge	Job		169		169	
2" x 4", 24" O.C. System						
Demolish	L.F.		.45		.45	Cost includes material and labor to
Install	L.F.	.41	1.13		1.54	install 2" x 4" fireblocks in wood frame
Demolish and Install	L.F.	.41	1.58		1.99	walls per L.F. of wall to be blocked.
Reinstall	L.F.		.90		.90	
Minimum Charge	Job		169		169	
2" x 6", 16" O.C. System						
Demolish	L.F.		.67		.67	Cost includes material and labor to
Install	L.F.	.65	1.13		1.78	install 2" x 6" fireblocks in wood frame
Demolish and Install	L.F.	.65	1.80		2.45	walls per L.F. of wall to be blocked.
Reinstall	L.F.		.90		.90	
Minimum Charge	Job		169		169	
2" x 6", 24" O.C. System						
Demolish	L.F.		.67		.67	Cost includes material and labor to
Install	L.F.	.65	1.13		1.78	install 2" x 6" fireblocks in wood frame
Demolish and Install	L.F.	.65	1.80		2.45	walls per L.F. of wall to be blocked.
Reinstall	L.F.		.90		.90	
Minimum Charge	Job		169		169	
Bracing						
1" x 3"						
Demolish	L.F.		.26		.26	Cost includes material and labor to
Install	L.F.	.31	2.11		2.42	install 1" x 3" let-in bracing.
Demolish and Install	L.F.	.31	2.37		2.68	
Minimum Charge	Job		169		169	

Wall Framing System		Unit	Material	Labor	Equip.	Total	Specification
1″ x 4″							
	Demolish	L.F.		.26		.26	Cost includes material and labor to
	Install	L.F.	.39	2.11		2.50	install 1″ x 4″ let-in bracing.
	Demolish and Install	L.F.	.39	2.37		2.76	
	Minimum Charge	Job		169		169	
1″ x 6″							
	Demolish	L.F.		.26		.26	Cost includes material and labor to
	Install	L.F.	.61	2.25		2.86	install 1″ x 6″ let-in bracing.
	Demolish and Install	L.F.	.61	2.51		3.12	
	Minimum Charge	Job		169		169	
2″ x 3″							
	Demolish	L.F.		.34		.34	Cost includes material and labor to
	Install	L.F.	.42	2.25		2.67	install 2″ x 3″ let-in bracing.
	Demolish and Install	L.F.	.42	2.59		3.01	
	Reinstall	L.F.		1.80		1.80	
	Minimum Charge	Job		169		169	
2″ x 4″							
	Demolish	L.F.		.34		.34	Cost includes material and labor to
	Install	L.F.	.41	2.25		2.66	install 2″ x 4″ let-in bracing.
	Demolish and Install	L.F.	.41	2.59		3	
	Reinstall	L.F.		1.80		1.80	
	Minimum Charge	Job		169		169	
2″ x 6″							
	Demolish	L.F.		.34		.34	Cost includes material and labor to
	Install	L.F.	.65	2.41		3.06	install 2″ x 6″ let-in bracing.
	Demolish and Install	L.F.	.65	2.75		3.40	
	Reinstall	L.F.		1.93		1.93	
	Minimum Charge	Job		169		169	
2″ x 8″							
	Demolish	L.F.		.34		.34	Cost includes material and labor to
	Install	L.F.	.94	2.41		3.35	install 2″ x 8″ let-in bracing.
	Demolish and Install	L.F.	.94	2.75		3.69	
	Reinstall	L.F.		1.93		1.93	
	Minimum Charge	Job		169		169	
Earthquake Strapping							
	Install	Ea.	2.17	2.11		4.28	Includes labor and material to replace
	Minimum Charge	Job		169		169	an earthquake strap.
Hurricane Clips							
	Install	Ea.	1.90	2.33		4.23	Includes labor and material to install a
	Minimum Charge	Job		169		169	hurricane clip.
Stud							
2″ x 4″							
	Demolish	L.F.		.27		.27	Cost includes material and labor to
	Install	L.F.	.41	.61		1.02	install 2″ x 4″ wall stud per L.F. of stud.
	Demolish and Install	L.F.	.41	.88		1.29	
	Reinstall	L.F.		.49		.49	
	Minimum Charge	Job		169		169	
2″ x 6″							
	Demolish	L.F.		.34		.34	Cost includes material and labor to
	Install	L.F.	.65	.68		1.33	install 2″ x 6″ stud per L.F. of stud.
	Demolish and Install	L.F.	.65	1.02		1.67	
	Reinstall	L.F.		.54		.54	
	Minimum Charge	Job		169		169	

Rough Frame / Structure

Wall Framing System

	Unit	Material	Labor	Equip.	Total	Specification
Plates						
2" x 4"						
Demolish	L.F.		.24		.24	Cost includes material and labor to
Install	L.F.	.41	.85		1.26	install 2" x 4" plate per L.F. of plate.
Demolish and Install	L.F.	.41	1.09		1.50	
Reinstall	L.F.		.68		.68	
Minimum Charge	Job		169		169	
2" x 6"						
Demolish	L.F.		.26		.26	Cost includes material and labor to
Install	L.F.	.65	.90		1.55	install 2" x 6" plate per L.F. of plate.
Demolish and Install	L.F.	.65	1.16		1.81	
Reinstall	L.F.		.72		.72	
Minimum Charge	Job		169		169	
Headers						
Demolish	L.F.		.24		.24	Includes material and labor to install
Install	L.F.	.65	1.88		2.53	header over wall openings and
Demolish and Install	L.F.	.65	2.12		2.77	around floor, ceiling and roof
Clean	L.F.	.01	.30		.31	openings or flush beam.
Paint	L.F.	.09	1.19		1.28	
Minimum Charge	Job		169		169	
2" x 8"						
Demolish	L.F.		.26		.26	Includes material and labor to install
Install	L.F.	.94	1.99		2.93	header over wall openings and
Demolish and Install	L.F.	.94	2.25		3.19	around floor, ceiling and roof
Clean	L.F.	.02	.31		.33	openings or flush beam.
Paint	L.F.	.12	1.60		1.72	
Minimum Charge	Job		169		169	
2" x 10"						
Demolish	L.F.		.26		.26	Includes material and labor to install
Install	L.F.	1.42	2.11		3.53	header over wall openings and
Demolish and Install	L.F.	1.42	2.37		3.79	around floor, ceiling and roof
Clean	L.F.	.02	.31		.33	openings or flush beam.
Paint	L.F.	.15	1.98		2.13	
Minimum Charge	Job		169		169	
2" x 12"						
Demolish	L.F.		.27		.27	Includes material and labor to install
Install	L.F.	1.76	2.25		4.01	header over wall openings and
Demolish and Install	L.F.	1.76	2.52		4.28	around floor, ceiling and roof
Clean	L.F.	.03	.31		.34	openings or flush beam.
Paint	L.F.	.19	2.37		2.56	
Minimum Charge	Job		169		169	
4" x 8"						
Demolish	L.F.		.48		.48	Includes material and labor to install
Install	L.F.	2.85	2.60		5.45	header over wall openings and
Demolish and Install	L.F.	2.85	3.08		5.93	around floor, ceiling and roof
Clean	L.F.	.02	.31		.33	openings or flush beam.
Paint	L.F.	.12	1.60		1.72	
Minimum Charge	Job		169		169	
4" x 10"						
Demolish	L.F.		.51		.51	Includes material and labor to install
Install	L.F.	3.59	2.82		6.41	header over wall openings and
Demolish and Install	L.F.	3.59	3.33		6.92	around floor, ceiling and roof
Clean	L.F.	.02	.31		.33	openings or flush beam.
Paint	L.F.	.15	1.98		2.13	
Minimum Charge	Job		169		169	

Rough Frame / Structure

Wall Framing System

Wall Framing System	Unit	Material	Labor	Equip.	Total	Specification
4" x 12"						
Demolish	L.F.		.54		.54	Includes material and labor to install
Install	L.F.	4.44	3.56		8	header over wall openings and
Demolish and Install	L.F.	4.44	4.10		8.54	around floor, ceiling and roof
Clean	L.F.	.03	.31		.34	openings or flush beam.
Paint	L.F.	.19	2.37		2.56	
Minimum Charge	Job		169		169	
6" x 8"						
Demolish	L.F.		.48		.48	Includes material and labor to install
Install	L.F.	6.65	1.88		8.53	header over wall openings and
Demolish and Install	L.F.	6.65	2.36		9.01	around floor, ceiling and roof
Clean	L.F.	.02	.31		.33	openings or flush beam.
Paint	L.F.	.12	1.60		1.72	
Minimum Charge	Job		169		169	
6" x 10"						
Demolish	L.F.		.51		.51	Includes material and labor to install
Install	L.F.	12	4.10		16.10	header over wall openings and
Demolish and Install	L.F.	12	4.61		16.61	around floor, ceiling and roof
Clean	L.F.	.02	.31		.33	openings or flush beam.
Paint	L.F.	.15	1.98		2.13	
Minimum Charge	Job		169		169	
6" x 12"						
Demolish	L.F.		.54		.54	Includes material and labor to install
Install	L.F.	10.55	4.83		15.38	header over wall openings and
Demolish and Install	L.F.	10.55	5.37		15.92	around floor, ceiling and roof
Clean	L.F.	.03	.31		.34	openings or flush beam.
Paint	L.F.	.19	.82		1.01	
Minimum Charge	Job		169		169	

Rough-in Opening

Rough-in Opening	Unit	Material	Labor	Equip.	Total	Specification
Door w / 2" x 4" Lumber						
3' Wide						
Demolish	Ea.		8.40		8.40	Includes material and labor to install
Install	Ea.	17.35	10.55		27.90	header, double studs each side,
Demolish and Install	Ea.	17.35	18.95		36.30	cripples, blocking and nails, up to 3'
Reinstall	Ea.		8.45		8.45	opening in 2" x 4" stud wall 8' high.
Minimum Charge	Job		169		169	
4' Wide						
Demolish	Ea.		8.40		8.40	Includes material and labor to install
Install	Ea.	18.65	10.55		29.20	header, double studs each side,
Demolish and Install	Ea.	18.65	18.95		37.60	cripples, blocking and nails, up to 4'
Reinstall	Ea.		8.45		8.45	opening in 2" x 4" stud wall 8' high.
Minimum Charge	Job		169		169	
5' Wide						
Demolish	Ea.		8.40		8.40	Includes material and labor to install
Install	Ea.	23	10.55		33.55	header, double studs each side,
Demolish and Install	Ea.	23	18.95		41.95	cripples, blocking and nails, up to 5'
Reinstall	Ea.		8.45		8.45	opening in 2" x 4" stud wall 8' high.
Minimum Charge	Job		169		169	
6' Wide						
Demolish	Ea.		8.40		8.40	Includes material and labor to install
Install	Ea.	25	10.55		35.55	header, double studs each side,
Demolish and Install	Ea.	25	18.95		43.95	cripples, blocking and nails, up to 6'
Reinstall	Ea.		8.45		8.45	opening in 2" x 4" stud wall 8' high.
Minimum Charge	Job		169		169	

Rough Frame / Structure

Rough-in Opening

	Unit	Material	Labor	Equip.	Total	Specification
8' Wide						
Demolish	Ea.		8.95		8.95	Includes material and labor to install
Install	Ea.	36.50	11.25		47.75	header, double studs each side,
Demolish and Install	Ea.	36.50	20.20		56.70	cripples, blocking and nails, up to 8'
Reinstall	Ea.		9.01		9.01	opening in 2" x 4" stud wall 8' high.
Minimum Charge	Job		169		169	
10' Wide						
Demolish	Ea.		8.95		8.95	Includes material and labor to install
Install	Ea.	49	11.25		60.25	header, double studs each side,
Demolish and Install	Ea.	49	20.20		69.20	cripples, blocking and nails, up to 10'
Reinstall	Ea.		9.01		9.01	opening in 2" x 4" stud wall 8' high.
Minimum Charge	Job		169		169	
12' Wide						
Demolish	Ea.		8.95		8.95	Includes material and labor to install
Install	Ea.	72	11.25		83.25	header, double studs each side,
Demolish and Install	Ea.	72	20.20		92.20	cripples, blocking and nails, up to 12'
Reinstall	Ea.		9.01		9.01	opening in 2" x 4" stud wall 8' high.
Minimum Charge	Job		169		169	

Door w / 2" x 6" Lumber

	Unit	Material	Labor	Equip.	Total	Specification
3' Wide						
Demolish	Ea.		8.40		8.40	Includes material and labor to install
Install	Ea.	25.50	10.55		36.05	header, double studs each side,
Demolish and Install	Ea.	25.50	18.95		44.45	cripples, blocking and nails, up to 3'
Reinstall	Ea.		8.45		8.45	opening in 2" x 6" stud wall 8' high.
Minimum Charge	Job		169		169	
4' Wide						
Demolish	Ea.		8.40		8.40	Includes material and labor to install
Install	Ea.	26.50	10.55		37.05	header, double studs each side,
Demolish and Install	Ea.	26.50	18.95		45.45	cripples, blocking and nails, up to 4'
Reinstall	Ea.		8.45		8.45	opening in 2" x 6" stud wall 8' high.
Minimum Charge	Job		169		169	
5' Wide						
Demolish	Ea.		8.40		8.40	Includes material and labor to install
Install	Ea.	31	10.55		41.55	header, double studs each side,
Demolish and Install	Ea.	31	18.95		49.95	cripples, blocking and nails up to 5'
Reinstall	Ea.		8.45		8.45	opening in 2" x 6" stud wall 8' high.
Minimum Charge	Job		169		169	
6' Wide						
Demolish	Ea.		8.40		8.40	Includes material and labor to install
Install	Ea.	32.50	10.55		43.05	header, double studs each side,
Demolish and Install	Ea.	32.50	18.95		51.45	cripples, blocking and nails up to 6'
Reinstall	Ea.		8.45		8.45	opening in 2" x 6" stud wall 8' high.
Minimum Charge	Job		169		169	
8' Wide						
Demolish	Ea.		8.95		8.95	Includes material and labor to install
Install	Ea.	44	11.25		55.25	header, double studs each side,
Demolish and Install	Ea.	44	20.20		64.20	cripples, blocking and nails up to 8'
Reinstall	Ea.		9.01		9.01	opening in 2" x 6" stud wall 8' high.
Minimum Charge	Job		169		169	
10' Wide						
Demolish	Ea.		8.95		8.95	Includes material and labor to install
Install	Ea.	56.50	11.25		67.75	header, double studs each side,
Demolish and Install	Ea.	56.50	20.20		76.70	cripples, blocking and nails up to 10'
Reinstall	Ea.		9.01		9.01	opening in 2" x 6" stud wall 8' high.
Minimum Charge	Job		169		169	
12' Wide						
Demolish	Ea.		8.95		8.95	Includes material and labor to install
Install	Ea.	80	11.25		91.25	header, double studs each side,
Demolish and Install	Ea.	80	20.20		100.20	cripples, blocking and nails, up to 12'
Reinstall	Ea.		9.01		9.01	opening in 2" x 6" stud wall 8' high.
Minimum Charge	Job		169		169	

Rough Frame / Structure

Rough-in Opening	Unit	Material	Labor	Equip.	Total	Specification
Window w / 2″ x 4″ Lumber						
2′ Wide						
Demolish	Ea.		11.20		11.20	Includes material and labor to install
Install	Ea.	18.20	14.10		32.30	header, double studs each side of
Demolish and Install	Ea.	18.20	25.30		43.50	opening, cripples, blocking, nails and
Reinstall	Ea.		11.27		11.27	sub-sills for opening up to 2′ in a 2″ x
Minimum Charge	Job		169		169	4″ stud wall 8′ high.
3′ Wide						
Demolish	Ea.		11.20		11.20	Includes material and labor to install
Install	Ea.	21.50	14.10		35.60	header, double studs each side of
Demolish and Install	Ea.	21.50	25.30		46.80	opening, cripples, blocking, nails and
Reinstall	Ea.		11.27		11.27	sub-sills for opening up to 3′ in a 2″ x
Minimum Charge	Job		169		169	4″ stud wall 8′ high.
4′ Wide						
Demolish	Ea.		11.20		11.20	Includes material and labor to install
Install	Ea.	23.50	14.10		37.60	header, double studs each side of
Demolish and Install	Ea.	23.50	25.30		48.80	opening, cripples, blocking, nails and
Reinstall	Ea.		11.27		11.27	sub-sills for opening up to 4′ in a 2″ x
Minimum Charge	Job		169		169	4″ stud wall 8′ high.
5′ Wide						
Demolish	Ea.		11.20		11.20	Includes material and labor to install
Install	Ea.	28	14.10		42.10	header, double studs each side of
Demolish and Install	Ea.	28	25.30		53.30	opening, cripples, blocking, nails and
Reinstall	Ea.		11.27		11.27	sub-sills for opening up to 5′ in a 2″ x
Minimum Charge	Job		169		169	4″ stud wall 8′ high.
6′ Wide						
Demolish	Ea.		11.20		11.20	Includes material and labor to install
Install	Ea.	31	14.10		45.10	header, double studs each side of
Demolish and Install	Ea.	31	25.30		56.30	opening, cripples, blocking, nails and
Reinstall	Ea.		11.27		11.27	sub-sills for opening up to 6′ in a 2″ x
Minimum Charge	Job		169		169	4″ stud wall 8′ high.
7′ Wide						
Demolish	Ea.		11.20		11.20	Includes material and labor to install
Install	Ea.	41.50	14.10		55.60	header, double studs each side of
Demolish and Install	Ea.	41.50	25.30		66.80	opening, cripples, blocking, nails and
Reinstall	Ea.		11.27		11.27	sub-sills for opening up to 7′ in a 2″ x
Minimum Charge	Job		169		169	4″ stud wall 8′ high.
8′ Wide						
Demolish	Ea.		12.20		12.20	Includes material and labor to install
Install	Ea.	45	15.35		60.35	header, double studs each side of
Demolish and Install	Ea.	45	27.55		72.55	opening, cripples, blocking, nails and
Reinstall	Ea.		12.29		12.29	sub-sills for opening up to 8′ in a 2″ x
Minimum Charge	Job		169		169	4″ stud wall 8′ high.
10′ Wide						
Demolish	Ea.		12.20		12.20	Includes material and labor to install
Install	Ea.	59	15.35		74.35	header, double studs each side of
Demolish and Install	Ea.	59	27.55		86.55	opening, cripples, blocking, nails and
Reinstall	Ea.		12.29		12.29	sub-sills for opening up to 10′ in a 2″
Minimum Charge	Job		169		169	x 4″ stud wall 8′ high.

Rough-in Opening		Unit	Material	Labor	Equip.	Total	Specification
12' Wide							
	Demolish	Ea.		12.20		12.20	Includes material and labor to install
	Install	Ea.	84	15.35		99.35	header, double studs each side of
	Demolish and Install	Ea.	84	27.55		111.55	opening, cripples, blocking, nails and
	Reinstall	Ea.		12.29		12.29	sub-sills for opening up to 12' in a 2"
	Minimum Charge	Job		169		169	x 4" stud wall 8' high.
Window w / 2" x 6" Lumber							
2' Wide							
	Demolish	Ea.		11.20		11.20	Includes material and labor to install
	Install	Ea.	28	14.10		42.10	header, double studs each side of
	Demolish and Install	Ea.	28	25.30		53.30	opening, cripples, blocking, nails and
	Reinstall	Ea.		11.27		11.27	sub-sills for opening up to 2' in a 2" x
	Minimum Charge	Job		169		169	6" stud wall 8' high.
3' Wide							
	Demolish	Ea.		11.20		11.20	Includes material and labor to install
	Install	Ea.	31.50	14.10		45.60	header, double studs each side of
	Demolish and Install	Ea.	31.50	25.30		56.80	opening, cripples, blocking, nails and
	Reinstall	Ea.		11.27		11.27	sub-sills for opening up to 3' in a 2" x
	Minimum Charge	Job		169		169	6" stud wall 8' high.
4' Wide							
	Demolish	Ea.		11.20		11.20	Includes material and labor to install
	Install	Ea.	34	14.10		48.10	header, double studs each side of
	Demolish and Install	Ea.	34	25.30		59.30	opening, cripples, blocking, nails and
	Reinstall	Ea.		11.27		11.27	sub-sills for opening up to 4' in a 2" x
	Minimum Charge	Job		169		169	6" stud wall 8' high.
5' Wide							
	Demolish	Ea.		11.20		11.20	Includes material and labor to install
	Install	Ea.	38.50	14.10		52.60	header, double studs each side of
	Demolish and Install	Ea.	38.50	25.30		63.80	opening, cripples, blocking, nails and
	Reinstall	Ea.		11.27		11.27	sub-sills for opening up to 5' in a 2" x
	Minimum Charge	Job		169		169	6" stud wall 8' high.
6' Wide							
	Demolish	Ea.		11.20		11.20	Includes material and labor to install
	Install	Ea.	42.50	14.10		56.60	header, double studs each side of
	Demolish and Install	Ea.	42.50	25.30		67.80	opening, cripples, blocking, nails and
	Reinstall	Ea.		11.27		11.27	sub-sills for opening up to 6' in a 2" x
	Minimum Charge	Job		169		169	6" stud wall 8' high.
7' Wide							
	Demolish	Ea.		11.20		11.20	Includes material and labor to install
	Install	Ea.	53.50	14.10		67.60	header, double studs each side of
	Demolish and Install	Ea.	53.50	25.30		78.80	opening, cripples, blocking, nails and
	Reinstall	Ea.		11.27		11.27	sub-sills for opening up to 7' in a 2" x
	Minimum Charge	Job		169		169	6" stud wall 8' high.
8' Wide							
	Demolish	Ea.		12.20		12.20	Includes material and labor to install
	Install	Ea.	58	15.35		73.35	header, double studs each side of
	Demolish and Install	Ea.	58	27.55		85.55	opening, cripples, blocking, nails and
	Reinstall	Ea.		12.29		12.29	sub-sills for opening up to 8' in a 2" x
	Minimum Charge	Job		169		169	6" stud wall 8' high.

Rough Frame / Structure

Rough-in Opening

	Unit	Material	Labor	Equip.	Total	Specification
10' Wide						
Demolish	Ea.		12.20		12.20	Includes material and labor to install
Install	Ea.	72.50	15.35		87.85	header, double studs each side of
Demolish and Install	Ea.	72.50	27.55		100.05	opening, cripples, blocking, nails and
Reinstall	Ea.		12.29		12.29	sub-sills for opening up to 10' in a 2"
Minimum Charge	Job		169		169	x 6" stud wall 8' high.
12' Wide						
Demolish	Ea.		12.20		12.20	Includes material and labor to install
Install	Ea.	99	15.35		114.35	header, double studs each side of
Demolish and Install	Ea.	99	27.55		126.55	opening, cripples, blocking, nails and
Reinstall	Ea.		12.29		12.29	sub-sills for opening up to 12' in a 2"
Minimum Charge	Job		169		169	x 6" stud wall 8' high.

Glue-Laminated Beams

	Unit	Material	Labor	Equip.	Total	Specification
3-1/2" x 6"						
Demolish	L.F.		2.02		2.02	Includes material and labor to install
Install	L.F.	5.05	2.53	1.17	8.75	glue-laminated wood beam.
Demolish and Install	L.F.	5.05	4.55	1.17	10.77	
Reinstall	L.F.		2.02	.93	2.95	
Clean	L.F.	.11	.30		.41	
Paint	L.F.	.29	.87		1.16	
Minimum Charge	Job		169		169	
3-1/2" x 9"						
Demolish	L.F.		2.02		2.02	Includes material and labor to install
Install	L.F.	6.80	2.53	1.17	10.50	glue-laminated wood beam.
Demolish and Install	L.F.	6.80	4.55	1.17	12.52	
Reinstall	L.F.		2.02	.93	2.95	
Clean	L.F.	.14	.31		.45	
Paint	L.F.	.36	1.13		1.49	
Minimum Charge	Job		169		169	
3-1/2" x 12"						
Demolish	L.F.		3.04		3.04	Includes material and labor to install
Install	L.F.	9.10	2.53	1.17	12.80	glue-laminated wood beam.
Demolish and Install	L.F.	9.10	5.57	1.17	15.84	
Reinstall	L.F.		2.02	.93	2.95	
Clean	L.F.	.18	.32		.50	
Paint	L.F.	.45	1.41		1.86	
Minimum Charge	Job		169		169	
3-1/2" x 15"						
Demolish	L.F.		3.04		3.04	Includes material and labor to install
Install	L.F.	11.35	2.61	1.21	15.17	glue-laminated wood beam.
Demolish and Install	L.F.	11.35	5.65	1.21	18.21	
Reinstall	L.F.		2.09	.97	3.06	
Clean	L.F.	.21	.33		.54	
Paint	L.F.	.53	1.65		2.18	
Minimum Charge	Job		169		169	
3-1/2" x 18"						
Demolish	L.F.		3.80		3.80	Includes material and labor to install
Install	L.F.	15.25	2.61	1.21	19.07	glue-laminated wood beam.
Demolish and Install	L.F.	15.25	6.41	1.21	22.87	
Reinstall	L.F.		2.09	.97	3.06	
Clean	L.F.	.24	.35		.59	
Paint	L.F.	.62	1.91		2.53	
Minimum Charge	Job		169		169	

Glue-Laminated Beams		Unit	Material	Labor	Equip.	Total	Specification
5-1/8″ x 6″							
	Demolish	L.F.		3.04		3.04	Includes material and labor to install
	Install	L.F.	7.45	2.53	1.17	11.15	glue-laminated wood beam.
	Demolish and Install	L.F.	7.45	5.57	1.17	14.19	
	Reinstall	L.F.		2.02	.93	2.95	
	Clean	L.F.	.12	.30		.42	
	Paint	L.F.	.30	.96		1.26	
	Minimum Charge	Job		169		169	
5-1/8″ x 9″							
	Demolish	L.F.		3.80		3.80	Includes material and labor to install
	Install	L.F.	11.15	2.53	1.17	14.85	glue-laminated wood beam.
	Demolish and Install	L.F.	11.15	6.33	1.17	18.65	
	Reinstall	L.F.		2.02	.93	2.95	
	Clean	L.F.	.15	.31		.46	
	Paint	L.F.	.40	1.22		1.62	
	Minimum Charge	Job		169		169	
5-1/8″ x 12″							
	Demolish	L.F.		4.52		4.52	Includes material and labor to install
	Install	L.F.	14.85	2.53	1.17	18.55	glue-laminated wood beam.
	Demolish and Install	L.F.	14.85	7.05	1.17	23.07	
	Reinstall	L.F.		2.02	.93	2.95	
	Clean	L.F.	.19	.32		.51	
	Paint	L.F.	.48	1.48		1.96	
	Minimum Charge	Job		169		169	
5-1/8″ x 18″							
	Demolish	L.F.		7.60		7.60	Includes material and labor to install
	Install	L.F.	22.50	2.61	1.21	26.32	glue-laminated wood beam.
	Demolish and Install	L.F.	22.50	10.21	1.21	33.92	
	Reinstall	L.F.		2.09	.97	3.06	
	Clean	L.F.	.25	.35		.60	
	Paint	L.F.	.56	2		2.56	
	Minimum Charge	Job		169		169	
6-3/4″ x 12″							
	Demolish	L.F.		4.52		4.52	Includes material and labor to install
	Install	L.F.	19.60	2.61	1.21	23.42	glue-laminated wood beam.
	Demolish and Install	L.F.	19.60	7.13	1.21	27.94	
	Reinstall	L.F.		2.09	.97	3.06	
	Clean	L.F.	.21	.33		.54	
	Paint	L.F.	.53	1.65		2.18	
	Minimum Charge	Job		169		169	
6-3/4″ x 15″							
	Demolish	L.F.		7.60		7.60	Includes material and labor to install
	Install	L.F.	24	2.61	1.21	27.82	glue-laminated wood beam.
	Demolish and Install	L.F.	24	10.21	1.21	35.42	
	Reinstall	L.F.		2.09	.97	3.06	
	Clean	L.F.	.24	.35		.59	
	Paint	L.F.	.62	1.91		2.53	
	Minimum Charge	Job		169		169	
6-3/4″ x 18″							
	Demolish	L.F.		7.60		7.60	Includes material and labor to install
	Install	L.F.	29	2.71	1.25	32.96	glue-laminated wood beam.
	Demolish and Install	L.F.	29	10.31	1.25	40.56	
	Reinstall	L.F.		2.17	1	3.17	
	Clean	L.F.	.28	.36		.64	
	Paint	L.F.	.70	2.18		2.88	
	Minimum Charge	Job		169		169	

Rough Frame / Structure

Glue-Laminated Beams		Unit	Material	Labor	Equip.	Total	Specification
Hardware							
5-1/4″ Glue-Lam Seat							
	Install	Ea.	62.50	1.88		64.38	Includes labor and material to install
	Minimum Charge	Job		169		169	beam hangers.
6-3/4″ Glue-Lam Seat							
	Install	Ea.	65	1.88		66.88	Includes labor and material to install
	Minimum Charge	Job		169		169	beam hangers.
8-3/4″ Glue-Lam Seat							
	Install	Ea.	74.50	1.88		76.38	Includes labor and material to install
	Minimum Charge	Job		169		169	beam hangers.
Earthquake Strapping							
	Install	Ea.	2.17	2.11		4.28	Includes labor and material to replace
	Minimum Charge	Job		169		169	an earthquake strap.
Hurricane Clips							
	Install	Ea.	1.90	2.33		4.23	Includes labor and material to install a
	Minimum Charge	Job		169		169	hurricane clip.

Metal Stud Framing		Unit	Material	Labor	Equip.	Total	Specification
16″ O.C. System							
4″, 16 Ga.							
	Demolish	S.F.		.99		.99	Includes material and labor to install
	Install	S.F.	1.17	1.04		2.21	load bearing cold rolled metal stud
	Demolish and Install	S.F.	1.17	2.03		3.20	walls, to 10′ high, including studs, top
	Reinstall	S.F.		.83		.83	and bottom track and screws.
	Minimum Charge	Job		169		169	
6″, 16 Ga.							
	Demolish	S.F.		.99		.99	Includes material and labor to install
	Install	S.F.	1.46	1.06		2.52	load bearing cold rolled metal stud
	Demolish and Install	S.F.	1.46	2.05		3.51	walls, to 10′ high, including studs, top
	Reinstall	S.F.		.85		.85	and bottom track and screws.
	Minimum Charge	Job		169		169	
4″, 25 Ga.							
	Demolish	S.F.		.69		.69	Includes material and labor to install
	Install	S.F.	.36	.71		1.07	cold rolled metal stud walls, to 10′
	Demolish and Install	S.F.	.36	1.40		1.76	high, including studs, top and bottom
	Reinstall	S.F.		.57		.57	track and screws.
	Minimum Charge	Job		169		169	
6″, 25 Ga.							
	Demolish	S.F.		.69		.69	Includes material and labor to install
	Install	S.F.	.53	.72		1.25	cold rolled metal stud walls, to 10′
	Demolish and Install	S.F.	.53	1.41		1.94	high, including studs, top and bottom
	Reinstall	S.F.		.58		.58	track and screws.
	Minimum Charge	Job		169		169	
24″ O.C. System							
4″, 25 Ga.							
	Demolish	S.F.		.59		.59	Includes material and labor to install
	Install	S.F.	.28	.46		.74	cold rolled metal stud walls, to 10′
	Demolish and Install	S.F.	.28	1.05		1.33	high, including studs, top and bottom
	Reinstall	S.F.		.37		.37	track and screws.
	Minimum Charge	Job		169		169	

Rough Frame / Structure

Metal Stud Framing

	Unit	Material	Labor	Equip.	Total	Specification
6", 25 Ga.						
Demolish	S.F.		.59		.59	Includes material and labor to install
Install	S.F.	.40	.47		.87	cold rolled metal stud walls, to 10'
Demolish and Install	S.F.	.40	1.06		1.46	high, including studs, top and bottom
Reinstall	S.F.		.37		.37	track and screws.
Minimum Charge	Job		169		169	
4", 16 Ga.						
Demolish	L.F.		6.70		6.70	Includes material and labor to install
Install	S.F.	.85	.75		1.60	load bearing cold rolled metal stud
Demolish and Install	S.F.	.85	7.45		8.30	walls, to 10' high, including studs, top
Reinstall	S.F.		.60		.60	and bottom track and screws.
Minimum Charge	Job		169		169	
6", 16 Ga.						
Demolish	L.F.		6.70		6.70	Includes material and labor to install
Install	S.F.	1.07	.77		1.84	load bearing cold rolled metal stud
Demolish and Install	S.F.	1.07	7.47		8.54	walls, to 10' high, including studs, top
Reinstall	S.F.		.61		.61	and bottom track and screws.
Minimum Charge	Job		169		169	

Metal Joist

	Unit	Material	Labor	Equip.	Total	Specification
Bar Joist						
18K9						
Demolish	L.F.		.80	.36	1.16	Includes material, labor and
Install	L.F.	7.45	1.85	.88	10.18	equipment to install open web joist.
Demolish and Install	L.F.	7.45	2.65	1.24	11.34	
Reinstall	L.F.		1.48	.71	2.19	
Minimum Charge	Job		1500	700	2200	
16K6						
Demolish	L.F.		.80	.36	1.16	Includes material, labor and
Install	L.F.	4.52	1.25	.07	5.84	equipment to install open web joist.
Demolish and Install	L.F.	4.52	2.05	.43	7	
Reinstall	L.F.		1	.06	1.06	
Minimum Charge	Job		207		207	

Exterior Sheathing

	Unit	Material	Labor	Equip.	Total	Specification
CDX Plywood						
5/16"						
Demolish	S.F.		.46		.46	Cost includes material and labor to
Install	S.F.	.68	.42		1.10	install 5/16" CDX plywood sheathing.
Demolish and Install	S.F.	.68	.88		1.56	
Minimum Charge	Job		169		169	
3/8"						
Demolish	S.F.		.47		.47	Cost includes material and labor to
Install	S.F.	.51	.56		1.07	install 3/8" CDX plywood sheathing.
Demolish and Install	S.F.	.51	1.03		1.54	
Minimum Charge	Job		169		169	
1/2"						
Demolish	S.F.		.48		.48	Cost includes material and labor to
Install	S.F.	.67	.60		1.27	install 1/2" CDX plywood sheathing.
Demolish and Install	S.F.	.67	1.08		1.75	
Minimum Charge	Job		169		169	

Rough Frame / Structure

Exterior Sheathing

Exterior Sheathing	Unit	Material	Labor	Equip.	Total	Specification
5/8"						
Demolish	S.F.		.49		.49	Cost includes material and labor to
Install	S.F.	.73	.64		1.37	install 5/8" CDX plywood sheathing.
Demolish and Install	S.F.	.73	1.13		1.86	
Minimum Charge	Job		169		169	
3/4"						
Demolish	S.F.		.50		.50	Cost includes material and labor to
Install	S.F.	.90	.69		1.59	install 3/4" CDX plywood sheathing.
Demolish and Install	S.F.	.90	1.19		2.09	
Minimum Charge	Job		169		169	
OSB						
1/2"						
Demolish	S.F.		.48		.48	Cost includes material and labor to
Install	S.F.	.68	.26		.94	install 4' x 8' x 1/2" OSB sheathing.
Demolish and Install	S.F.	.68	.74		1.42	
Minimum Charge	Job		169		169	
5/8"						
Demolish	S.F.		.49		.49	Cost includes material and labor to
Install	S.F.	.86	.27		1.13	install 4' x 8' x 5/8" OSB sheathing.
Demolish and Install	S.F.	.86	.76		1.62	
Minimum Charge	Job		169		169	
Vapor Barrier						
Black Paper						
Install	S.F.	.03	.09		.12	Includes material and labor to install
Minimum Charge	Job		169		169	#15 felt building paper.

Plywood Sheathing

Plywood Sheathing	Unit	Material	Labor	Equip.	Total	Specification
Finish Plywood						
5/16"						
Demolish	S.F.		.46		.46	Includes material and labor to install
Install	S.F.	.86	.21		1.07	exterior 5/16" AC plywood on walls.
Demolish and Install	S.F.	.86	.67		1.53	
Minimum Charge	Job		169		169	
3/8"						
Demolish	S.F.		.47		.47	Includes material and labor to install
Install	S.F.	.86	.28		1.14	exterior 3/8" AC plywood on walls.
Demolish and Install	S.F.	.86	.75		1.61	
Minimum Charge	Job		169		169	
1/2"						
Demolish	S.F.		.48		.48	Includes material and labor to install
Install	S.F.	.96	.30		1.26	exterior 1/2" AC plywood on walls.
Demolish and Install	S.F.	.96	.78		1.74	
Minimum Charge	Job		169		169	
5/8"						
Demolish	S.F.		.49		.49	Includes material and labor to install
Install	S.F.	1.34	.30		1.64	exterior 5/8" AC plywood on walls.
Demolish and Install	S.F.	1.34	.79		2.13	
Reinstall	S.F.		.30		.30	
Minimum Charge	Job		169		169	
3/4"						
Demolish	S.F.		.50		.50	Includes material and labor to install
Install	S.F.	1.45	.30		1.75	exterior 3/4" AC plywood on walls.
Demolish and Install	S.F.	1.45	.80		2.25	
Reinstall	S.F.		.30		.30	
Minimum Charge	Job		169		169	

Stairs

Stairs	Unit	Material	Labor	Equip.	Total	Specification
Job-built						
Treads and Risers						
Demolish	Ea.		33.50		33.50	Includes material and labor to install
Install	Ea.	485	42.50		527.50	three 2" x 12" stringers, treads and
Demolish and Install	Ea.	485	76		561	risers of 3/4" CDX plywood, installed
Clean	Flight	3.28	21.50		24.78	in a straight or "L" shaped run
Paint	Ea.	.09	59.50		59.59	including carpet.
Minimum Charge	Job		169		169	
Landing						
Demolish	S.F.		.27		.27	Includes material and labor to install
Install	S.F.	49.50	2.35		51.85	landing framing, 3/4" CDX plywood
Demolish and Install	S.F.	49.50	2.62		52.12	surface and carpet.
Clean	S.F.		.30		.30	
Paint	S.F.	.28	.73		1.01	
Minimum Charge	Job		169		169	

Ceiling Framing

Ceiling Framing	Unit	Material	Labor	Equip.	Total	Specification
16" O.C. System						
2" x 6" Joists						
Demolish	S.F.		.41		.41	Cost includes material and labor to
Install	S.F.	.50	.56		1.06	install 2" x 6" joists including end and
Demolish and Install	S.F.	.50	.97		1.47	header joist installed 16" O.C. Does
Reinstall	S.F.		.44		.44	not include beams, ledger strips,
Minimum Charge	Job		169		169	blocking or bridging.
2" x 8" Joists						
Demolish	S.F.		.43		.43	Cost includes material and labor to
Install	S.F.	.72	.68		1.40	install 2" x 8" joists including end and
Demolish and Install	S.F.	.72	1.11		1.83	header joist installed 16" O.C. Does
Reinstall	S.F.		.55		.55	not include beams, ledger strips,
Minimum Charge	Job		169		169	blocking or bridging.
2" x 10" Joists						
Demolish	S.F.		.44		.44	Cost includes material and labor to
Install	S.F.	1.08	.81		1.89	install 2" x 10" joists including end
Demolish and Install	S.F.	1.08	1.25		2.33	and header joist installed 16" O.C.
Reinstall	S.F.		.65		.65	Does not include beams, ledger strips,
Minimum Charge	Job		169		169	blocking or bridging.
2" x 12" Joists						
Demolish	S.F.		.46		.46	Cost includes material and labor to
Install	S.F.	1.34	.99		2.33	install 2" x 12" joists including end
Demolish and Install	S.F.	1.34	1.45		2.79	and header joist installed 16" O.C.
Reinstall	S.F.		.79		.79	Does not include beams, ledger strips,
Minimum Charge	Job		169		169	blocking or bridging.
Block / Bridge						
Demolish	Ea.		.84		.84	Includes material and labor to install
Install	Ea.	1.90	1.90		3.80	set of cross bridging or per block of
Demolish and Install	Ea.	1.90	2.74		4.64	solid bridging for 2" x 10" joists cut to
Minimum Charge	Job		169		169	size on site.

Rough Frame / Structure

Ceiling Framing		Unit	Material	Labor	Equip.	Total	Specification
24″ O.C. System							
2″ x 6″ Joists							
	Demolish	S.F.		.28		.28	Cost includes material and labor to
	Install	S.F.	.33	.37		.70	install 2″ x 6″ joists including end and
	Demolish and Install	S.F.	.33	.65		.98	header joist installed 24″ O.C. Does
	Reinstall	S.F.		.30		.30	not include beams, ledger strips,
	Minimum Charge	Job		169		169	blocking or bridging.
2″ x 8″ Joists							
	Demolish	S.F.		.29		.29	Cost includes material and labor to
	Install	S.F.	.48	.46		.94	install 2″ x 8″ joists including end and
	Demolish and Install	S.F.	.48	.75		1.23	header joist installed 24″ O.C. Does
	Reinstall	S.F.		.36		.36	not include beams, ledger strips,
	Minimum Charge	Job		169		169	blocking or bridging.
2″ x 10″ Joists							
	Demolish	S.F.		.29		.29	Cost includes material and labor to
	Install	S.F.	.73	.54		1.27	install 2″ x 10″ joists including end
	Demolish and Install	S.F.	.73	.83		1.56	and header joist installed 24″ O.C.
	Reinstall	S.F.		.43		.43	Does not include beams, ledger strips,
	Minimum Charge	Job		169		169	blocking or bridging.
2″ x 12″ Joists							
	Demolish	S.F.		.30		.30	Cost includes material and labor to
	Install	S.F.	.90	.66		1.56	install 2″ x 12″ joists including end
	Demolish and Install	S.F.	.90	.96		1.86	and header joist installed 24″ O.C.
	Reinstall	S.F.		.53		.53	Does not include beams, ledger strips,
	Minimum Charge	Job		169		169	blocking or bridging.
Block / Bridge							
	Demolish	Ea.		.84		.84	Includes material and labor to install
	Install	Ea.	1.90	1.90		3.80	set of cross bridging or per block of
	Demolish and Install	Ea.	1.90	2.74		4.64	solid bridging for 2″ x 10″ joists cut to
	Minimum Charge	Job		169		169	size on site.
Ledger Strips							
1″ x 2″							
	Demolish	L.F.		.22		.22	Cost includes material and labor to
	Install	L.F.	.21	.98		1.19	install 1″ x 2″ ledger strip nailed to the
	Demolish and Install	L.F.	.21	1.20		1.41	face of studs, beams or joist.
	Minimum Charge	Job		169		169	
1″ x 3″							
	Demolish	L.F.		.22		.22	Includes material and labor to install
	Install	L.F.	.51	1.13		1.64	up to 1″ x 4″ ledger strip nailed to the
	Demolish and Install	L.F.	.51	1.35		1.86	face of studs, beams or joist.
	Minimum Charge	Job		169		169	
1″ x 4″							
	Demolish	L.F.		.22		.22	Includes material and labor to install
	Install	L.F.	.51	1.13		1.64	up to 1″ x 4″ ledger strip nailed to the
	Demolish and Install	L.F.	.51	1.35		1.86	face of studs, beams or joist.
	Minimum Charge	Job		169		169	
2″ x 2″							
	Demolish	L.F.		.24		.24	Cost includes material and labor to
	Install	L.F.	.42	1.02		1.44	install 2″ x 2″ ledger strip nailed to the
	Demolish and Install	L.F.	.42	1.26		1.68	face of studs, beams or joist.
	Minimum Charge	Job		169		169	
2″ x 4″							
	Demolish	L.F.		.27		.27	Cost includes material and labor to
	Install	L.F.	.41	1.35		1.76	install 2″ x 4″ ledger strip nailed to the
	Demolish and Install	L.F.	.41	1.62		2.03	face of studs, beams or joist.
	Reinstall	L.F.		1.08		1.08	
	Minimum Charge	Job		169		169	

48

Rough Frame / Structure

Ceiling Framing

Ceiling Framing		Unit	Material	Labor	Equip.	Total	Specification
Ledger Boards							
	Demolish	L.F.		.27		.27	Cost includes material and labor to
	Install	L.F.	1.82	1.88		3.70	install 2" x 4" ledger board fastened to
	Demolish and Install	L.F.	1.82	2.15		3.97	a wall, joists or studs.
	Reinstall	L.F.		1.50		1.50	
	Minimum Charge	Job		169		169	
4" x 6"							
	Demolish	L.F.		.45		.45	Cost includes material and labor to
	Install	L.F.	3.55	2.11		5.66	install 4" x 6" ledger board fastened to
	Demolish and Install	L.F.	3.55	2.56		6.11	a wall, joists or studs.
	Reinstall	L.F.		1.69		1.69	
	Minimum Charge	Job		169		169	
4" x 8"							
	Demolish	L.F.		.60		.60	Cost includes material and labor to
	Install	L.F.	4.35	2.11		6.46	install 4" x 8" ledger board fastened to
	Demolish and Install	L.F.	4.35	2.71		7.06	a wall, joists or studs.
	Reinstall	L.F.		1.69		1.69	
	Minimum Charge	Job		169		169	
Earthquake Strapping							
	Install	Ea.	2.17	2.11		4.28	Includes labor and material to replace
	Minimum Charge	Job		169		169	an earthquake strap.
Hurricane Clips							
	Install	Ea.	1.90	2.33		4.23	Includes labor and material to install a
	Minimum Charge	Job		169		169	hurricane clip.

Roof Framing

Roof Framing		Unit	Material	Labor	Equip.	Total	Specification
16" O.C. System							
2" x 4" Rafters							
	Demolish	S.F.		.61		.61	Cost includes material and labor to
	Install	S.F.	.31	.30		.61	install 2" x 4", 16" O.C. rafter framing
	Demolish and Install	S.F.	.31	.91		1.22	for flat, shed or gable roofs, 25' span,
	Reinstall	S.F.		.24		.24	up to 5/12 slope, per S.F. of roof.
	Minimum Charge	Job		169		169	
2" x 6" Rafters							
	Demolish	S.F.		.64		.64	Cost includes material and labor to
	Install	S.F.	.48	.32		.80	install 2" x 6", 16" O.C. rafter framing
	Demolish and Install	S.F.	.48	.96		1.44	for flat, shed or gable roofs, 25' span,
	Reinstall	S.F.		.25		.25	up to 5/12 slope, per S.F. of roof.
	Minimum Charge	Job		169		169	
2" x 8" Rafters							
	Demolish	S.F.		.65		.65	Cost includes material and labor to
	Install	S.F.	.70	.34		1.04	install 2" x 8", 16" O.C. rafter framing
	Demolish and Install	S.F.	.70	.99		1.69	for flat, shed or gable roofs, 25' span,
	Reinstall	S.F.		.27		.27	up to 5/12 slope, per S.F. of roof.
	Minimum Charge	Job		169		169	
2" x 10" Rafters							
	Demolish	S.F.		.65		.65	Cost includes material and labor to
	Install	S.F.	1.07	.51		1.58	install 2" x 10", 16" O.C. rafter
	Demolish and Install	S.F.	1.07	1.16		2.23	framing for flat, shed or gable roofs,
	Reinstall	S.F.		.41		.41	25' span, up to 5/12 slope, per S.F.
	Minimum Charge	Job		169		169	of roof.

Rough Frame / Structure

Roof Framing		Unit	Material	Labor	Equip.	Total	Specification
2" x 12" Rafters							
	Demolish	S.F.		.66		.66	Cost includes material and labor to
	Install	S.F.	1.32	.56		1.88	install 2" x 12", 16" O.C. rafter
	Demolish and Install	S.F.	1.32	1.22		2.54	framing for flat, shed or gable roofs,
	Reinstall	S.F.		.45		.45	25' span, up to 5/12 slope, per S.F.
	Minimum Charge	Job		169		169	of roof.
24" O.C. System							
2" x 4" Rafters							
	Demolish	S.F.		.46		.46	Cost includes material and labor to
	Install	S.F.	.21	.23		.44	install 2" x 4", 24" O.C. rafter framing
	Demolish and Install	S.F.	.21	.69		.90	for flat, shed or gable roofs, 25' span,
	Reinstall	S.F.		.18		.18	up to 5/12 slope, per S.F. of roof.
	Minimum Charge	Job		169		169	
2" x 6" Rafters							
	Demolish	S.F.		.48		.48	Cost includes material and labor to
	Install	S.F.	.33	.24		.57	install 2" x 6", 24" O.C. rafter framing
	Demolish and Install	S.F.	.33	.72		1.05	for flat, shed or gable roofs, 25' span,
	Reinstall	S.F.		.19		.19	up to 5/12 slope, per S.F. of roof.
	Minimum Charge	Job		169		169	
2" x 8" Rafters							
	Demolish	S.F.		.49		.49	Cost includes material and labor to
	Install	S.F.	.47	.25		.72	install 2" x 8", 24" O.C. rafter framing
	Demolish and Install	S.F.	.47	.74		1.21	for flat, shed or gable roofs, 25' span,
	Reinstall	S.F.		.20		.20	up to 5/12 slope, per S.F. of roof.
	Minimum Charge	Job		169		169	
2" x 10" Rafters							
	Demolish	S.F.		.49		.49	Cost includes material and labor to
	Install	S.F.	.72	.38		1.10	install 2" x 10", 24" O.C. rafter
	Demolish and Install	S.F.	.72	.87		1.59	framing for flat, shed or gable roofs,
	Reinstall	S.F.		.31		.31	25' span, up to 5/12 slope, per S.F.
	Minimum Charge	Job		169		169	of roof.
2" x 12" Rafters							
	Demolish	S.F.		.50		.50	Cost includes material and labor to
	Install	S.F.	.88	.42		1.30	install 2" x 12", 24" O.C. rafter
	Demolish and Install	S.F.	.88	.92		1.80	framing for flat, shed or gable roofs,
	Reinstall	S.F.		.34		.34	25' span, up to 5/12 slope, per S.F.
	Minimum Charge	Job		169		169	of roof.
Rafter							
2" x 4"							
	Demolish	L.F.		.62		.62	Cost includes material and labor to
	Install	L.F.	.41	.98		1.39	install 2" x 4" rafters for flat, shed or
	Demolish and Install	L.F.	.41	1.60		2.01	gable roofs, up to 5/12 slope, 25'
	Reinstall	L.F.		.78		.78	span, per L.F.
	Minimum Charge	Job		169		169	
2" x 6"							
	Demolish	L.F.		.63		.63	Cost includes material and labor to
	Install	L.F.	.65	.85		1.50	install 2" x 6" rafters for flat, shed or
	Demolish and Install	L.F.	.65	1.48		2.13	gable roofs, up to 5/12 slope, 25'
	Reinstall	L.F.		.68		.68	span, per L.F.
	Minimum Charge	Job		169		169	
2" x 8"							
	Demolish	L.F.		.64		.64	Cost includes material and labor to
	Install	L.F.	.94	.90		1.84	install 2" x 8" rafters for flat, shed or
	Demolish and Install	L.F.	.94	1.54		2.48	gable roofs, up to 5/12 slope, 25'
	Reinstall	L.F.		.72		.72	span, per L.F.
	Minimum Charge	Job		169		169	

Rough Frame / Structure

Roof Framing		Unit	Material	Labor	Equip.	Total	Specification
2″ x 10″							
	Demolish	L.F.		.65		.65	Cost includes material and labor to
	Install	L.F.	1.42	1.37		2.79	install 2″ x 10″ rafters for flat, shed or
	Demolish and Install	L.F.	1.42	2.02		3.44	gable roofs, up to 5/12 slope, 25′
	Reinstall	L.F.		1.09		1.09	span, per L.F.
	Minimum Charge	Job		169		169	
2″ x 12″							
	Demolish	L.F.		.66		.66	Cost includes material and labor to
	Install	L.F.	1.76	1.49		3.25	install 2″ x 12″ rafters for flat, shed or
	Demolish and Install	L.F.	1.76	2.15		3.91	gable roofs, up to 5/12 slope, 25′
	Reinstall	L.F.		1.19		1.19	span, per L.F.
	Minimum Charge	Job		169		169	
2″x 4″ Valley / Jack							
	Demolish	L.F.		.62		.62	Includes material and labor to install
	Install	L.F.	.65	1.42		2.07	up to 2″ x 6″ valley/jack rafters for
	Demolish and Install	L.F.	.65	2.04		2.69	flat, shed or gable roofs, up to 5/12
	Reinstall	L.F.		1.14		1.14	slope, 25′ span, per L.F.
	Minimum Charge	Job		169		169	
2″x 6″ Valley / Jack							
	Demolish	L.F.		.63		.63	Includes material and labor to install
	Install	L.F.	.65	1.42		2.07	up to 2″ x 6″ valley/jack rafters for
	Demolish and Install	L.F.	.65	2.05		2.70	flat, shed or gable roofs, up to 5/12
	Reinstall	L.F.		1.14		1.14	slope, 25′ span, per L.F.
	Minimum Charge	Job		169		169	
Ridgeboard							
2″ x 4″							
	Demolish	L.F.		.60		.60	Cost includes material and labor to
	Install	L.F.	.41	.61		1.02	install 2″ x 4″ ridgeboard for flat, shed
	Demolish and Install	L.F.	.41	1.21		1.62	or gable roofs, up to 5/12 slope, 25′
	Reinstall	L.F.		.49		.49	span, per L.F.
	Minimum Charge	Job		169		169	
2″ x 6″							
	Demolish	L.F.		.61		.61	Cost includes material and labor to
	Install	L.F.	.65	1.35		2	install 2″ x 6″ ridgeboard for flat, shed
	Demolish and Install	L.F.	.65	1.96		2.61	or gable roofs, up to 5/12 slope, 25′
	Reinstall	L.F.		1.08		1.08	span, per L.F.
	Minimum Charge	Job		169		169	
2″ x 8″							
	Demolish	L.F.		.63		.63	Cost includes material and labor to
	Install	L.F.	.94	1.50		2.44	install 2″ x 8″ ridgeboard for flat, shed
	Demolish and Install	L.F.	.94	2.13		3.07	or gable roofs, up to 5/12 slope, 25′
	Reinstall	L.F.		1.20		1.20	span, per L.F.
	Minimum Charge	Job		169		169	
2″ x 10″							
	Demolish	L.F.		.65		.65	Cost includes material and labor to
	Install	L.F.	1.42	1.69		3.11	install 2″ x 10″ ridgeboard for flat,
	Demolish and Install	L.F.	1.42	2.34		3.76	shed or gable roofs, up to 5/12
	Reinstall	L.F.		1.35		1.35	slope, 25′ span, per L.F.
	Minimum Charge	Job		169		169	
2″ x 12″							
	Demolish	L.F.		.67		.67	Cost includes material and labor to
	Install	L.F.	1.76	.97		2.73	install 2″ x 12″ ridgeboard for flat,
	Demolish and Install	L.F.	1.76	1.64		3.40	shed or gable roofs, up to 5/12
	Reinstall	L.F.		.77		.77	slope, 25′ span, per L.F.
	Minimum Charge	Job		169		169	

Rough Frame / Structure

Roof Framing		Unit	Material	Labor	Equip.	Total	Specification
Collar Beam							
1″ x 6″							
	Demolish	L.F.		.47		.47	Cost includes material and labor to
	Install	L.F.	.61	.42		1.03	install 1″ x 6″ collar beam or tie for
	Demolish and Install	L.F.	.61	.89		1.50	roof framing.
	Reinstall	L.F.		.34		.34	
	Minimum Charge	Job		169		169	
2″ x 6″							
	Demolish	L.F.		.55		.55	Cost includes material and labor to
	Install	L.F.	.65	.42		1.07	install 2″ x 6″ collar beam or tie for
	Demolish and Install	L.F.	.65	.97		1.62	roof framing.
	Reinstall	L.F.		.34		.34	
	Minimum Charge	Job		169		169	
Purlins							
	Demolish	L.F.		.54		.54	Cost includes material and labor to
	Install	L.F.	.65	.38		1.03	install 2″ x 6″ purlins below roof
	Demolish and Install	L.F.	.65	.92		1.57	rafters.
	Reinstall	L.F.		.30		.30	
	Minimum Charge	Job		169		169	
2″ x 8″							
	Demolish	L.F.		.54		.54	Cost includes material and labor to
	Install	L.F.	.94	.38		1.32	install 2″ x 8″ purlins below roof
	Demolish and Install	L.F.	.94	.92		1.86	rafters.
	Reinstall	L.F.		.30		.30	
	Minimum Charge	Job		169		169	
2″ x 10″							
	Demolish	L.F.		.55		.55	Cost includes material and labor to
	Install	L.F.	1.42	.38		1.80	install 2″ x 10″ purlins below roof
	Demolish and Install	L.F.	1.42	.93		2.35	rafters.
	Reinstall	L.F.		.31		.31	
	Minimum Charge	Job		169		169	
2″ x 12″							
	Demolish	L.F.		.55		.55	Cost includes material and labor to
	Install	L.F.	1.76	.39		2.15	install 2″ x 12″ purlins below roof
	Demolish and Install	L.F.	1.76	.94		2.70	rafters.
	Reinstall	L.F.		.31		.31	
	Minimum Charge	Job		169		169	
4″ x 6″							
	Demolish	L.F.		.55		.55	Cost includes material and labor to
	Install	L.F.	2.06	.39		2.45	install 4″ x 6″ purlins below roof
	Demolish and Install	L.F.	2.06	.94		3	rafters.
	Reinstall	L.F.		.31		.31	
	Minimum Charge	Job		169		169	
4″ x 8″							
	Demolish	L.F.		.56		.56	Cost includes material and labor to
	Install	L.F.	2.85	.40		3.25	install 4″ x 8″ purlins below roof
	Demolish and Install	L.F.	2.85	.96		3.81	rafters.
	Reinstall	L.F.		.32		.32	
	Minimum Charge	Job		169		169	
Ledger Board							
2″ x 6″							
	Demolish	L.F.		.27		.27	Cost includes material and labor to
	Install	L.F.	.65	1.52		2.17	install 2″ x 6″ ledger board nailed to
	Demolish and Install	L.F.	.65	1.79		2.44	the face of a rafter.
	Reinstall	L.F.		1.22		1.22	
	Minimum Charge	Job		169		169	

Rough Frame / Structure

Roof Framing		Unit	Material	Labor	Equip.	Total	Specification
4″ x 6″							
	Demolish	L.F.		.45		.45	Cost includes material and labor to
	Install	L.F.	2.06	2.24		4.30	install 4″ x 6″ ledger board nailed to
	Demolish and Install	L.F.	2.06	2.69		4.75	the face of a rafter.
	Reinstall	L.F.		1.79		1.79	
	Minimum Charge	Job		169		169	
4″ x 8″							
	Demolish	L.F.		.60		.60	Cost includes material and labor to
	Install	L.F.	2.85	2.58		5.43	install 4″ x 8″ ledger board nailed to
	Demolish and Install	L.F.	2.85	3.18		6.03	the face of a rafter.
	Reinstall	L.F.		2.06		2.06	
	Minimum Charge	Job		169		169	
Outriggers							
2″ x 4″							
	Demolish	L.F.		.18		.18	Cost includes material and labor to
	Install	L.F.	.41	1.35		1.76	install 2″ x 4″ outrigger rafters for flat,
	Demolish and Install	L.F.	.41	1.53		1.94	shed or gabled roofs, up to 5/12
	Reinstall	L.F.		1.08		1.08	slope, per L.F.
	Minimum Charge	Job		169		169	
2″ x 6″							
	Demolish	L.F.		.28		.28	Cost includes material and labor to
	Install	L.F.	.65	1.35		2	install 2″ x 6″ outrigger rafters for flat,
	Demolish and Install	L.F.	.65	1.63		2.28	shed or gabled roofs, up to 5/12
	Reinstall	L.F.		1.08		1.08	slope, per L.F.
	Minimum Charge	Job		169		169	
Lookout Rafter							
2″ x 4″							
	Demolish	L.F.		.62		.62	Cost includes material and labor to
	Install	L.F.	.41	.56		.97	install 2″ x 4″ lookout rafters for flat,
	Demolish and Install	L.F.	.41	1.18		1.59	shed or gabled roofs, up to 5/12
	Reinstall	L.F.		.45		.45	slope, per L.F.
	Minimum Charge	Job		169		169	
2″ x 6″							
	Demolish	L.F.		.63		.63	Cost includes material and labor to
	Install	L.F.	.65	.57		1.22	install 2″ x 6″ lookout rafters for flat,
	Demolish and Install	L.F.	.65	1.20		1.85	shed or gabled roofs, up to 5/12
	Reinstall	L.F.		.46		.46	slope, per L.F.
	Minimum Charge	Job		169		169	
Fly Rafter							
2″ x 4″							
	Demolish	L.F.		.62		.62	Cost includes material and labor to
	Install	L.F.	.41	.56		.97	install 2″ x 4″ fly rafters for flat, shed
	Demolish and Install	L.F.	.41	1.18		1.59	or gabled roofs, up to 5/12 slope,
	Reinstall	L.F.		.45		.45	per L.F.
	Minimum Charge	Job		169		169	
2″ x 6″							
	Demolish	L.F.		.63		.63	Cost includes material and labor to
	Install	L.F.	.65	.57		1.22	install 2″ x 6″ fly rafters for flat, shed
	Demolish and Install	L.F.	.65	1.20		1.85	or gabled roofs, up to 5/12 slope,
	Reinstall	L.F.		.46		.46	per L.F.
	Minimum Charge	Job		169		169	

Rough Frame / Structure

Residential Trusses		Unit	Material	Labor	Equip.	Total	Specification
W or Fink Truss							
24' Span							
	Demolish	Ea.		25.50	11.80	37.30	Includes material, labor and
	Install	Ea.	60	25	11.65	96.65	equipment to install wood gang-nailed
	Demolish and Install	Ea.	60	50.50	23.45	133.95	residential truss, up to 24' span.
	Reinstall	Ea.		25.10	11.67	36.77	
	Minimum Charge	Job		169		169	
28' Span							
	Demolish	Ea.		27	12.55	39.55	Includes material, labor and
	Install	Ea.	70.50	28.50	13.20	112.20	equipment to install wood gang-nailed
	Demolish and Install	Ea.	70.50	55.50	25.75	151.75	residential truss, up to 28' span.
	Reinstall	Ea.		28.42	13.22	41.64	
	Minimum Charge	Job		169		169	
32' Span							
	Demolish	Ea.		30	13.90	43.90	Includes material, labor and
	Install	Ea.	110	30	14	154	equipment to install wood gang-nailed
	Demolish and Install	Ea.	110	60	27.90	197.90	residential truss, up to 32' span.
	Reinstall	Ea.		30.12	14.01	44.13	
	Minimum Charge	Job		169		169	
36' Span							
	Demolish	Ea.		32.50	14.95	47.45	Includes material, labor and
	Install	Ea.	147	32.50	15.25	194.75	equipment to install wood gang-nailed
	Demolish and Install	Ea.	147	65	30.20	242.20	residential truss, up to 36' span.
	Reinstall	Ea.		32.74	15.23	47.97	
	Minimum Charge	Job		169		169	
40' Span							
	Demolish	Ea.		35	16.20	51.20	Includes material, labor and
	Install	Ea.	118	35	16.30	169.30	equipment to install wood gang-nailed
	Demolish and Install	Ea.	118	70	32.50	220.50	residential truss, up to 40' span.
	Reinstall	Ea.		35.02	16.29	51.31	
	Minimum Charge	Job		169		169	
Gable End							
24' Span							
	Demolish	Ea.		25.50	11.80	37.30	Includes material, labor and
	Install	Ea.	116	27	12.50	155.50	equipment to install wood gang-nailed
	Demolish and Install	Ea.	116	52.50	24.30	192.80	residential gable end truss, up to 24'
	Reinstall	Ea.		26.89	12.51	39.40	span.
	Clean	Ea.	1.36	32		33.36	
	Paint	Ea.	3.52	39.50		43.02	
	Minimum Charge	Job		169		169	
28' Span							
	Demolish	Ea.		27	12.55	39.55	Includes material, labor and
	Install	Ea.	110	32	14.90	156.90	equipment to install wood gang-nailed
	Demolish and Install	Ea.	110	59	27.45	196.45	residential gable end truss, up to 30'
	Reinstall	Ea.		32.04	14.90	46.94	span.
	Clean	Ea.	1.43	44		45.43	
	Paint	Ea.	3.67	54		57.67	
	Minimum Charge	Job		169		169	
32' Span							
	Demolish	Ea.		30	13.90	43.90	Includes material, labor and
	Install	Ea.	123	32.50	15.25	170.75	equipment to install wood gang-nailed
	Demolish and Install	Ea.	123	62.50	29.15	214.65	residential gable end truss, up to 32'
	Reinstall	Ea.		32.74	15.23	47.97	span.
	Clean	Ea.	1.49	57.50		58.99	
	Paint	Ea.	3.84	70.50		74.34	
	Minimum Charge	Job		169		169	

Rough Frame / Structure

Residential Trusses

	Unit	Material	Labor	Equip.	Total	Specification
36' Span						
Demolish	Ea.		32.50	14.95	47.45	Includes material, labor and
Install	Ea.	146	37.50	17.50	201	equipment to install wood gang-nailed
Demolish and Install	Ea.	146	70	32.45	248.45	residential gable end truss, up to 36'
Reinstall	Ea.		37.65	17.51	55.16	span.
Clean	Ea.	1.56	72.50		74.06	
Paint	Ea.	4.03	89		93.03	
Minimum Charge	Job		169		169	
40' Span						
Demolish	Ea.		35	16.20	51.20	Includes material, labor and
Install	Ea.	164	39.50	18.45	221.95	equipment to install wood gang-nailed
Demolish and Install	Ea.	164	74.50	34.65	273.15	residential gable end truss, up to 40'
Reinstall	Ea.		39.63	18.43	58.06	span.
Clean	Ea.	4.22	89.50		93.72	
Paint	Ea.	4.22	110		114.22	
Minimum Charge	Job		169		169	
Truss Hardware						
5-1/4″ Glue-Lam Seat						
Install	Ea.	62.50	1.88		64.38	Includes labor and material to install
Minimum Charge	Job		169		169	beam hangers.
6-3/4″ Glue-Lam Seat						
Install	Ea.	65	1.88		66.88	Includes labor and material to install
Minimum Charge	Job		169		169	beam hangers.
8-3/4″ Glue-Lam Seat						
Install	Ea.	74.50	1.88		76.38	Includes labor and material to install
Minimum Charge	Job		169		169	beam hangers.
2″ x 4″ Joist Hanger						
Install	Ea.	.77	1.93		2.70	Includes labor and material to install
Minimum Charge	Job		169		169	joist and beam hangers.
2″ x 12″ Joist Hanger						
Install	Ea.	.88	2.05		2.93	Cost includes labor and material to
Minimum Charge	Job		169		169	install 2″ x 12″ joist hanger.

Commercial Trusses

	Unit	Material	Labor	Equip.	Total	Specification
Engineered Lumber, Truss / Joist						
9-1/2″						
Demolish	S.F.		.44		.44	Includes material and labor to install
Install	S.F.	1.79	.34		2.13	engineered lumber truss / joists per
Demolish and Install	S.F.	1.79	.78		2.57	S.F. of floor area based on joists 16″
Reinstall	S.F.		.27		.27	O.C. Beams, supports and bridging
Minimum Charge	Job		169		169	are not included.
11-7/8″						
Demolish	S.F.		.46		.46	Includes material and labor to install
Install	S.F.	1.94	.35		2.29	engineered lumber truss / joists per
Demolish and Install	S.F.	1.94	.81		2.75	S.F. of floor area based on joists 16″
Reinstall	S.F.		.28		.28	O.C. Beams, supports and bridging
Minimum Charge	Job		169		169	are not included.
14″						
Demolish	S.F.		.47		.47	Includes material and labor to install
Install	S.F.	2.23	.37		2.60	engineered lumber truss / joists per
Demolish and Install	S.F.	2.23	.84		3.07	S.F. of floor area based on joists 16″
Reinstall	S.F.		.30		.30	O.C. Beams, supports and bridging
Minimum Charge	Job		169		169	are not included.

55

Rough Frame / Structure

Commercial Trusses

		Unit	Material	Labor	Equip.	Total	Specification
16"							Includes material and labor to install
	Demolish	S.F.		.49		.49	engineered lumber truss / joists per
	Install	S.F.	2.52	.39		2.91	S.F. of floor area based on joists 16"
	Demolish and Install	S.F.	2.52	.88		3.40	O.C. Beams, supports and bridging
	Reinstall	S.F.		.31		.31	are not included.
	Minimum Charge	Job		169		169	
Bowstring Truss							
100' Clear Span							Includes material, labor and
	Demolish	SF Flr.		.45	.32	.77	equipment to install bow string truss
	Install	SF Flr.	6.65	.38	.26	7.29	system up to 100' clear span per S.F.
	Demolish and Install	SF Flr.	6.65	.83	.58	8.06	of floor area.
	Minimum Charge	Job		169		169	
120' Clear Span							Includes material, labor and
	Demolish	SF Flr.		.50	.36	.86	equipment to install bow string truss
	Install	SF Flr.	7.15	.42	.29	7.86	system up to 120' clear span per S.F.
	Demolish and Install	SF Flr.	7.15	.92	.65	8.72	of floor area.
	Minimum Charge	Job		169		169	

Roof Sheathing

		Unit	Material	Labor	Equip.	Total	Specification
Plywood							
3/8"							Cost includes material and labor to
	Demolish	S.F.		.35		.35	install 3/8" CDX plywood roof
	Install	S.F.	.51	.44		.95	sheathing.
	Demolish and Install	S.F.	.51	.79		1.30	
	Reinstall	S.F.		.44		.44	
	Minimum Charge	Job		169		169	
1/2"							Cost includes material and labor to
	Demolish	S.F.		.35		.35	install 1/2" CDX plywood roof
	Install	S.F.	.67	.48		1.15	sheathing.
	Demolish and Install	S.F.	.67	.83		1.50	
	Minimum Charge	Job		169		169	
5/8"							Cost includes material and labor to
	Demolish	S.F.		.35		.35	install 5/8" CDX plywood roof
	Install	S.F.	.73	.52		1.25	sheathing.
	Demolish and Install	S.F.	.73	.87		1.60	
	Minimum Charge	Job		169		169	
3/4"							Cost includes material and labor to
	Demolish	S.F.		.35		.35	install 3/4" CDX plywood roof
	Install	S.F.	.90	.56		1.46	sheathing.
	Demolish and Install	S.F.	.90	.91		1.81	
	Minimum Charge	Job		169		169	
OSB							
1/2"							Cost includes material and labor to
	Demolish	S.F.		.35		.35	install 4' x 8' x 1/2" OSB sheathing.
	Install	S.F.	.68	.26		.94	
	Demolish and Install	S.F.	.68	.61		1.29	
	Minimum Charge	Job		169		169	
5/8"							Cost includes material and labor to
	Demolish	S.F.		.35		.35	install 4' x 8' x 5/8" OSB sheathing.
	Install	S.F.	.86	.27		1.13	
	Demolish and Install	S.F.	.86	.62		1.48	
	Minimum Charge	Job		169		169	

Rough Frame / Structure

Roof Sheathing

	Unit	Material	Labor	Equip.	Total	Specification
Plank						
Demolish	S.F.		.35		.35	Cost includes material and labor to
Install	S.F.	1.22	.93		2.15	install 1″ x 6″ or 1″ x 8″ utility T&G
Demolish and Install	S.F.	1.22	1.28		2.50	board sheathing laid diagonal.
Reinstall	S.F.		.93		.93	
Minimum Charge	Job		169		169	
Skip-type						
1″ x 4″						
Demolish	S.F.		.54		.54	Includes material and labor to install
Install	S.F.	.48	.28		.76	sheathing board material 1″ x 4″, 7″
Demolish and Install	S.F.	.48	.82		1.30	O.C.
Minimum Charge	Job		169		169	
1″ x 6″						
Demolish	S.F.		.54		.54	Includes material and labor to install
Install	S.F.	.61	.23		.84	sheathing board material 1″ x 6″, 7″
Demolish and Install	S.F.	.61	.77		1.38	or 9″ O.C.
Minimum Charge	Job		169		169	

Wood Deck

	Unit	Material	Labor	Equip.	Total	Specification
Treated						
Demolish	S.F.		1.12		1.12	Includes material and labor to install
Install	S.F.	4.55	8.45		13	unfinished deck using 2″ x 4″ treated
Demolish and Install	S.F.	4.55	9.57		14.12	decking, with 2″ x 6″ double beams
Reinstall	S.F.		6.76		6.76	24″ O.C., 4″ x 4″ posts 5′ O.C. set in
Clean	S.F.	.07	.25		.32	concrete.
Paint	S.F.	.06	.23		.29	
Minimum Charge	Job		169		169	
Redwood						
Demolish	S.F.		1.12		1.12	Includes material and labor to install
Install	S.F.	16.50	8.45		24.95	unfinished deck using 2″ x 4″ select
Demolish and Install	S.F.	16.50	9.57		26.07	redwood decking, with 2″ x 6″ double
Reinstall	S.F.		6.76		6.76	beams 24″ O.C., 4″ x 4″ posts 5′
Clean	S.F.	.07	.25		.32	O.C. set in concrete.
Paint	S.F.	.06	.23		.29	
Minimum Charge	Job		169		169	
Cedar						
Demolish	S.F.		1.12		1.12	Includes material and labor to install
Install	S.F.	7.60	8.45		16.05	unfinished deck using 2″ x 4″ select
Demolish and Install	S.F.	7.60	9.57		17.17	cedar decking, with 2″ x 6″ double
Reinstall	S.F.		6.76		6.76	beams 24″ O.C., 4″ x 4″ posts 5′
Clean	S.F.	.07	.25		.32	O.C. set in concrete.
Paint	S.F.	.06	.23		.29	
Minimum Charge	Job		169		169	
Stairs						
Demolish	Ea.		325		325	Includes material and labor to install
Install	Ea.	132	580		712	wood steps. 5 - 7 risers included.
Demolish and Install	Ea.	132	905		1037	
Reinstall	Ea.		465.66		465.66	
Clean	Flight	3.28	21.50		24.78	
Paint	Flight	10.35	39.50		49.85	
Minimum Charge	Job		169		169	

Rough Frame / Structure

Wood Deck

	Unit	Material	Labor	Equip.	Total	Specification
Railing						
Demolish	L.F.		4.13		4.13	Includes material and labor to install
Install	L.F.	16.95	15.35		32.30	porch rail with balusters.
Demolish and Install	L.F.	16.95	19.48		36.43	
Clean	L.F.	.66	1.51		2.17	
Paint	L.F.	.53	2.47		3	
Minimum Charge	Job		169		169	
Bench Seating						
Demolish	L.F.		4.47		4.47	Includes material and labor to install
Install	L.F.	2.18	16.90		19.08	pressure treated bench seating.
Demolish and Install	L.F.	2.18	21.37		23.55	
Reinstall	L.F.		13.52		13.52	
Clean	L.F.	.13	1.21		1.34	
Paint	L.F.	.51	2.37		2.88	
Minimum Charge	Job		169		169	
Privacy Wall						
Demolish	SF Flr.		1.07		1.07	Includes material, labor and
Install	SF Flr.	.47	.56	.11	1.14	equipment to install privacy wall.
Demolish and Install	SF Flr.	.47	1.63	.11	2.21	
Reinstall	SF Flr.		.45	.09	.54	
Clean	SF Flr.	.07	.25		.32	
Paint	S.F.	.06	.23		.29	
Minimum Charge	Job		169		169	

Porch

	Unit	Material	Labor	Equip.	Total	Specification
Treated						
Demolish	S.F.		2.24		2.24	Includes material and labor to install
Install	S.F.	5.60	16.90		22.50	floor and roof frame and decking
Demolish and Install	S.F.	5.60	19.14		24.74	using 2" x 4" treated decking, with 2"
Reinstall	S.F.		13.52		13.52	x 6" double beams 24" O.C., 4" x 4"
Clean	S.F.	.13	.50		.63	posts 5' O.C. set in concrete.
Paint	S.F.	.17	.59		.76	
Minimum Charge	Job		169		169	
Redwood						
Demolish	S.F.		2.24		2.24	Includes material and labor to install
Install	S.F.	17.80	16.90		34.70	floor and roof frame and decking
Demolish and Install	S.F.	17.80	19.14		36.94	using 2" x 4" redwood decking, with
Reinstall	S.F.		13.52		13.52	2" x 6" double beams 24" O.C., 4" x
Clean	S.F.	.13	.50		.63	4" posts 5' O.C. set in concrete.
Paint	S.F.	.17	.59		.76	
Minimum Charge	Job		169		169	
Cedar						
Demolish	S.F.		2.24		2.24	Includes material and labor to install
Install	S.F.	8.95	16.90		25.85	floor and roof frame and cedar
Demolish and Install	S.F.	8.95	19.14		28.09	decking using 2" x 4" select cedar,
Reinstall	S.F.		13.52		13.52	with 2" x 6" double beams 24" O.C.,
Clean	S.F.	.13	.50		.63	4" x 4" posts 5' O.C. set in concrete.
Paint	S.F.	.17	.59		.76	
Minimum Charge	Job		169		169	
Stairs						
Demolish	Ea.		325		325	Includes material and labor to install
Install	Ea.	132	580		712	wood steps. 5 - 7 risers included.
Demolish and Install	Ea.	132	905		1037	
Reinstall	Ea.		465.66		465.66	
Clean	Flight	3.28	21.50		24.78	
Paint	Flight	10.35	39.50		49.85	
Minimum Charge	Job		169		169	

Rough Frame / Structure

Porch	Unit	Material	Labor	Equip.	Total	Specification
Railing						
Demolish	L.F.		4.13		4.13	Includes material and labor to install
Install	L.F.	16.95	15.35		32.30	porch rail with balusters.
Demolish and Install	L.F.	16.95	19.48		36.43	
Clean	L.F.	.66	1.89		2.55	
Paint	L.F.	.53	2.47		3	
Minimum Charge	Job		169		169	
Bench Seating						
Demolish	L.F.		4.47		4.47	Includes material and labor to install
Install	L.F.	2.18	16.90		19.08	pressure treated bench seating.
Demolish and Install	L.F.	2.18	21.37		23.55	
Reinstall	L.F.		13.52		13.52	
Clean	L.F.	.13	1.21		1.34	
Paint	L.F.	.51	2.37		2.88	
Minimum Charge	Job		169		169	
Privacy Wall						
Demolish	SF Flr.		1.07		1.07	Includes material, labor and
Install	SF Flr.	.47	.56	.11	1.14	equipment to install privacy wall.
Demolish and Install	SF Flr.	.47	1.63	.11	2.21	
Reinstall	SF Flr.		.45	.09	.54	
Clean	SF Flr.	.07	.25		.32	
Paint	S.F.	.06	.23		.29	
Minimum Charge	Job		169		169	
Wood Board Ceiling						
Demolish	S.F.		.45		.45	Includes material and labor to install
Install	S.F.	1.97	1.50		3.47	wood board on ceiling.
Demolish and Install	S.F.	1.97	1.95		3.92	
Clean	S.F.	.02	.14		.16	
Paint	S.F.	.15	.37		.52	
Minimum Charge	Job		169		169	
Re-screen						
Install	S.F.	.59	1.35		1.94	Includes labor and material to
Minimum Charge	Job		169		169	re-screen wood frame.

Exterior Trim

Exterior Trim

Exterior Trim		Unit	Material	Labor	Equip.	Total	Specification
Stock Lumber							
1″ x 2″							
	Demolish	L.F.		.22		.22	Cost includes material and labor to
	Install	L.F.	.40	1.02		1.42	install 1″ x 2″ pine trim.
	Demolish and Install	L.F.	.40	1.24		1.64	
	Clean	L.F.	.01	.16		.17	
	Paint	L.F.	.02	.46		.48	
	Minimum Charge	Job		169		169	
1″ x 3″							
	Demolish	L.F.		.22		.22	Cost includes material and labor to
	Install	L.F.	.31	1.13		1.44	install 1″ x 3″ pine trim.
	Demolish and Install	L.F.	.31	1.35		1.66	
	Clean	L.F.	.01	.16		.17	
	Paint	L.F.	.02	.46		.48	
	Minimum Charge	Job		169		169	
1″ x 4″							
	Demolish	L.F.		.22		.22	Cost includes material and labor to
	Install	L.F.	.78	1.35		2.13	install 1″ x 4″ pine trim.
	Demolish and Install	L.F.	.78	1.57		2.35	
	Clean	L.F.	.01	.16		.17	
	Paint	L.F.	.02	.46		.48	
	Minimum Charge	Job		169		169	
1″ x 6″							
	Demolish	L.F.		.22		.22	Cost includes material and labor to
	Install	L.F.	1.36	1.35		2.71	install 1″ x 6″ pine trim.
	Demolish and Install	L.F.	1.36	1.57		2.93	
	Clean	L.F.	.01	.16		.17	
	Paint	L.F.	.07	.46		.53	
	Minimum Charge	Job		169		169	
1″ x 8″							
	Demolish	L.F.		.24		.24	Cost includes material and labor to
	Install	L.F.	1.58	1.69		3.27	install 1″ x 8″ pine trim.
	Demolish and Install	L.F.	1.58	1.93		3.51	
	Reinstall	L.F.		1.35		1.35	
	Clean	L.F.	.02	.17		.19	
	Paint	L.F.	.07	.46		.53	
	Minimum Charge	Job		169		169	

Exterior Trim

Exterior Trim		Unit	Material	Labor	Equip.	Total	Specification
1″ x 10″							
	Demolish	L.F.		.37		.37	Cost includes material and labor to
	Install	L.F.	1.85	1.88		3.73	install 1″ x 10″ pine trim.
	Demolish and Install	L.F.	1.85	2.25		4.10	
	Minimum Charge	Job		169		169	
1″ x 12″							
	Demolish	L.F.		.37		.37	Cost includes material and labor to
	Install	L.F.	2.08	1.88		3.96	install 1″ x 12″ pine trim.
	Demolish and Install	L.F.	2.08	2.25		4.33	
	Minimum Charge	Job		169		169	

Fascia		Unit	Material	Labor	Equip.	Total	Specification
Aluminum							
	Demolish	L.F.		.47		.47	Includes material and labor to install
	Install	S.F.	1.36	1.61		2.97	aluminum fascia.
	Demolish and Install	S.F.	1.36	2.08		3.44	
	Reinstall	S.F.		1.29		1.29	
	Clean	L.F.		.30		.30	
	Minimum Charge	Job		169		169	
Vinyl							
	Demolish	L.F.		.47		.47	Cost includes material and labor to
	Install	L.F.	2.88	1.93		4.81	install 6″ vinyl fascia with 12″ vinyl
	Demolish and Install	L.F.	2.88	2.40		5.28	soffit and J-channel.
	Reinstall	L.F.		1.55		1.55	
	Clean	L.F.		.30		.30	
	Minimum Charge	Job		169		169	
Plywood							
	Demolish	L.F.		.46		.46	Cost includes material and labor to
	Install	L.F.	1.23	.75		1.98	install 12″ wide plywood fascia.
	Demolish and Install	L.F.	1.23	1.21		2.44	
	Reinstall	L.F.		.60		.60	
	Clean	L.F.	.03	.20		.23	
	Paint	L.F.	.14	.74		.88	
	Minimum Charge	Job		169		169	
Cedar							
	Demolish	L.F.		.46		.46	Cost includes material and labor to
	Install	L.F.	3.18	.75		3.93	install 12″ wide cedar fascia.
	Demolish and Install	L.F.	3.18	1.21		4.39	
	Reinstall	L.F.		.60		.60	
	Clean	L.F.	.03	.20		.23	
	Paint	L.F.	.14	.74		.88	
	Minimum Charge	Job		169		169	
Redwood							
1″ x 6″							
	Demolish	L.F.		.36		.36	Cost includes material and labor to
	Install	L.F.	1.65	1.35		3	install 1″ x 6″ redwood fascia board.
	Demolish and Install	L.F.	1.65	1.71		3.36	
	Reinstall	L.F.		1.08		1.08	
	Minimum Charge	Job		169		169	
1″ x 8″							
	Demolish	L.F.		.40		.40	Cost includes material and labor to
	Install	L.F.	3.60	1.50		5.10	install 1″ x 8″ redwood fascia board.
	Demolish and Install	L.F.	3.60	1.90		5.50	
	Reinstall	L.F.		1.20		1.20	
	Minimum Charge	Job		169		169	

Exterior Trim

Fascia		Unit	Material	Labor	Equip.	Total	Specification
2" x 6"							
	Demolish	L.F.		.36		.36	Cost includes material and labor to
	Install	L.F.	2.54	1.50		4.04	install 2" x 6" redwood fascia board.
	Demolish and Install	L.F.	2.54	1.86		4.40	
	Reinstall	L.F.		1.20		1.20	
	Minimum Charge	Job		169		169	
2" x 8"							
	Demolish	L.F.		.40		.40	Cost includes material and labor to
	Install	L.F.	3.44	1.69		5.13	install 2" x 8" redwood fascia board.
	Demolish and Install	L.F.	3.44	2.09		5.53	
	Reinstall	L.F.		1.35		1.35	
	Minimum Charge	Job		169		169	

Hem-fir Std & Better

		Unit	Material	Labor	Equip.	Total	Specification
2" x 6"							
	Demolish	L.F.		.36		.36	Cost includes material and labor to
	Install	L.F.	.65	2.70		3.35	install 2" x 6" hem-fir exterior trim
	Demolish and Install	L.F.	.65	3.06		3.71	board.
	Reinstall	L.F.		2.16		2.16	
	Clean	L.F.	.02	.20		.22	
	Paint	L.F.	.14	.74		.88	
	Minimum Charge	Job		169		169	
2" x 8"							
	Demolish	L.F.		.40		.40	Cost includes material and labor to
	Install	L.F.	.94	3		3.94	install 2" x 8" hem-fir exterior trim
	Demolish and Install	L.F.	.94	3.40		4.34	board.
	Reinstall	L.F.		2.40		2.40	
	Clean	L.F.	.02	.20		.22	
	Paint	L.F.	.14	.74		.88	
	Minimum Charge	Job		169		169	
2" x 10"							
	Demolish	L.F.		.46		.46	Cost includes material and labor to
	Install	L.F.	1.42	3.76		5.18	install 2" x 10" hem-fir exterior trim
	Demolish and Install	L.F.	1.42	4.22		5.64	board.
	Reinstall	L.F.		3		3	
	Clean	L.F.	.03	.22		.25	
	Paint	L.F.	.14	.74		.88	
	Minimum Charge	Job		169		169	

Barge Rafter

		Unit	Material	Labor	Equip.	Total	Specification
2" x 6"							
	Demolish	L.F.		.72		.72	Cost includes material and labor to
	Install	L.F.	.65	.38		1.03	install 2" x 6" barge rafters.
	Demolish and Install	L.F.	.65	1.10		1.75	
	Reinstall	L.F.		.30		.30	
	Clean	L.F.	.02	.20		.22	
	Paint	L.F.	.14	.74		.88	
	Minimum Charge	Job		169		169	
2" x 8"							
	Demolish	L.F.		.73		.73	Cost includes material and labor to
	Install	L.F.	.94	.40		1.34	install 2" x 8" barge rafters.
	Demolish and Install	L.F.	.94	1.13		2.07	
	Reinstall	L.F.		.32		.32	
	Clean	L.F.	.02	.20		.22	
	Paint	L.F.	.14	.74		.88	
	Minimum Charge	Job		169		169	

Exterior Trim

Fascia

Fascia	Unit	Material	Labor	Equip.	Total	Specification
2″ x 10″						
Demolish	L.F.		.74		.74	Cost includes material and labor to
Install	L.F.	1.42	.41		1.83	install 2″ x 10″ barge rafters.
Demolish and Install	L.F.	1.42	1.15		2.57	
Reinstall	L.F.		.33		.33	
Clean	L.F.	.03	.22		.25	
Paint	L.F.	.14	.74		.88	
Minimum Charge	Job		169		169	

Soffit

Soffit	Unit	Material	Labor	Equip.	Total	Specification
Aluminum						
12″ w / Fascia						
Demolish	L.F.		.47		.47	Cost includes material and labor to
Install	L.F.	3.32	1.93		5.25	install 12″ aluminum soffit with 6″
Demolish and Install	L.F.	3.32	2.40		5.72	fascia and J-channel.
Reinstall	L.F.		1.55		1.55	
Clean	L.F.		.30		.30	
Paint	L.F.	.18	.74		.92	
Minimum Charge	Job		169		169	
18″ w / Fascia						
Demolish	L.F.		.47		.47	Cost includes material and labor to
Install	L.F.	3.96	2.25		6.21	install 18″ aluminum soffit with 6″
Demolish and Install	L.F.	3.96	2.72		6.68	fascia and J-channel.
Reinstall	L.F.		1.80		1.80	
Clean	L.F.		.30		.30	
Paint	L.F.	.18	.74		.92	
Minimum Charge	Job		169		169	
24″ w / Fascia						
Demolish	L.F.		.47		.47	Cost includes material and labor to
Install	L.F.	4.63	2.50		7.13	install 24″ aluminum soffit with 6″
Demolish and Install	L.F.	4.63	2.97		7.60	fascia and J-channel.
Reinstall	L.F.		2		2	
Clean	L.F.		.44		.44	
Paint	L.F.	.18	.74		.92	
Minimum Charge	Job		169		169	
Vinyl						
Demolish	L.F.		.47		.47	Cost includes material and labor to
Install	L.F.	2.88	1.93		4.81	install 6″ vinyl fascia with 12″ vinyl
Demolish and Install	L.F.	2.88	2.40		5.28	soffit and J-channel.
Reinstall	L.F.		1.55		1.55	
Clean	L.F.		.30		.30	
Paint	L.F.	.18	.74		.92	
Minimum Charge	Job		169		169	
Plywood						
Demolish	L.F.		.52		.52	Cost includes material and labor to
Install	L.F.	1.23	.97		2.20	install 12″ wide plywood soffit.
Demolish and Install	L.F.	1.23	1.49		2.72	
Reinstall	L.F.		.77		.77	
Clean	L.F.	.03	.20		.23	
Paint	L.F.	.14	.74		.88	
Minimum Charge	Job		169		169	

Exterior Trim

Soffit		Unit	Material	Labor	Equip.	Total	Specification
Redwood							
	Demolish	L.F.		.52		.52	Cost includes material and labor to
	Install	L.F.	3.30	.97		4.27	install 12″ wide redwood soffit.
	Demolish and Install	L.F.	3.30	1.49		4.79	
	Reinstall	L.F.		.77		.77	
	Clean	L.F.	.03	.20		.23	
	Paint	L.F.	.14	.74		.88	
	Minimum Charge	Job		169		169	
Cedar							
	Demolish	L.F.		.52		.52	Cost includes material and labor to
	Install	L.F.	3.18	.97		4.15	install 12″ wide cedar soffit.
	Demolish and Install	L.F.	3.18	1.49		4.67	
	Reinstall	L.F.		.77		.77	
	Clean	L.F.	.03	.20		.23	
	Paint	L.F.	.14	.74		.88	
	Minimum Charge	Job		169		169	
Douglas Fir							
	Demolish	L.F.		.52		.52	Cost includes material and labor to
	Install	L.F.	2.08	.97		3.05	install 12″ wide Douglas fir soffit.
	Demolish and Install	L.F.	2.08	1.49		3.57	
	Reinstall	L.F.		.77		.77	
	Clean	L.F.	.03	.20		.23	
	Paint	L.F.	.14	.74		.88	
	Minimum Charge	Job		169		169	

65

Roofing and Flashing

Tile Roofing

Metal Flashing	Unit	Material	Labor	Equip.	Total	Specification
Galvanized						
6″						
Demolish	L.F.		.18		.18	Includes material and labor to install
Install	L.F.	.51	.78		1.29	flat galvanized steel flashing, 6″ wide.
Demolish and Install	L.F.	.51	.96		1.47	
Paint	L.F.	.10	.37		.47	
Minimum Charge	Job		155		155	
14″						
Demolish	L.F.		.18		.18	Includes material and labor to install
Install	L.F.	1.19	.78		1.97	flat galvanized steel flashing, 14″
Demolish and Install	L.F.	1.19	.96		2.15	wide.
Paint	L.F.	.10	.37		.47	
Minimum Charge	Job		155		155	
20″						
Demolish	L.F.		.18		.18	Includes material and labor to install
Install	L.F.	1.68	.78		2.46	flat galvanized steel flashing, 20″
Demolish and Install	L.F.	1.68	.96		2.64	wide.
Paint	L.F.	.10	.37		.47	
Minimum Charge	Job		155		155	
Aluminum						
6″						
Demolish	L.F.		.18		.18	Includes material and labor to install
Install	L.F.	.30	.78		1.08	flat aluminum flashing, 6″ wide.
Demolish and Install	L.F.	.30	.96		1.26	
Paint	L.F.	.10	.37		.47	
Minimum Charge	Job		155		155	
14″						
Demolish	L.F.		.18		.18	Includes material and labor to install
Install	L.F.	.61	.78		1.39	flat aluminum flashing, 14″ wide.
Demolish and Install	L.F.	.61	.96		1.57	
Paint	L.F.	.10	.37		.47	
Minimum Charge	Job		155		155	
20″						
Demolish	L.F.		.18		.18	Includes material and labor to install
Install	L.F.	.78	.78		1.56	flat aluminum flashing, 20″ wide.
Demolish and Install	L.F.	.78	.96		1.74	
Paint	L.F.	.10	.37		.47	
Minimum Charge	Job		155		155	

Roofing

Metal Flashing	Unit	Material	Labor	Equip.	Total	Specification
Copper						
6"						
Demolish	L.F.		.18		.18	Includes material and labor to install
Install	L.F.	1.80	1.03		2.83	flat copper flashing, 6" wide.
Demolish and Install	L.F.	1.80	1.21		3.01	
Minimum Charge	Job		187		187	
14"						
Demolish	L.F.		.18		.18	Includes material and labor to install
Install	L.F.	4.18	1.03		5.21	flat copper flashing, 14" wide.
Demolish and Install	L.F.	4.18	1.21		5.39	
Minimum Charge	Job		187		187	
18"						
Demolish	L.F.		.18		.18	Includes material and labor to install
Install	L.F.	6	1.03		7.03	flat copper flashing, 18" wide.
Demolish and Install	L.F.	6	1.21		7.21	
Minimum Charge	Job		187		187	
Aluminum Valley						
Demolish	L.F.		.18		.18	Includes material and labor to install
Install	L.F.	2.73	1.99		4.72	aluminum valley flashing, .024" thick.
Demolish and Install	L.F.	2.73	2.17		4.90	
Paint	L.F.	.10	.37		.47	
Minimum Charge	Job		169		169	
Gravel Stop						
Demolish	L.F.		.16		.16	Includes material and labor to install
Install	L.F.	4.71	2.58		7.29	aluminum gravel stop, .050" thick, 4"
Demolish and Install	L.F.	4.71	2.74		7.45	high, mill finish.
Reinstall	L.F.		2.06		2.06	
Paint	L.F.	.15	.37		.52	
Minimum Charge	Job		169		169	
Drip Edge						
Demolish	L.F.		.16		.16	Includes material and labor to install
Install	L.F.	.33	.85		1.18	aluminum drip edge, .016" thick, 5"
Demolish and Install	L.F.	.33	1.01		1.34	wide, mill finish.
Reinstall	L.F.		.68		.68	
Paint	L.F.	.15	.37		.52	
Minimum Charge	Job		169		169	
End Wall						
Demolish	L.F.		.83		.83	Includes material and labor to install
Install	L.F.	1.77	1.99		3.76	aluminum end wall flashing, .024"
Demolish and Install	L.F.	1.77	2.82		4.59	thick.
Minimum Charge	Job		169		169	
Side Wall						
Demolish	L.F.		.16		.16	Includes material and labor to install
Install	L.F.	1.78	1.99		3.77	aluminum side wall flashing, .024"
Demolish and Install	L.F.	1.78	2.15		3.93	thick.
Minimum Charge	Job		169		169	
Coping and Wall Cap						
Demolish	L.F.		1.12		1.12	Includes material and labor to install
Install	L.F.	1.74	2.14		3.88	aluminum flashing, mill finish, 0.40
Demolish and Install	L.F.	1.74	3.26		5	thick, 12" wide.
Clean	L.F.	.08	.76		.84	
Minimum Charge	Job		155		155	

Roofing

Roof Shingles	Unit	Material	Labor	Equip.	Total	Specification
Underlayment						
15#						
Demolish	Sq.		8.95		8.95	Includes material and labor to install
Install	Sq.	3.75	4.84		8.59	#15 felt underlayment.
Demolish and Install	Sq.	3.75	13.79		17.54	
Minimum Charge	Job		155		155	
30#						
Demolish	Sq.		8.95		8.95	Includes material and labor to install
Install	Sq.	7.90	5.35		13.25	#30 felt underlayment.
Demolish and Install	Sq.	7.90	14.30		22.20	
Minimum Charge	Job		155		155	
Self Adhering						
Demolish	Sq.		22.50		22.50	Includes material and labor to install
Install	Sq.	50.50	14.10		64.60	self adhering ice barrier roofing
Demolish and Install	Sq.	50.50	36.60		87.10	underlayment.
Minimum Charge	Job		155		155	

Rolled Roofing	Unit	Material	Labor	Equip.	Total	Specification
90 lb						
Demolish	Sq.		22.50		22.50	Cost includes material and labor to
Install	Sq.	18.20	20.50		38.70	install 90 lb mineral surface rolled
Demolish and Install	Sq.	18.20	43		61.20	roofing.
Minimum Charge	Job		155		155	
110 lb						
Demolish	Sq.		22.50		22.50	Cost includes material and labor to
Install	Sq.	48	20.50		68.50	install 110 lb mineral surface rolled
Demolish and Install	Sq.	48	43		91	roofing.
Minimum Charge	Job		155		155	
140 lb						
Demolish	Sq.		22.50		22.50	Cost includes material and labor to
Install	Sq.	48	31		79	install 140 lb mineral surface rolled
Demolish and Install	Sq.	48	53.50		101.50	roofing.
Minimum Charge	Job		155		155	

Composition Shingles	Unit	Material	Labor	Equip.	Total	Specification
Fiberglass (3-tab)						
225 lb, 20 Year						
Demolish	Sq.		24.50		24.50	Cost includes material and labor to
Install	Sq.	36	60.50		96.50	install 225 lb, 20 year fiberglass
Demolish and Install	Sq.	36	85		121	shingles.
Minimum Charge	Job		155		155	
300 lb, 25 Year						
Demolish	Sq.		24.50		24.50	Cost includes material and labor to
Install	Sq.	28.50	60.50		89	install 300 lb, 25 year fiberglass
Demolish and Install	Sq.	28.50	85		113.50	shingles.
Minimum Charge	Job		155		155	
Emergency Repairs						
Install	Ea.	72	155		227	Includes labor and material for
Minimum Charge	Job		155		155	emergency repairs.

Roofing

Composition Shingles	Unit	Material	Labor	Equip.	Total	Specification
Architectural						
Laminated, 25 Year						
Demolish	Sq.		24.50		24.50	Cost includes material and labor to
Install	Sq.	51	60.50		111.50	install 25 year fiberglass roof shingles.
Demolish and Install	Sq.	51	85		136	
Minimum Charge	Job		155		155	
Laminated, 30 Year						
Demolish	Sq.		24.50		24.50	Cost includes material and labor to
Install	Sq.	51	60.50		111.50	install 30 year fiberglass roof shingles.
Demolish and Install	Sq.	51	85		136	
Minimum Charge	Job		155		155	
Laminated, 40 Year						
Demolish	Sq.		24.50		24.50	Cost includes material and labor to
Install	Sq.	73.50	60.50		134	install 40 year fiberglass roof shingles.
Demolish and Install	Sq.	73.50	85		158.50	
Minimum Charge	Job		155		155	
Laminated Shake, 40 Year						
Demolish	Sq.		24.50		24.50	Cost includes material and labor to
Install	Sq.	83	80.50		163.50	install 40 year fiberglass roof shakes.
Demolish and Install	Sq.	83	105		188	
Minimum Charge	Job		155		155	
Emergency Repairs						
Install	Ea.	72	155		227	Includes labor and material for
Minimum Charge	Job		155		155	emergency repairs.
Roof Jack						
Demolish	Ea.		4.88		4.88	Includes material and labor to install
Install	Ea.	13.80	34		47.80	residential roof jack, w/bird screen,
Demolish and Install	Ea.	13.80	38.88		52.68	backdraft damper, 3" & 4" dia. round
Paint	Ea.	.06	3.71		3.77	duct.
Minimum Charge	Job		184		184	
Ridge Vent						
Demolish	L.F.		.87		.87	Includes material and labor to install a
Install	L.F.	2.93	2.41		5.34	mill finish aluminum ridge vent strip.
Demolish and Install	L.F.	2.93	3.28		6.21	
Minimum Charge	Job		169		169	
Shingle Molding						
1" x 2"						
Demolish	L.F.		.17		.17	Cost includes material and labor to
Install	L.F.	.40	1.02		1.42	install 1" x 2" pine trim.
Demolish and Install	L.F.	.40	1.19		1.59	
Paint	L.F.	.02	.46		.48	
Minimum Charge	Job		169		169	
1" x 3"						
Demolish	L.F.		.17		.17	Cost includes material and labor to
Install	L.F.	.44	1.17		1.61	install 1" x 3" pine trim.
Demolish and Install	L.F.	.44	1.34		1.78	
Paint	L.F.	.02	.46		.48	
Minimum Charge	Job		169		169	
1" x 4"						
Demolish	L.F.		.17		.17	Cost includes material and labor to
Install	L.F.	.78	1.35		2.13	install 1" x 4" pine trim.
Demolish and Install	L.F.	.78	1.52		2.30	
Paint	L.F.	.02	.46		.48	
Minimum Charge	Job		169		169	

Roofing

Composition Shingles		Unit	Material	Labor	Equip.	Total	Specification
Add for Steep Pitch	Install	Sq.		19.40		19.40	Includes labor and material added for 2nd story or steep roofs.
Add for Additional Story	Install	Sq.		19.40		19.40	Includes labor and material added for 2nd story or steep roofs.

Wood Shingles		Unit	Material	Labor	Equip.	Total	Specification
Cedar Wood Shingles							
16″ Red Label (#2)							
	Demolish	Sq.		59.50		59.50	Includes material and labor to install
	Install	Sq.	121	135		256	wood shingles/shakes.
	Demolish and Install	Sq.	121	194.50		315.50	
	Minimum Charge	Job		169		169	
16″ Blue Label (#1)							
	Demolish	Sq.		59.50		59.50	Includes material and labor to install
	Install	Sq.	206	104		310	wood shingles/shakes.
	Demolish and Install	Sq.	206	163.50		369.50	
	Minimum Charge	Job		169		169	
18″ Red Label (#2)							
	Demolish	Sq.		59.50		59.50	Includes material and labor to install
	Install	Sq.	231	86.50		317.50	wood shingles/shakes.
	Demolish and Install	Sq.	231	146		377	
	Minimum Charge	Job		169		169	
18″ Blue Label (#1)							
	Demolish	Sq.		59.50		59.50	Includes material and labor to install
	Install	Sq.	185	94.50		279.50	wood shingles/shakes.
	Demolish and Install	Sq.	185	154		339	
	Minimum Charge	Job		169		169	
Replace Shingle							
	Install	Ea.	13.20	8.45		21.65	Includes labor and material to replace
	Minimum Charge	Job		169		169	individual wood shingle/shake. Cost is per shingle.
Cedar Shake							
18″ Medium Handsplit							
	Demolish	Sq.		59.50		59.50	Includes material and labor to install
	Install	Sq.	98.50	130		228.50	wood shingles/shakes.
	Demolish and Install	Sq.	98.50	189.50		288	
	Minimum Charge	Job		169		169	
18″ Heavy Handsplit							
	Demolish	Sq.		59.50		59.50	Includes material and labor to install
	Install	Sq.	98.50	144		242.50	wood shingles/shakes.
	Demolish and Install	Sq.	98.50	203.50		302	
	Minimum Charge	Job		169		169	
24″ Heavy Handsplit							
	Demolish	Sq.		59.50		59.50	Includes material and labor to install
	Install	Sq.	128	116		244	wood shingles/shakes.
	Demolish and Install	Sq.	128	175.50		303.50	
	Minimum Charge	Job		169		169	
24″ Medium Handsplit							
	Demolish	Sq.		59.50		59.50	Includes material and labor to install
	Install	Sq.	128	104		232	wood shingles/shakes.
	Demolish and Install	Sq.	128	163.50		291.50	
	Minimum Charge	Job		169		169	

Roofing

Wood Shingles

Wood Shingles		Unit	Material	Labor	Equip.	Total	Specification
24″ Straight Split							
	Demolish	Sq.		59.50		59.50	Includes material and labor to install
	Install	Sq.	156	154		310	wood shingles/shakes.
	Demolish and Install	Sq.	156	213.50		369.50	
	Minimum Charge	Job		169		169	
24″ Tapersplit							
	Demolish	Sq.		59.50		59.50	Includes material and labor to install
	Install	Sq.	156	154		310	wood shingles/shakes.
	Demolish and Install	Sq.	156	213.50		369.50	
	Minimum Charge	Job		169		169	
18″ Straight Split							
	Demolish	Sq.		59.50		59.50	Includes material and labor to install
	Install	Sq.	132	154		286	wood shingles/shakes.
	Demolish and Install	Sq.	132	213.50		345.50	
	Minimum Charge	Job		169		169	
Replace Shingle							
	Install	Ea.	13.20	8.45		21.65	Includes labor and material to replace
	Minimum Charge	Job		169		169	individual wood shingle/shake. Cost is per shingle.

Slate Tile Roofing

Slate Tile Roofing		Unit	Material	Labor	Equip.	Total	Specification
Unfading Green							
	Demolish	Sq.		53.50		53.50	Includes material and labor to install
	Install	Sq.	400	177		577	clear Vermont slate tile.
	Demolish and Install	Sq.	400	230.50		630.50	
	Reinstall	Sq.		141.71		141.71	
	Minimum Charge	Job		155		155	
Unfading Purple							
	Demolish	Sq.		53.50		53.50	Includes material and labor to install
	Install	Sq.	430	177		607	clear Vermont slate tile.
	Demolish and Install	Sq.	430	230.50		660.50	
	Reinstall	Sq.		141.71		141.71	
	Minimum Charge	Job		155		155	
Replace Slate Tiles							
	Install	Ea.	6.25	16.30		22.55	Includes labor and material to replace
	Minimum Charge	Job		155		155	individual slate tiles. Cost is per tile.
Variegated Purple							
	Demolish	Sq.		53.50		53.50	Includes material and labor to install
	Install	Sq.	450	177		627	clear Vermont slate tile.
	Demolish and Install	Sq.	450	230.50		680.50	
	Reinstall	Sq.		141.71		141.71	
	Minimum Charge	Job		155		155	
Unfading Grey / Black							
	Demolish	Sq.		53.50		53.50	Includes material and labor to install
	Install	Sq.	395	177		572	clear Vermont slate tile.
	Demolish and Install	Sq.	395	230.50		625.50	
	Reinstall	Sq.		141.71		141.71	
	Minimum Charge	Job		155		155	
Unfading Red							
	Demolish	Sq.		53.50		53.50	Includes material and labor to install
	Install	Sq.	1300	177		1477	clear Vermont slate tile.
	Demolish and Install	Sq.	1300	230.50		1530.50	
	Reinstall	Sq.		141.71		141.71	
	Minimum Charge	Job		155		155	

Roofing

Slate Tile Roofing

	Unit	Material	Labor	Equip.	Total	Specification
Weathering Black						Includes material and labor to install clear Pennsylvania slate tile.
Demolish	Sq.		53.50		53.50	
Install	Sq.	525	177		702	
Demolish and Install	Sq.	525	230.50		755.50	
Reinstall	Sq.		141.71		141.71	
Minimum Charge	Job		155		155	
Weathering Green						Includes material and labor to install clear Vermont slate tile.
Demolish	Sq.		53.50		53.50	
Install	Sq.	385	177		562	
Demolish and Install	Sq.	385	230.50		615.50	
Reinstall	Sq.		141.71		141.71	
Minimum Charge	Job		155		155	

Flat Clay Tile Roofing

	Unit	Material	Labor	Equip.	Total	Specification
Glazed						Includes material and labor to install tile roofing. Cost based on terra cotta red flat clay tile and includes ridge and rake tiles.
Demolish	Sq.		59.50		59.50	
Install	Sq.	510	124		634	
Demolish and Install	Sq.	510	183.50		693.50	
Reinstall	Sq.		99.20		99.20	
Minimum Charge	Job		155		155	
Terra Cotta Red						Includes material and labor to install tile roofing. Cost based on terra cotta red flat clay tile and includes ridge and rake tiles.
Demolish	Sq.		59.50		59.50	
Install	Sq.	510	124		634	
Demolish and Install	Sq.	510	183.50		693.50	
Reinstall	Sq.		99.20		99.20	
Minimum Charge	Job		155		155	

Mission Tile Roofing

	Unit	Material	Labor	Equip.	Total	Specification
Glazed Red						Includes material and labor to install tile roofing. Cost based on glazed red tile and includes birdstop, ridge and rake tiles.
Demolish	Sq.		59.50		59.50	
Install	Sq.	735	207		942	
Demolish and Install	Sq.	735	266.50		1001.50	
Reinstall	Sq.		165.33		165.33	
Minimum Charge	Job		155		155	
Unglazed Red						Includes material and labor to install tile roofing. Cost based on unglazed red tile and includes birdstop, ridge and rake tiles.
Demolish	Sq.		59.50		59.50	
Install	Sq.	735	207		942	
Demolish and Install	Sq.	735	266.50		1001.50	
Reinstall	Sq.		165.33		165.33	
Minimum Charge	Job		155		155	
Glazed Blue						Includes material and labor to install tile roofing. Cost based on glazed blue tile and includes birdstop, ridge and rake tiles.
Demolish	Sq.		59.50		59.50	
Install	Sq.	735	207		942	
Demolish and Install	Sq.	735	266.50		1001.50	
Reinstall	Sq.		165.33		165.33	
Minimum Charge	Job		155		155	
Glazed White						Includes material and labor to install tile roofing. Cost based on glazed white tile and includes birdstop, ridge and rake tiles.
Demolish	Sq.		59.50		59.50	
Install	Sq.	735	207		942	
Demolish and Install	Sq.	735	266.50		1001.50	
Reinstall	Sq.		165.33		165.33	
Minimum Charge	Job		155		155	

Mission Tile Roofing	Unit	Material	Labor	Equip.	Total	Specification
Unglazed White						Includes material and labor to install
Demolish	Sq.		59.50		59.50	tile roofing. Cost based on unglazed
Install	Sq.	735	207		942	white tile and includes birdstop, ridge
Demolish and Install	Sq.	735	266.50		1001.50	and rake tiles.
Reinstall	Sq.		165.33		165.33	
Minimum Charge	Job		155		155	
Color Blend						Includes material and labor to install
Demolish	Sq.		59.50		59.50	tile roofing. Cost based on color blend
Install	Sq.	735	207		942	tile and includes birdstop, ridge and
Demolish and Install	Sq.	735	266.50		1001.50	rake tiles.
Reinstall	Sq.		165.33		165.33	
Minimum Charge	Job		155		155	
Accessories / Extras						
Add for Cap Furring Strips						Includes labor and material to install
Install	Sq.	80.50	194		274.50	vertically placed wood furring strips under cap tiles.
Add for Tile Adhesive						Includes material and labor for
Install	Sq.	13.75	11.30		25.05	adhesive material placed between tiles.
Add for Wire Attachment						Includes labor and material to install
Install	Sq.	13.60	135		148.60	single wire anchors for attaching tile.
Add for Braided Runners						Includes labor and material to install
Install	Sq.	24	129		153	braided wire runners for secure attachment in high wind areas.
Add for Stainless Steel Nails						Includes material only for stainless
Install	Sq.	18.75			18.75	steel nails.
Add for Brass Nails						
Install	Sq.	25.50			25.50	Includes material only for brass nails.
Add for Hurricane Clips						Includes material and labor for
Install	Sq.	16.45	182		198.45	Hurricane or wind clips for high wind areas.
Add for Copper Nails						
Install	Sq.	25.50			25.50	Includes material only for copper nails.
Add for Vertical Furring Strips						Includes labor and material to install
Install	Sq.	15.85	36		51.85	vertically placed wood furring strips under tiles.
Add for Horizontal Furring Strips						Includes labor and material to install
Install	Sq.	10.20	16.15		26.35	horizontally placed wood furring strips under tiles.

Roofing

Spanish Tile Roofing		Unit	Material	Labor	Equip.	Total	Specification
Glazed Red							
	Demolish	Sq.		59.50		59.50	Includes material and labor to install
	Install	Sq.	510	129		639	tile roofing. Cost based on glazed red
	Demolish and Install	Sq.	510	188.50		698.50	tile and includes birdstop, ridge and
	Reinstall	Sq.		103.33		103.33	rake tiles.
	Minimum Charge	Job		155		155	
Glazed White							
	Demolish	Sq.		59.50		59.50	Includes material and labor to install
	Install	Sq.	510	129		639	tile roofing. Cost based on glazed
	Demolish and Install	Sq.	510	188.50		698.50	white tile and includes birdstop, ridge
	Reinstall	Sq.		103.33		103.33	and rake tiles.
	Minimum Charge	Job		155		155	
Glazed Blue							
	Demolish	Sq.		59.50		59.50	Includes material and labor to install
	Install	Sq.	510	129		639	tile roofing. Cost based on glazed
	Demolish and Install	Sq.	510	188.50		698.50	blue tile and includes birdstop, ridge
	Reinstall	Sq.		103.33		103.33	and rake tiles.
	Minimum Charge	Job		155		155	
Unglazed Red							
	Demolish	Sq.		59.50		59.50	Includes material and labor to install
	Install	Sq.	510	129		639	tile roofing. Cost based on unglazed
	Demolish and Install	Sq.	510	188.50		698.50	red tile and includes birdstop, ridge
	Reinstall	Sq.		103.33		103.33	and rake tiles.
	Minimum Charge	Job		155		155	
Unglazed White							
	Demolish	Sq.		59.50		59.50	Includes material and labor to install
	Install	Sq.	510	129		639	tile roofing. Cost based on unglazed
	Demolish and Install	Sq.	510	188.50		698.50	white tile and includes birdstop, ridge
	Reinstall	Sq.		103.33		103.33	and rake tiles.
	Minimum Charge	Job		155		155	
Color Blend							
	Demolish	Sq.		59.50		59.50	Includes material and labor to install
	Install	Sq.	510	129		639	tile roofing. Cost based on color blend
	Demolish and Install	Sq.	510	188.50		698.50	tile and includes birdstop, ridge and
	Reinstall	Sq.		103.33		103.33	rake tiles.
	Minimum Charge	Job		155		155	
Accessories / Extras							
Add for Vertical Furring Strips							
	Install	Sq.	15.85	36		51.85	Includes labor and material to install vertically placed wood furring strips under tiles.
Add for Wire Attachment							
	Install	Sq.	11.35	103		114.35	Includes labor and material to install single wire ties.
Add for Braided Runners							
	Install	Sq.	20	88.50		108.50	Includes labor and material to install braided wire runners for secure attachment in high wind areas.
Add for Stainless Steel Nails							
	Install	Sq.	15.85			15.85	Includes material only for stainless steel nails.
Add for Copper Nails							
	Install	Sq.	21			21	Includes material only for copper nails.

Roofing

Spanish Tile Roofing		Unit	Material	Labor	Equip.	Total	Specification
Add for Brass Nails							
	Install	Sq.	21			21	Includes material only for brass nails.
Add for Hurricane Clips							
	Install	Sq.	12.60	107		119.60	Includes material and labor to install hurricane or wind clips for high wind areas.
Add for Mortar Set Tiles							
	Install	Sq.	39	155		194	Includes material and labor to install setting tiles by using mortar material placed between tiles.
Add for Tile Adhesive							
	Install	Sq.	12.75	10.05		22.80	Includes material and labor for adhesive placed between tiles.
Add for Cap Furring Strips							
	Install	Sq.	80.50	194		274.50	Includes labor and material to install vertically placed wood furring strips under cap tiles.
Add for Horizontal Furring Strips							
	Install	Sq.	10.20	16.15		26.35	Includes labor and material to install horizontally placed wood furring strips under tiles.

Concrete Tile Roofing		Unit	Material	Labor	Equip.	Total	Specification
Corrugated Red / Brown							
	Demolish	Sq.		59.50		59.50	Includes material and labor to install tile roofing. Cost includes birdstop, booster, ridge and rake tiles.
	Install	Sq.	124	230		354	
	Demolish and Install	Sq.	124	289.50		413.50	
	Reinstall	Sq.		183.70		183.70	
	Minimum Charge	Job		155		155	
Corrugated Gray							
	Demolish	Sq.		59.50		59.50	Includes material and labor to install tile roofing. Cost includes birdstop, booster, ridge and rake tiles.
	Install	Sq.	124	230		354	
	Demolish and Install	Sq.	124	289.50		413.50	
	Reinstall	Sq.		183.70		183.70	
	Minimum Charge	Job		155		155	
Corrugated Bright Red							
	Demolish	Sq.		59.50		59.50	Includes material and labor to install tile roofing. Cost includes birdstop, booster, ridge and rake tiles.
	Install	Sq.	124	230		354	
	Demolish and Install	Sq.	124	289.50		413.50	
	Reinstall	Sq.		183.70		183.70	
	Minimum Charge	Job		155		155	
Corrugated Black							
	Demolish	Sq.		59.50		59.50	Includes material and labor to install tile roofing. Cost includes birdstop, booster, ridge and rake tiles.
	Install	Sq.	124	230		354	
	Demolish and Install	Sq.	124	289.50		413.50	
	Reinstall	Sq.		183.70		183.70	
	Minimum Charge	Job		155		155	
Corrugated Green							
	Demolish	Sq.		59.50		59.50	Includes material and labor to install tile roofing. Cost includes birdstop, booster, ridge and rake tiles.
	Install	Sq.	124	230		354	
	Demolish and Install	Sq.	124	289.50		413.50	
	Reinstall	Sq.		183.70		183.70	
	Minimum Charge	Job		155		155	

Concrete Tile Roofing		Unit	Material	Labor	Equip.	Total	Specification
Corrugated Blue							
	Demolish	Sq.		59.50		59.50	Includes material and labor to install
	Install	Sq.	124	230		354	tile roofing. Cost includes birdstop,
	Demolish and Install	Sq.	124	289.50		413.50	booster, ridge and rake tiles.
	Reinstall	Sq.		183.70		183.70	
	Minimum Charge	Job		155		155	
Flat Natural Gray							
	Demolish	Sq.		59.50		59.50	Includes material and labor to install
	Install	Sq.	108	230		338	tile roofing. Cost includes birdstop,
	Demolish and Install	Sq.	108	289.50		397.50	booster, ridge and rake tiles.
	Reinstall	Sq.		183.70		183.70	
	Minimum Charge	Job		155		155	
Flat Red / Brown							
	Demolish	Sq.		59.50		59.50	Includes material and labor to install
	Install	Sq.	108	230		338	tile roofing. Cost includes birdstop,
	Demolish and Install	Sq.	108	289.50		397.50	booster, ridge and rake tiles.
	Reinstall	Sq.		183.70		183.70	
	Minimum Charge	Job		155		155	
Flat Bright Red							
	Demolish	Sq.		59.50		59.50	Includes material and labor to install
	Install	Sq.	108	230		338	tile roofing. Cost includes birdstop,
	Demolish and Install	Sq.	108	289.50		397.50	booster, ridge and rake tiles.
	Reinstall	Sq.		183.70		183.70	
	Minimum Charge	Job		155		155	
Flat Green							
	Demolish	Sq.		59.50		59.50	Includes material and labor to install
	Install	Sq.	108	230		338	tile roofing. Cost includes birdstop,
	Demolish and Install	Sq.	108	289.50		397.50	booster, ridge and rake tiles.
	Minimum Charge	Job		155		155	
Flat Blue							
	Demolish	Sq.		59.50		59.50	Includes material and labor to install
	Install	Sq.	108	230		338	tile roofing. Cost includes birdstop,
	Demolish and Install	Sq.	108	289.50		397.50	booster, ridge and rake tiles.
	Reinstall	Sq.		183.70		183.70	
	Minimum Charge	Job		155		155	
Flat Black							
	Demolish	Sq.		59.50		59.50	Includes material and labor to install
	Install	Sq.	108	230		338	tile roofing. Cost includes birdstop,
	Demolish and Install	Sq.	108	289.50		397.50	booster, ridge and rake tiles.
	Reinstall	Sq.		183.70		183.70	
	Minimum Charge	Job		155		155	

Aluminum Sheet		Unit	Material	Labor	Equip.	Total	Specification
Corrugated							
.019" Thick Natural							
	Demolish	S.F.		.64		.64	Includes material and labor to install
	Install	S.F.	.99	1.56		2.55	aluminum sheet metal roofing.
	Demolish and Install	S.F.	.99	2.20		3.19	
	Reinstall	S.F.		1.25		1.25	
	Minimum Charge	Job		375		375	
.016" Thick Natural							
	Demolish	S.F.		.64		.64	Includes material and labor to install
	Install	S.F.	.76	1.56		2.32	aluminum sheet metal roofing.
	Demolish and Install	S.F.	.76	2.20		2.96	
	Reinstall	S.F.		1.25		1.25	
	Minimum Charge	Job		375		375	

Roofing

Aluminum Sheet	Unit	Material	Labor	Equip.	Total	Specification
.016″ Colored Finish						
Demolish	S.F.		.64		.64	Includes material and labor to install
Install	S.F.	1.10	1.56		2.66	aluminum sheet metal roofing.
Demolish and Install	S.F.	1.10	2.20		3.30	
Reinstall	S.F.		1.25		1.25	
Minimum Charge	Job		375		375	
.019″ Colored Finish						
Demolish	S.F.		.64		.64	Includes material and labor to install
Install	S.F.	1.22	1.56		2.78	aluminum sheet metal roofing.
Demolish and Install	S.F.	1.22	2.20		3.42	
Reinstall	S.F.		1.25		1.25	
Minimum Charge	Job		375		375	

Ribbed

	Unit	Material	Labor	Equip.	Total	Specification
.019″ Thick Natural						
Demolish	S.F.		.64		.64	Includes material and labor to install
Install	S.F.	.99	1.56		2.55	aluminum sheet metal roofing.
Demolish and Install	S.F.	.99	2.20		3.19	
Reinstall	S.F.		1.25		1.25	
Minimum Charge	Job		375		375	
.016″ Thick Natural						
Demolish	S.F.		.64		.64	Includes material and labor to install
Install	S.F.	.76	1.56		2.32	aluminum sheet metal roofing.
Demolish and Install	S.F.	.76	2.20		2.96	
Reinstall	S.F.		1.25		1.25	
Minimum Charge	Job		375		375	
.050″ Colored Finish						
Demolish	S.F.		.64		.64	Includes material and labor to install
Install	S.F.	3.41	1.56		4.97	aluminum sheet metal roofing.
Demolish and Install	S.F.	3.41	2.20		5.61	
Reinstall	S.F.		1.25		1.25	
Minimum Charge	Job		375		375	
.032″ Thick Natural						
Demolish	S.F.		.64		.64	Includes material and labor to install
Install	S.F.	1.90	1.56		3.46	aluminum sheet metal roofing.
Demolish and Install	S.F.	1.90	2.20		4.10	
Reinstall	S.F.		1.25		1.25	
Minimum Charge	Job		375		375	
.040″ Thick Natural						
Demolish	S.F.		.64		.64	Includes material and labor to install
Install	S.F.	2.33	1.56		3.89	aluminum sheet metal roofing.
Demolish and Install	S.F.	2.33	2.20		4.53	
Reinstall	S.F.		1.25		1.25	
Minimum Charge	Job		375		375	
.050″ Thick Natural						
Demolish	S.F.		.64		.64	Includes material and labor to install
Install	S.F.	2.83	1.56		4.39	aluminum sheet metal roofing.
Demolish and Install	S.F.	2.83	2.20		5.03	
Reinstall	S.F.		1.25		1.25	
Minimum Charge	Job		375		375	
.016″ Colored Finish						
Demolish	S.F.		.64		.64	Includes material and labor to install
Install	S.F.	1.10	1.56		2.66	aluminum sheet metal roofing.
Demolish and Install	S.F.	1.10	2.20		3.30	
Reinstall	S.F.		1.25		1.25	
Minimum Charge	Job		375		375	

Roofing

Aluminum Sheet

Aluminum Sheet	Unit	Material	Labor	Equip.	Total	Specification
.019″ Colored Finish						
Demolish	S.F.		.64		.64	Includes material and labor to install
Install	S.F.	1.22	1.56		2.78	aluminum sheet metal roofing.
Demolish and Install	S.F.	1.22	2.20		3.42	
Reinstall	S.F.		1.25		1.25	
Minimum Charge	Job		375		375	
.032″ Colored Finish						
Demolish	S.F.		.64		.64	Includes material and labor to install
Install	S.F.	2.38	1.56		3.94	aluminum sheet metal roofing.
Demolish and Install	S.F.	2.38	2.20		4.58	
Reinstall	S.F.		1.25		1.25	
Minimum Charge	Job		375		375	
.040″ Colored Finish						
Demolish	S.F.		.64		.64	Includes material and labor to install
Install	S.F.	2.84	1.56		4.40	aluminum sheet metal roofing.
Demolish and Install	S.F.	2.84	2.20		5.04	
Reinstall	S.F.		1.25		1.25	
Minimum Charge	Job		375		375	

Fiberglass Sheet

Fiberglass Sheet	Unit	Material	Labor	Equip.	Total	Specification
12 Ounce Corrugated						
Demolish	S.F.		1.14		1.14	Cost includes material and labor to
Install	S.F.	3.55	1.23		4.78	install 12 ounce corrugated fiberglass
Demolish and Install	S.F.	3.55	2.37		5.92	sheet roofing.
Reinstall	S.F.		.99		.99	
Minimum Charge	Job		155		155	
8 Ounce Corrugated						
Demolish	S.F.		1.14		1.14	Cost includes material and labor to
Install	S.F.	2.46	1.23		3.69	install 8 ounce corrugated fiberglass
Demolish and Install	S.F.	2.46	2.37		4.83	roofing.
Reinstall	S.F.		.99		.99	
Minimum Charge	Job		155		155	

Galvanized Steel

Galvanized Steel	Unit	Material	Labor	Equip.	Total	Specification
Corrugated						
24 Gauge						
Demolish	S.F.		.64		.64	Includes material and labor to install
Install	S.F.	2.08	1.30		3.38	galvanized steel sheet metal roofing.
Demolish and Install	S.F.	2.08	1.94		4.02	Cost based on 24 gauge corrugated
Reinstall	S.F.		1.04		1.04	or ribbed steel roofing.
Minimum Charge	Job		155		155	
26 Gauge						
Demolish	S.F.		.64		.64	Includes material and labor to install
Install	S.F.	1.68	1.23		2.91	galvanized steel sheet metal roofing.
Demolish and Install	S.F.	1.68	1.87		3.55	Cost based on 26 gauge corrugated
Reinstall	S.F.		.99		.99	or ribbed steel roofing.
Minimum Charge	Job		155		155	
28 Gauge						
Demolish	S.F.		.64		.64	Includes material and labor to install
Install	S.F.	1.21	1.17		2.38	galvanized steel sheet metal roofing.
Demolish and Install	S.F.	1.21	1.81		3.02	Cost based on 28 gauge corrugated
Reinstall	S.F.		.94		.94	or ribbed steel roofing.
Minimum Charge	Job		155		155	

Roofing

Galvanized Steel	Unit	Material	Labor	Equip.	Total	Specification
30 Gauge						
Demolish	S.F.		.64		.64	Includes material and labor to install
Install	S.F.	1.16	1.12		2.28	galvanized steel sheet metal roofing.
Demolish and Install	S.F.	1.16	1.76		2.92	Cost based on 30 gauge corrugated
Reinstall	S.F.		.90		.90	or ribbed steel roofing.
Minimum Charge	Job		155		155	
Ribbed						
24 Gauge						
Demolish	S.F.		.64		.64	Includes material and labor to install
Install	S.F.	2.08	1.30		3.38	galvanized steel sheet metal roofing.
Demolish and Install	S.F.	2.08	1.94		4.02	Cost based on 24 gauge corrugated
Reinstall	S.F.		1.04		1.04	or ribbed steel roofing.
Minimum Charge	Job		155		155	
26 Gauge						
Demolish	S.F.		.64		.64	Includes material and labor to install
Install	S.F.	1.68	1.23		2.91	galvanized steel sheet metal roofing.
Demolish and Install	S.F.	1.68	1.87		3.55	Cost based on 26 gauge corrugated
Reinstall	S.F.		.99		.99	or ribbed steel roofing.
Minimum Charge	Job		155		155	
28 Gauge						
Demolish	S.F.		.64		.64	Includes material and labor to install
Install	S.F.	1.21	1.17		2.38	galvanized steel sheet metal roofing.
Demolish and Install	S.F.	1.21	1.81		3.02	Cost based on 28 gauge corrugated
Reinstall	S.F.		.94		.94	or ribbed steel roofing.
Minimum Charge	Job		155		155	
30 Gauge						
Demolish	S.F.		.64		.64	Includes material and labor to install
Install	S.F.	1.16	1.12		2.28	galvanized steel sheet metal roofing.
Demolish and Install	S.F.	1.16	1.76		2.92	Cost based on 30 gauge corrugated
Reinstall	S.F.		.90		.90	or ribbed steel roofing.
Minimum Charge	Job		155		155	

Standing Seam	Unit	Material	Labor	Equip.	Total	Specification
Copper						
16 Ounce						
Demolish	Sq.		64		64	Includes material and labor to install
Install	Sq.	470	288		758	copper standing seam metal roofing.
Demolish and Install	Sq.	470	352		822	Cost based on 16 ounce copper
Reinstall	Sq.		230.15		230.15	roofing.
Minimum Charge	Job		187		187	
18 Ounce						
Demolish	Sq.		64		64	Includes material and labor to install
Install	Sq.	525	310		835	copper standing seam metal roofing.
Demolish and Install	Sq.	525	374		899	Cost based on 18 ounce copper
Reinstall	Sq.		249.33		249.33	roofing.
Minimum Charge	Job		187		187	
20 Ounce						
Demolish	Sq.		64		64	Includes material and labor to install
Install	Sq.	560	340		900	copper standing seam metal roofing.
Demolish and Install	Sq.	560	404		964	Cost based on 20 ounce copper
Reinstall	Sq.		272		272	roofing.

Roofing

Standing Seam

Standing Seam	Unit	Material	Labor	Equip.	Total	Specification
Stainless Steel						
26 Gauge						
Demolish	Sq.		64		64	Cost includes material and labor to
Install	Sq.	545	325		870	install 26 gauge stainless steel
Demolish and Install	Sq.	545	389		934	standing seam metal roofing.
Reinstall	Sq.		260.17		260.17	
Minimum Charge	Job		187		187	
28 Gauge						
Demolish	Sq.		64		64	Includes material and labor to install
Install	Sq.	440	310		750	standing seam metal roofing. Cost
Demolish and Install	Sq.	440	374		814	based on 28 gauge stainless steel
Reinstall	Sq.		249.33		249.33	roofing.
Minimum Charge	Job		187		187	

Elastomeric Roofing

Elastomeric Roofing	Unit	Material	Labor	Equip.	Total	Specification
Neoprene						
1/16"						
Demolish	Sq.		22		22	Includes material and labor to install
Install	Sq.	243	148	27	418	neoprene membrane, fully adhered,
Demolish and Install	Sq.	243	170	27	440	1/16" thick.
Minimum Charge	Job		710	80.50	790.50	
GRM Membrane						
Demolish	Sq.		22		22	Includes material and labor to install
Install	Sq.	246	236	27	509	coal tar based elastomeric roofing
Demolish and Install	Sq.	246	258	27	531	membrane, polyester fiber reinforced.
Minimum Charge	Job		710	80.50	790.50	
Cant Strips						
3"						
Demolish	L.F.		.69		.69	Includes material and labor to install
Install	L.F.	.45	.95		1.40	cants 3" x 3", treated timber, cut
Demolish and Install	L.F.	.45	1.64		2.09	diagonally.
Minimum Charge	Job		155		155	
4"						
Demolish	L.F.		.69		.69	Includes material and labor to install
Install	L.F.	1.52	.95		2.47	cants 4" x 4", treated timber, cut
Demolish and Install	L.F.	1.52	1.64		3.16	diagonally.
Minimum Charge	Job		155		155	
Polyvinyl Chloride (PVC)						
45 mil (loose laid)						
Demolish	Sq.		22		22	Cost includes material and labor to
Install	Sq.	92.50	28	3.15	123.65	install 45 mil PVC elastomeric roofing,
Demolish and Install	Sq.	92.50	50	3.15	145.65	loose-laid with stone.
Minimum Charge	Job		710	80.50	790.50	
48 mil (loose laid)						
Demolish	Sq.		22		22	Cost includes material and labor to
Install	Sq.	107	28	3.15	138.15	install 48 mil PVC elastomeric roofing,
Demolish and Install	Sq.	107	50	3.15	160.15	loose-laid with stone.
Minimum Charge	Job		710	80.50	790.50	
60 mil (loose laid)						
Demolish	Sq.		22		22	Cost includes material and labor to
Install	Sq.	99	28	3.15	130.15	install 60 mil PVC elastomeric roofing,
Demolish and Install	Sq.	99	50	3.15	152.15	loose-laid with stone.
Minimum Charge	Job		710	80.50	790.50	

Roofing

Elastomeric Roofing	Unit	Material	Labor	Equip.	Total	Specification
45 mil (attached)						
Demolish	Sq.		32.50		32.50	Cost includes material and labor to
Install	Sq.	81	54.50	6.20	141.70	install 45 mil PVC elastomeric roofing
Demolish and Install	Sq.	81	87	6.20	174.20	fully attached to roof deck.
Minimum Charge	Job		710	80.50	790.50	
48 mil (attached)						
Demolish	Sq.		32.50		32.50	Cost includes material and labor to
Install	Sq.	131	54.50	6.20	191.70	install 48 mil PVC elastomeric roofing
Demolish and Install	Sq.	131	87	6.20	224.20	fully attached to roof deck.
Minimum Charge	Job		710	80.50	790.50	
60 mil (attached)						
Demolish	Sq.		32.50		32.50	Cost includes material and labor to
Install	Sq.	122	54.50	6.20	182.70	install 60 mil PVC elastomeric roofing
Demolish and Install	Sq.	122	87	6.20	215.20	fully attached to roof deck.
Minimum Charge	Job		710	80.50	790.50	
CSPE Type						
45 mil (loose laid w / stone)						
Demolish	Sq.		22		22	Cost includes material and labor to
Install	Sq.	143	28	3.15	174.15	install 45 mil CSPE (chlorosulfonated
Demolish and Install	Sq.	143	50	3.15	196.15	polyethylene-hypalon) roofing material
Minimum Charge	Job		710	80.50	790.50	loose laid on deck and ballasted with
						stone.
45 mil (attached at seams)						
Demolish	Sq.		53		53	Cost includes material and labor to
Install	Sq.	138	40.50	4.59	183.09	install 45 mil CSPE (chlorosulfonated
Demolish and Install	Sq.	138	93.50	4.59	236.09	polyethylene-hypalon) roofing material
Minimum Charge	Job		710	80.50	790.50	attached to deck at seams with batten
						strips.
45 mil (fully attached)						
Demolish	Sq.		32.50		32.50	Cost includes material and labor to
Install	Sq.	135	54.50	6.20	195.70	install 45 mil CSPE (chlorosulfonated
Demolish and Install	Sq.	135	87	6.20	228.20	polyethylene-hypalon) roofing material
Minimum Charge	Job		710	80.50	790.50	fully attached to deck.
EPDM Type						
45 mil (attached at seams)						
Demolish	Sq.		53		53	Cost includes material and labor to
Install	Sq.	59	40.50	4.59	104.09	install 45 mil EPDM (ethylene
Demolish and Install	Sq.	59	93.50	4.59	157.09	propylene diene monomer) roofing
Minimum Charge	Job		710	80.50	790.50	material attached to deck at seams.
45 mil (fully attached)						
Demolish	Sq.		32.50		32.50	Cost includes material and labor to
Install	Sq.	84.50	54.50	6.20	145.20	install 45 mil EPDM (ethylene
Demolish and Install	Sq.	84.50	87	6.20	177.70	propylene diene monomer) roofing
Minimum Charge	Job		710	80.50	790.50	material fully attached to deck.
45 mil (loose laid w / stone)						
Demolish	Sq.		22		22	Cost includes material and labor to
Install	Sq.	67	28	3.15	98.15	install 45 mil EPDM (ethylene
Demolish and Install	Sq.	67	50	3.15	120.15	propylene diene monomer) roofing
Minimum Charge	Job		710	80.50	790.50	material loose laid on deck and
						ballasted with stones (approx. 1/2 ton
						of stone per square).
60 mil (attached at seams)						
Demolish	Sq.		53		53	Cost includes material and labor to
Install	Sq.	73	40.50	4.59	118.09	install 60 mil EPDM (ethylene
Demolish and Install	Sq.	73	93.50	4.59	171.09	propylene diene monomer) roofing
Minimum Charge	Job		710	80.50	790.50	material attached to deck at seams.

Roofing

Elastomeric Roofing

	Unit	Material	Labor	Equip.	Total	Specification
60 mil (fully attached)						Cost includes material and labor to
Demolish	Sq.		32.50		32.50	install 60 mil EPDM (ethylene
Install	Sq.	99	54.50	6.20	159.70	propylene diene monomer) roofing
Demolish and Install	Sq.	99	87	6.20	192.20	material fully attached to deck.
Minimum Charge	Job		710	80.50	790.50	
60 mil (loose laid w / stone)						Cost includes material and labor to
Demolish	Sq.		22		22	install 60 mil EPDM (ethylene
Install	Sq.	83	28	3.15	114.15	propylene diene monomer) roofing
Demolish and Install	Sq.	83	50	3.15	136.15	material loose laid on deck and
Minimum Charge	Job		710	80.50	790.50	ballasted with stones (approx. 1/2 ton of stone per square).
Silicone-urethane Foam						
Prep Roof for Foam System						Includes labor and material to removal
Install	Sq.	5.25	124		129.25	of any gravel and prepare existing
Minimum Charge	Job		710	80.50	790.50	built-up roof for application of silicone-urethane foam system.

Modified Bitumen Roofing

	Unit	Material	Labor	Equip.	Total	Specification
Fully Attached						
120 mil						
Demolish	Sq.		53		53	Cost includes material and labor to
Install	Sq.	84	6.90	1.26	92.16	install 120 mil modified bitumen
Demolish and Install	Sq.	84	59.90	1.26	145.16	roofing, fully attached to deck by
Minimum Charge	Job		155		155	torch.
150 mil						
Demolish	Sq.		53		53	Cost includes material and labor to
Install	Sq.	121	16.95	3.09	141.04	install 150 mil modified bitumen
Demolish and Install	Sq.	121	69.95	3.09	194.04	roofing, fully attached to deck by
Minimum Charge	Job		1025	185	1210	torch.
160 mil						
Demolish	Sq.		53		53	Cost includes material and labor to
Install	Sq.	97	17.70	3.22	117.92	install 160 mil modified bitumen
Demolish and Install	Sq.	97	70.70	3.22	170.92	roofing, fully attached to deck by
Minimum Charge	Job		1025	185	1210	torch.

Built-up Roofing

	Unit	Material	Labor	Equip.	Total	Specification
Emergency Repairs						
Install	Ea.	154	1450	265	1869	Minimum charge for work requiring
Minimum Charge	Job		1025	185	1210	hot asphalt.
Asphalt						
3-ply Asphalt						
Demolish	Sq.		55		55	Cost includes material and labor to
Install	Sq.	40.50	85	15.45	140.95	install 3-ply asphalt built-up roof
Demolish and Install	Sq.	40.50	140	15.45	195.95	including one 30 lb and two 15 lb
Minimum Charge	Job		1025	185	1210	layers of felt.

Built-up Roofing		Unit	Material	Labor	Equip.	Total	Specification
4-ply Asphalt							Cost includes material and labor to
	Demolish	Sq.		65.50		65.50	install 4-ply asphalt built-up roof
	Install	Sq.	54.50	92.50	16.85	163.85	including one 30 lb and three 15 lb
	Demolish and Install	Sq.	54.50	158	16.85	229.35	layers of felt.
	Minimum Charge	Job		1025	185	1210	
5-ply Asphalt							Cost includes material and labor to
	Demolish	Sq.		86		86	install 5-ply asphalt built-up roof
	Install	Sq.	78.50	102	18.55	199.05	including one 30 lb and four 15 lb
	Demolish and Install	Sq.	78.50	188	18.55	285.05	layers of felt.
	Minimum Charge	Job		1025	185	1210	
Components / Accessories							
Flood Coat							Includes labor and material to install
	Install	Sq.	13.55	47.50		61.05	flood coat.
R / R Gravel and Flood Coat							
	Install	Sq.	18.45	155		173.45	Includes labor and material to install flood coat and gravel.
Aluminum UV Coat							
	Install	Sq.	15.55	177	20	212.55	Includes labor and material to install aluminum-based UV coating.
Gravel Coat							
	Demolish	Sq.		27.50		27.50	Includes labor and material to install
	Install	Sq.	4.91	26		30.91	gravel coat.
	Demolish and Install	Sq.	4.91	53.50		58.41	
Mineral Surface Cap Sheet							Cost includes labor and material to
	Install	Sq.	21	5.35		26.35	install 50 # mineral surfaced cap sheet.
Fibrated Aluminum UV Coating							
	Install	Sq.	17.70	47.50		65.20	Includes labor and material to install fibered aluminum-based UV coating.
Impregnated Walk							
1"							
	Install	S.F.	2.51	1.65		4.16	Cost includes labor and material to install 1" layer of asphalt coating.
1/2"							
	Install	S.F.	1.69	1.31		3	Cost includes labor and material to install 1/2" layer of asphalt coating.
3/4"							
	Install	S.F.	2.10	1.38		3.48	Cost includes labor and material to install 3/4" layer of asphalt coating.

Exterior Doors and Windows

| Doors | Roof Window | Double Hung Window |

Exterior Type Door	Unit	Material	Labor	Equip.	Total	Specification
Entry						
Flush (1-3/4")						
Demolish	Ea.		31.50		31.50	Cost includes material and labor to
Install	Ea.	219	17.80		236.80	install 3' x 6' 8" pre-hung 1-3/4"
Demolish and Install	Ea.	219	49.30		268.30	door, including casing and stop,
Reinstall	Ea.		17.79		17.79	hinges, jamb, aluminum sill,
Clean	Ea.	3.28	10.05		13.33	weatherstripped.
Paint	Ea.	8.05	42.50		50.55	
Minimum Charge	Job		169		169	
Raised Panel						
Demolish	Ea.		31.50		31.50	Cost includes material and labor to
Install	Ea.	390	18.80		408.80	install 3' x 6' 8" pre-hung 1-3/4"
Demolish and Install	Ea.	390	50.30		440.30	door, jamb, including casing and
Reinstall	Ea.		18.78		18.78	stop, hinges, aluminum sill,
Clean	Ea.	3.28	10.05		13.33	weatherstripping.
Paint	Ea.	8.05	42.50		50.55	
Minimum Charge	Job		169		169	
Entry w / Lights						
Good Quality						
Demolish	Ea.		31.50		31.50	Cost includes material and labor to
Install	Ea.	365	19.90		384.90	install 3' x 6' 8" pre-hung 1-3/4"
Demolish and Install	Ea.	365	51.40		416.40	door, solid wood lower panel, with up
Reinstall	Ea.		19.88		19.88	to 9-light glass panels, casing, sill,
Clean	Ea.	3.28	10.05		13.33	weatherstripped.
Paint	Ea.	5.70	19.75		25.45	
Minimum Charge	Job		169		169	
Better Quality						
Demolish	Ea.		31.50		31.50	Cost includes material and labor to
Install	Ea.	400	19.90		419.90	install 3' x 6' 8" pre-hung 1-3/4" Ash
Demolish and Install	Ea.	400	51.40		451.40	door, solid wood, raised panel door
Reinstall	Ea.		19.88		19.88	without glass panels.
Clean	Ea.	3.28	10.05		13.33	
Paint	Ea.	8.05	42.50		50.55	
Minimum Charge	Job		169		169	
Premium Quality						
Demolish	Ea.		31.50		31.50	Cost includes material and labor to
Install	Ea.	645	19.90		664.90	install 3' x 6' 8" pre-hung 1-3/4" Oak
Demolish and Install	Ea.	645	51.40		696.40	door, solid wood, raised panel door
Reinstall	Ea.		19.88		19.88	without glass panels.
Clean	Ea.	3.28	10.05		13.33	
Paint	Ea.	8.05	42.50		50.55	
Minimum Charge	Job		169		169	

Exterior Type Door		Unit	Material	Labor	Equip.	Total	Specification
Custom Quality							
	Demolish	Ea.		31.50		31.50	Cost includes material and labor to
	Install	Ea.	1325	19.90		1344.90	install 3' x 6' 8" pre-hung 1-3/4"
	Demolish and Install	Ea.	1325	51.40		1376.40	hand-made, solid wood, raised panel
	Reinstall	Ea.		19.88		19.88	door with custom stained glass.
	Clean	Ea.	.22	10.05		10.27	
	Paint	Ea.	2.20	99		101.20	
	Minimum Charge	Job		169		169	
Miami Style							
	Demolish	Ea.		31.50		31.50	Cost includes material and labor to
	Install	Ea.	207	16.90		223.90	install 3' x 6' 8" pre-hung 1-3/4"
	Demolish and Install	Ea.	207	48.40		255.40	lauan door of solid wood.
	Reinstall	Ea.		16.90		16.90	
	Clean	Ea.	.22	10.05		10.27	
	Paint	Ea.	8.05	42.50		50.55	
	Minimum Charge	Job		169		169	
French Style							
1-Light							
	Demolish	Ea.		31.50		31.50	Cost includes material and labor to
	Install	Ea.	365	24		389	install 3' x 6' 8" pre-hung 1-3/4"
	Demolish and Install	Ea.	365	55.50		420.50	french style door with 1 light.
	Reinstall	Ea.		24.14		24.14	
	Clean	Ea.	2.05	15.10		17.15	
	Paint	Ea.	9.30	74		83.30	
	Minimum Charge	Job		169		169	
5-Light							
	Demolish	Ea.		31.50		31.50	Cost includes material and labor to
	Install	Ea.	365	24		389	install 3' x 6' 8" pre-hung 1-3/4"
	Demolish and Install	Ea.	365	55.50		420.50	french style door with 5 lights.
	Reinstall	Ea.		24.14		24.14	
	Clean	Ea.	.45	20		20.45	
	Paint	Ea.	9.30	74		83.30	
	Minimum Charge	Job		169		169	
9-Light							
	Demolish	Ea.		31.50		31.50	Cost includes material and labor to
	Install	Ea.	350	24		374	install 3' x 6' 8" pre-hung 1-3/4"
	Demolish and Install	Ea.	350	55.50		405.50	french style door with 9 lights.
	Reinstall	Ea.		24.14		24.14	
	Clean	Ea.	.45	20		20.45	
	Paint	Ea.	9.30	74		83.30	
	Minimum Charge	Job		169		169	
10-Light							
	Demolish	Ea.		33.50		33.50	Cost includes material and labor to
	Install	Ea.	585	42.50		627.50	install 3' x 6' 8" pre-hung 1-3/4"
	Demolish and Install	Ea.	585	76		661	french style door with 10 lights.
	Reinstall	Ea.		42.25		42.25	
	Clean	Ea.	.45	20		20.45	
	Paint	Ea.	9.30	74		83.30	
	Minimum Charge	Job		169		169	
15-Light							
	Demolish	Ea.		33.50		33.50	Cost includes material and labor to
	Install	Ea.	655	48.50		703.50	install 3' x 6' 8" pre-hung 1-3/4"
	Demolish and Install	Ea.	655	82		737	french style door with 15 lights.
	Reinstall	Ea.		48.29		48.29	
	Clean	Ea.	.45	20		20.45	
	Paint	Ea.	9.30	74		83.30	
	Minimum Charge	Job		169		169	

Exterior Type Door		Unit	Material	Labor	Equip.	Total	Specification
Jamb							
Wood							
	Demolish	Ea.		18.80		18.80	Includes material and labor to install
	Install	Ea.	98.50	9.15		107.65	flat pine jamb with square cut heads
	Demolish and Install	Ea.	98.50	27.95		126.45	and rabbeted sides for 6' 8" high and
	Clean	Ea.	.10	6.05		6.15	3-9/16" door including trim sets for
	Paint	Ea.	5.70	19.75		25.45	both sides.
	Minimum Charge	Job		169		169	
Metal 6' 8" High							
	Demolish	Ea.		18.80		18.80	Cost includes material and labor to
	Install	Ea.	127	21		148	install 18 gauge hollow metal door
	Demolish and Install	Ea.	127	39.80		166.80	frame to fit a 4-1/2" jamb 6' 8" high
	Reinstall	Ea.		21.13		21.13	and up to 3' 6" opening.
	Clean	Ea.	.68	5.05		5.73	
	Paint	Ea.	1.07	5.95		7.02	
	Minimum Charge	Job		169		169	
Metal 7' High							
	Demolish	Ea.		18.80		18.80	Cost includes material and labor to
	Install	Ea.	129	22.50		151.50	install 18 gauge hollow metal door
	Demolish and Install	Ea.	129	41.30		170.30	frame to fit a 4-1/2" jamb 7' high and
	Reinstall	Ea.		22.53		22.53	up to 3' 6" opening.
	Clean	Ea.	.68	5.05		5.73	
	Paint	Ea.	1.07	5.95		7.02	
	Minimum Charge	Job		169		169	
Metal 8' High							
	Demolish	Ea.		18.80		18.80	Cost includes material and labor to
	Install	Ea.	156	28		184	install 18 gauge hollow metal door
	Demolish and Install	Ea.	156	46.80		202.80	frame to fit a 4-1/2" jamb 8' high and
	Reinstall	Ea.		28.17		28.17	up to 3' 6" opening.
	Clean	Ea.	.97	6.05		7.02	
	Paint	Ea.	1.20	6.60		7.80	
	Minimum Charge	Job		169		169	
Metal 9' High							
	Demolish	Ea.		18.80		18.80	Cost includes material and labor to
	Install	Ea.	179	34		213	install 18 gauge hollow metal door
	Demolish and Install	Ea.	179	52.80		231.80	frame to fit a 4-1/2" jamb 9' high and
	Reinstall	Ea.		33.80		33.80	up to 3' 6" opening.
	Clean	Ea.	.97	6.05		7.02	
	Paint	Ea.	1.20	6.60		7.80	
	Minimum Charge	Job		169		169	
Casing Trim							
Single Width							
	Demolish	Ea.		5.95		5.95	Cost includes material and labor to
	Install	Opng.	24	57.50		81.50	install 11/16" x 2-1/2" pine ranch
	Demolish and Install	Opng.	24	63.45		87.45	style casing for one side of a standard
	Reinstall	Opng.		45.83		45.83	door opening.
	Clean	Opng.	.46	2.42		2.88	
	Paint	Ea.	3.15	5.95		9.10	
	Minimum Charge	Job		169		169	
Double Width							
	Demolish	Ea.		7.10		7.10	Cost includes material and labor to
	Install	Opng.	28	67.50		95.50	install 11/16" x 2-1/2" pine ranch
	Demolish and Install	Opng.	28	74.60		102.60	style door casing for one side of a
	Reinstall	Opng.		54.08		54.08	double door opening.
	Clean	Opng.	.55	4.83		5.38	
	Paint	Ea.	3.15	6.75		9.90	
	Minimum Charge	Job		169		169	

Residential Metal Door

	Unit	Material	Labor	Equip.	Total	Specification
Pre-hung, Steel-clad						
2′ 8″ x 6′ 8″						
Demolish	Ea.		31.50		31.50	Cost includes material and labor to
Install	Ea.	238	42.50		280.50	install 2′ 6″ x 6′ 8″ pre-hung 1-3/4″
Demolish and Install	Ea.	238	74		312	24 gauge steel clad door with 4-1/2″
Reinstall	Ea.		42.25		42.25	wood frame, trims, hinges, aluminum
Clean	Ea.	3.28	10.05		13.33	sill, weatherstripped.
Paint	Ea.	9.30	42.50		51.80	
Minimum Charge	Job		169		169	
3′ x 6′ 8″						
Demolish	Ea.		31.50		31.50	Cost includes material and labor to
Install	Ea.	238	45		283	install 3′ x 6′ 8″ pre-hung 1-3/4″ 24
Demolish and Install	Ea.	238	76.50		314.50	gauge steel clad door with 4-1/2″
Reinstall	Ea.		45.07		45.07	wood frame, trims, hinges, aluminum
Clean	Ea.	3.28	10.05		13.33	sill, weatherstripped.
Paint	Ea.	9.30	42.50		51.80	
Minimum Charge	Job		169		169	

Metal Door

	Unit	Material	Labor	Equip.	Total	Specification
2′ 8″ x 6′ 8″, 18 Gauge						
Demolish	Ea.		6.70		6.70	Cost includes material and labor to
Install	Ea.	232	21		253	install 18 gauge 2′ 8″ x 6′ 8″ steel
Demolish and Install	Ea.	232	27.70		259.70	door only, 1-3/4″ thick, hinges.
Reinstall	Ea.		21.13		21.13	
Clean	Ea.	1.36	7.55		8.91	
Paint	Ea.	3.87	24.50		28.37	
Minimum Charge	Job		169		169	
3′ x 6′ 8″, 18 Gauge						
Demolish	Ea.		6.70		6.70	Cost includes material and labor to
Install	Ea.	226	22.50		248.50	install 18 gauge 3′ x 6′ 8″ steel door
Demolish and Install	Ea.	226	29.20		255.20	only, 1-3/4″ thick, hinges.
Reinstall	Ea.		22.53		22.53	
Clean	Ea.	1.36	7.55		8.91	
Paint	Ea.	3.87	24.50		28.37	
Minimum Charge	Job		169		169	
2′ 8″ x 7′, 18 Gauge						
Demolish	Ea.		6.70		6.70	Cost includes material and labor to
Install	Ea.	400	22.50		422.50	install 18 gauge 2′ 8″ x 7′ steel door
Demolish and Install	Ea.	400	29.20		429.20	only, 1-3/4″ thick, hinges.
Reinstall	Ea.		22.53		22.53	
Clean	Ea.	1.36	7.55		8.91	
Paint	Ea.	3.87	24.50		28.37	
Minimum Charge	Job		169		169	
3′ x 7′, 18 Gauge						
Demolish	Ea.		6.70		6.70	Cost includes material and labor to
Install	Ea.	385	24		409	install 18 gauge 3′ x 7′ steel door
Demolish and Install	Ea.	385	30.70		415.70	only, 1-3/4″ thick, hinges.
Reinstall	Ea.		24.14		24.14	
Clean	Ea.	1.36	7.55		8.91	
Paint	Ea.	3.87	24.50		28.37	
Minimum Charge	Job		169		169	
3′ 6″ x 7′, 18 Gauge						
Demolish	Ea.		6.70		6.70	Cost includes material and labor to
Install	Ea.	415	26		441	install 18 gauge 3′ 6″ x 7′ steel door
Demolish and Install	Ea.	415	32.70		447.70	only, 1-3/4″ thick, hinges.
Reinstall	Ea.		26		26	
Clean	Ea.	1.36	7.55		8.91	
Paint	Ea.	3.87	24.50		28.37	
Minimum Charge	Job		169		169	

Exterior Doors and Windows

Metal Door		Unit	Material	Labor	Equip.	Total	Specification
Casing Trim							
Single Width							
	Demolish	Ea.		5.95		5.95	Cost includes material and labor to
	Install	Opng.	24	57.50		81.50	install 11/16" x 2-1/2" pine ranch
	Demolish and Install	Opng.	24	63.45		87.45	style casing for one side of a standard
	Reinstall	Opng.		45.83		45.83	door opening.
	Clean	Opng.	.46	2.42		2.88	
	Paint	Ea.	3.15	5.95		9.10	
	Minimum Charge	Job		169		169	
Double Width							
	Demolish	Ea.		7.10		7.10	Cost includes material and labor to
	Install	Opng.	28	67.50		95.50	install 11/16" x 2-1/2" pine ranch
	Demolish and Install	Opng.	28	74.60		102.60	style door casing for one side of a
	Reinstall	Opng.		54.08		54.08	double door opening.
	Clean	Opng.	.55	4.83		5.38	
	Paint	Ea.	3.15	6.75		9.90	
	Minimum Charge	Job		169		169	

Solid Core Door		Unit	Material	Labor	Equip.	Total	Specification
Exterior Type							
Flush (1-3/4")							
	Demolish	Ea.		31.50		31.50	Cost includes material and labor to
	Install	Ea.	325	42.50		367.50	install 3' x 6' 8" pre-hung 1-3/4"
	Demolish and Install	Ea.	325	74		399	door, including casing and stop,
	Reinstall	Ea.		42.25		42.25	hinges, jamb, aluminum sill,
	Clean	Ea.	3.28	10.05		13.33	weatherstripped.
	Paint	Ea.	9.30	42.50		51.80	
	Minimum Charge	Job		169		169	
Raised Panel							
	Demolish	Ea.		31.50		31.50	Cost includes material and labor to
	Install	Ea.	390	18.80		408.80	install 3' x 6' 8" pre-hung 1-3/4"
	Demolish and Install	Ea.	390	50.30		440.30	door, jamb, including casing and
	Reinstall	Ea.		18.78		18.78	stop, hinges, aluminum sill,
	Clean	Ea.	3.28	10.05		13.33	weatherstripping.
	Paint	Ea.	9.30	99		108.30	
	Minimum Charge	Job		169		169	
Entry w / Lights							
	Demolish	Ea.		31.50		31.50	Cost includes material and labor to
	Install	Ea.	365	19.90		384.90	install 3' x 6' 8" pre-hung 1-3/4"
	Demolish and Install	Ea.	365	51.40		416.40	door, solid wood lower panel, with up
	Reinstall	Ea.		19.88		19.88	to 9-light glass panels, casing, sill,
	Clean	Ea.	3.28	10.05		13.33	weatherstripped.
	Paint	Ea.	2.20	99		101.20	
	Minimum Charge	Job		169		169	
Custom Entry							
	Demolish	Ea.		31.50		31.50	Includes material and labor to install
	Install	Ea.	1250	19.90		1269.90	prehung, fir, single glazed with lead
	Demolish and Install	Ea.	1250	51.40		1301.40	caming, 1-3/4" x 6' 8" x 3'.
	Reinstall	Ea.		19.88		19.88	
	Clean	Ea.	3.28	10.05		13.33	
	Paint	Ea.	3.87	24.50		28.37	
	Minimum Charge	Job		169		169	

Exterior Doors and Windows

Solid Core Door		Unit	Material	Labor	Equip.	Total	Specification
Dutch Style							
	Demolish	Ea.		31.50		31.50	Includes material and labor to install
	Install	Ea.	745	56.50		801.50	dutch door, pine, 1-3/4" x 6' 8" x 2'
	Demolish and Install	Ea.	745	88		833	8" wide.
	Reinstall	Ea.		56.33		56.33	
	Clean	Ea.	3.28	10.05		13.33	
	Paint	Ea.	2.20	99		101.20	
	Minimum Charge	Job		169		169	
Miami Style							
	Demolish	Ea.		31.50		31.50	Cost includes material and labor to
	Install	Ea.	207	16.90		223.90	install 3' x 6' 8" pre-hung 1-3/4"
	Demolish and Install	Ea.	207	48.40		255.40	lauan door of solid wood.
	Reinstall	Ea.		16.90		16.90	
	Clean	Ea.	3.28	10.05		13.33	
	Paint	Ea.	8.05	42.50		50.55	
	Minimum Charge	Job		169		169	
French Style							
	Demolish	Ea.		33.50		33.50	Cost includes material and labor to
	Install	Ea.	655	48.50		703.50	install 3' x 6' 8" pre-hung 1-3/4"
	Demolish and Install	Ea.	655	82		737	french style door with 15 lights.
	Reinstall	Ea.		48.29		48.29	
	Clean	Ea.	.22	10.05		10.27	
	Paint	Ea.	5.90	148		153.90	
	Minimum Charge	Job		169		169	
Double							
	Demolish	Ea.		22.50		22.50	Includes material and labor to install a
	Install	Ea.	1150	84.50		1234.50	pre-hung double entry door including
	Demolish and Install	Ea.	1150	107		1257	frame and exterior trim.
	Clean	Ea.	.12	34.50		34.62	
	Paint	Ea.	18.55	148		166.55	
	Minimum Charge	Job		84.50		84.50	
Commercial Grade							
3' x 6' 8"							
	Demolish	Ea.		31.50		31.50	Includes material and labor to install
	Install	Ea.	285	19.90		304.90	flush solid core 1-3/4" hardwood
	Demolish and Install	Ea.	285	51.40		336.40	veneer 3' x 6' 8' door only.
	Reinstall	Ea.		19.88		19.88	
	Clean	Ea.	3.28	10.05		13.33	
	Paint	Ea.	6.45	24.50		30.95	
	Minimum Charge	Job		169		169	
3' 6" x 6' 8"							
	Demolish	Ea.		31.50		31.50	Includes material and labor to install
	Install	Ea.	385	19.90		404.90	flush solid core 1-3/4" hardwood
	Demolish and Install	Ea.	385	51.40		436.40	veneer 3' 6" x 6' 8' door only.
	Reinstall	Ea.		19.88		19.88	
	Clean	Ea.	3.28	10.05		13.33	
	Paint	Ea.	6.45	24.50		30.95	
	Minimum Charge	Job		169		169	
3' x 7'							
	Demolish	Ea.		31.50		31.50	Includes material and labor to install
	Install	Ea.	305	48.50		353.50	flush solid core 1-3/4" hardwood
	Demolish and Install	Ea.	305	80		385	veneer 3' x 7' door only.
	Reinstall	Ea.		48.29		48.29	
	Clean	Ea.	3.28	10.05		13.33	
	Paint	Ea.	6.45	24.50		30.95	
	Minimum Charge	Job		169		169	

Exterior Doors and Windows

Solid Core Door		Unit	Material	Labor	Equip.	Total	Specification
3' 6" x 7'							
	Demolish	Ea.		31.50		31.50	Includes material and labor to install
	Install	Ea.	415	24		439	flush solid core 1-3/4" hardwood
	Demolish and Install	Ea.	415	55.50		470.50	veneer 3' 6" x 7' door only.
	Reinstall	Ea.		24.14		24.14	
	Clean	Ea.	3.28	10.05		13.33	
	Paint	Ea.	6.45	24.50		30.95	
	Minimum Charge	Job		169		169	
3' x 8'							
	Demolish	Ea.		31.50		31.50	Includes material and labor to install
	Install	Ea.	360	42.50		402.50	flush solid core 1-3/4" hardwood
	Demolish and Install	Ea.	360	74		434	veneer 3' x 8' door only.
	Reinstall	Ea.		42.25		42.25	
	Clean	Ea.	3.94	15.10		19.04	
	Paint	Ea.	6.45	24.50		30.95	
	Minimum Charge	Job		169		169	
3' 6" x 8'							
	Demolish	Ea.		31.50		31.50	Includes material and labor to install
	Install	Ea.	440	42.50		482.50	flush solid core 1-3/4" hardwood
	Demolish and Install	Ea.	440	74		514	veneer 3' 6" x 8' door only.
	Reinstall	Ea.		42.25		42.25	
	Clean	Ea.	3.94	15.10		19.04	
	Paint	Ea.	6.45	24.50		30.95	
	Minimum Charge	Job		169		169	
3' x 9'							
	Demolish	Ea.		31.50		31.50	Includes material and labor to install
	Install	Ea.	460	48.50		508.50	flush solid core 1-3/4" hardwood
	Demolish and Install	Ea.	460	80		540	veneer 3' x 9' door only.
	Reinstall	Ea.		48.29		48.29	
	Clean	Ea.	3.94	15.10		19.04	
	Paint	Ea.	6.45	24.50		30.95	
	Minimum Charge	Job		169		169	
Stain							
	Install	Ea.	6.10	33		39.10	Includes labor and material to stain
	Minimum Charge	Job		148		148	exterior type door and trim on both sides.
Door Only (SC)							
2' 4" x 6' 8"							
	Demolish	Ea.		6.70		6.70	Cost includes material and labor to
	Install	Ea.	193	18.80		211.80	install 2' 4" x 6' 8" lauan solid core
	Demolish and Install	Ea.	193	25.50		218.50	door slab.
	Reinstall	Ea.		18.78		18.78	
	Clean	Ea.	1.36	7.55		8.91	
	Paint	Ea.	2.41	37		39.41	
	Minimum Charge	Job		169		169	
2' 8" x 6' 8"							
	Demolish	Ea.		6.70		6.70	Cost includes material and labor to
	Install	Ea.	197	18.80		215.80	install 2' 8" x 6' 8" lauan solid core
	Demolish and Install	Ea.	197	25.50		222.50	door slab.
	Reinstall	Ea.		18.78		18.78	
	Clean	Ea.	1.36	7.55		8.91	
	Paint	Ea.	2.41	37		39.41	
	Minimum Charge	Job		169		169	

Exterior Doors and Windows

Solid Core Door		Unit	Material	Labor	Equip.	Total	Specification
3' x 6' 8"							
	Demolish	Ea.		6.70		6.70	Cost includes material and labor to
	Install	Ea.	205	19.90		224.90	install 3' x 6' 8" lauan solid core door
	Demolish and Install	Ea.	205	26.60		231.60	slab.
	Reinstall	Ea.		19.88		19.88	
	Clean	Ea.	1.36	7.55		8.91	
	Paint	Ea.	2.41	37		39.41	
	Minimum Charge	Job		169		169	
3' 6" x 6' 8"							
	Demolish	Ea.		6.70		6.70	Cost includes material and labor to
	Install	Ea.	232	21		253	install 3' 6" x 6' 8" lauan solid core
	Demolish and Install	Ea.	232	27.70		259.70	door slab.
	Reinstall	Ea.		21.13		21.13	
	Clean	Ea.	1.36	7.55		8.91	
	Paint	Ea.	2.41	37		39.41	
	Minimum Charge	Job		169		169	
Panel							
	Demolish	Ea.		6.70		6.70	Cost includes material and labor to
	Install	Ea.	174	40		214	install 3' x 6' 8" six (6) panel interior
	Demolish and Install	Ea.	174	46.70		220.70	door.
	Reinstall	Ea.		39.76		39.76	
	Clean	Ea.	1.64	7.55		9.19	
	Paint	Ea.	2.41	37		39.41	
	Minimum Charge	Job		169		169	
Casing Trim							
Single Width							
	Demolish	Ea.		5.95		5.95	Cost includes material and labor to
	Install	Opng.	24	57.50		81.50	install 11/16" x 2-1/2" pine ranch
	Demolish and Install	Opng.	24	63.45		87.45	style casing for one side of a standard
	Reinstall	Opng.		45.83		45.83	door opening.
	Clean	Opng.	.46	2.42		2.88	
	Paint	Ea.	3.15	5.95		9.10	
	Minimum Charge	Job		169		169	
Double Width							
	Demolish	Ea.		7.10		7.10	Cost includes material and labor to
	Install	Opng.	28	67.50		95.50	install 11/16" x 2-1/2" pine ranch
	Demolish and Install	Opng.	28	74.60		102.60	style door casing for one side of a
	Reinstall	Opng.		54.08		54.08	double door opening.
	Clean	Opng.	.55	4.83		5.38	
	Paint	Ea.	3.15	6.75		9.90	
	Minimum Charge	Job		169		169	

Components		Unit	Material	Labor	Equip.	Total	Specification
Hardware							
Doorknob							
	Demolish	Ea.		12.50		12.50	Includes material and labor to install
	Install	Ea.	18.20	21		39.20	residential passage lockset, keyless
	Demolish and Install	Ea.	18.20	33.50		51.70	bored type.
	Reinstall	Ea.		16.90		16.90	
	Clean	Ea.	.33	7.55		7.88	
	Minimum Charge	Job		169		169	

Exterior Doors and Windows

Components		Unit	Material	Labor	Equip.	Total	Specification
Doorknob w / Lock							
	Demolish	Ea.		12.50		12.50	Includes material and labor to install
	Install	Ea.	21	21		42	privacy lockset.
	Demolish and Install	Ea.	21	33.50		54.50	
	Reinstall	Ea.		16.90		16.90	
	Clean	Ea.	.33	7.55		7.88	
	Minimum Charge	Job		169		169	
Lever Handle							
	Demolish	Ea.		12.50		12.50	Includes material and labor to install
	Install	Ea.	28.50	34		62.50	residential passage lockset, keyless
	Demolish and Install	Ea.	28.50	46.50		75	bored type with lever handle.
	Reinstall	Ea.		27.04		27.04	
	Clean	Ea.	.33	7.55		7.88	
	Minimum Charge	Job		169		169	
Deadbolt							
	Demolish	Ea.		11.75		11.75	Includes material and labor to install a
	Install	Ea.	38.50	24		62.50	deadbolt.
	Demolish and Install	Ea.	38.50	35.75		74.25	
	Reinstall	Ea.		19.31		19.31	
	Clean	Ea.	.33	7.55		7.88	
	Minimum Charge	Job		169		169	
Grip Handle Entry							
	Demolish	Ea.		10.75		10.75	Includes material and labor to install
	Install	Ea.	127	37.50		164.50	residential keyed entry lock set with
	Demolish and Install	Ea.	127	48.25		175.25	grip type handle.
	Reinstall	Ea.		30.04		30.04	
	Clean	Ea.	.33	7.55		7.88	
	Minimum Charge	Job		169		169	
Panic Device							
	Demolish	Ea.		27		27	Includes material and labor to install
	Install	Ea.	365	56.50		421.50	touch bar, low profile, exit only.
	Demolish and Install	Ea.	365	83.50		448.50	
	Reinstall	Ea.		45.07		45.07	
	Clean	Ea.	.17	13.40		13.57	
	Minimum Charge	Job		169		169	
Closer							
	Demolish	Ea.		7.05		7.05	Includes material and labor to install
	Install	Ea.	176	56.50		232.50	pneumatic heavy duty closer for
	Demolish and Install	Ea.	176	63.55		239.55	exterior type doors.
	Reinstall	Ea.		45.07		45.07	
	Clean	Ea.	.17	10.05		10.22	
	Minimum Charge	Job		169		169	
Peep-hole							
	Demolish	Ea.		5.95		5.95	Includes material and labor to install
	Install	Ea.	15.50	10.55		26.05	peep-hole for door.
	Demolish and Install	Ea.	15.50	16.50		32	
	Reinstall	Ea.		8.45		8.45	
	Clean	Ea.	.02	2.67		2.69	
	Minimum Charge	Job		169		169	
Exit Sign							
	Demolish	Ea.		18.55		18.55	Includes material and labor to install a
	Install	Ea.	50.50	46		96.50	wall mounted interior electric exit sign.
	Demolish and Install	Ea.	50.50	64.55		115.05	
	Reinstall	Ea.		36.72		36.72	
	Clean	Ea.	.02	5.45		5.47	
	Minimum Charge	Job		169		169	

Exterior Doors and Windows

Components		Unit	Material	Labor	Equip.	Total	Specification
Kickplate							
	Demolish	Ea.		10.45		10.45	Includes material and labor to install
	Install	Ea.	22	22.50		44.50	aluminum kickplate, 10" x 28".
	Demolish and Install	Ea.	22	32.95		54.95	
	Reinstall	Ea.		18.03		18.03	
	Clean	Ea.	.33	7.55		7.88	
	Minimum Charge	Job		169		169	
Threshold							
	Install	L.F.	2.86	3.38		6.24	Includes labor and material to install
	Minimum Charge	Job		84.50		84.50	oak threshold.
Weatherstripping							
	Install	L.F.	.29	.85		1.14	Includes labor and material to install
	Minimum Charge	Job		84.50		84.50	rubber weatherstripping for door.
Metal Jamb							
6' 8" High							
	Demolish	Ea.		18.80		18.80	Cost includes material and labor to
	Install	Ea.	127	21		148	install 18 gauge hollow metal door
	Demolish and Install	Ea.	127	39.80		166.80	frame to fit a 4-1/2" jamb 6' 8" high
	Reinstall	Ea.		21.13		21.13	and up to 3' 6" opening.
	Clean	Ea.	.68	5.05		5.73	
	Paint	Ea.	1.07	5.95		7.02	
	Minimum Charge	Job		169		169	
7' High							
	Demolish	Ea.		18.80		18.80	Cost includes material and labor to
	Install	Ea.	129	22.50		151.50	install 18 gauge hollow metal door
	Demolish and Install	Ea.	129	41.30		170.30	frame to fit a 4-1/2" jamb 7' high and
	Reinstall	Ea.		22.53		22.53	up to 3' 6" opening.
	Clean	Ea.	.68	5.05		5.73	
	Paint	Ea.	1.07	5.95		7.02	
	Minimum Charge	Job		169		169	
8' High							
	Demolish	Ea.		18.80		18.80	Cost includes material and labor to
	Install	Ea.	156	28		184	install 18 gauge hollow metal door
	Demolish and Install	Ea.	156	46.80		202.80	frame to fit a 4-1/2" jamb 8' high and
	Reinstall	Ea.		28.17		28.17	up to 3' 6" opening.
	Clean	Ea.	.97	6.05		7.02	
	Paint	Ea.	1.20	6.60		7.80	
	Minimum Charge	Job		169		169	
9' High							
	Demolish	Ea.		18.80		18.80	Cost includes material and labor to
	Install	Ea.	179	34		213	install 18 gauge hollow metal door
	Demolish and Install	Ea.	179	52.80		231.80	frame to fit a 4-1/2" jamb 9' high and
	Reinstall	Ea.		33.80		33.80	up to 3' 6" opening.
	Clean	Ea.	.97	6.05		7.02	
	Paint	Ea.	1.20	6.60		7.80	
	Minimum Charge	Job		169		169	
Wood Jamb							
	Demolish	Ea.		23.50		23.50	Includes material and labor to install
	Install	Ea.	98.50	9.15		107.65	flat pine jamb with square cut heads
	Demolish and Install	Ea.	98.50	32.65		131.15	and rabbeted sides for 6' 8" high and
	Clean	Ea.	.10	6.05		6.15	3-9/16" door including trim sets for
	Paint	Ea.	5.70	19.75		25.45	both sides.
	Minimum Charge	Job		84.50		84.50	
Stain							
	Install	Ea.	4.22	24.50		28.72	Includes labor and material to stain
	Minimum Charge	Job		148		148	single door and trim on both sides.

Exterior Doors and Windows

Components		Unit	Material	Labor	Equip.	Total	Specification
Shave & Refit							
	Install	Ea.		28		28	Includes labor to shave and rework
	Minimum Charge	Job		84.50		84.50	door to fit opening at the job site.

Sliding Patio Door		Unit	Material	Labor	Equip.	Total	Specification
Aluminum							
6' Single Pane							
	Demolish	Ea.		47		47	Cost includes material and labor to
	Install	Ea.	430	169		599	install 6' x 6' 8" sliding anodized
	Demolish and Install	Ea.	430	216		646	aluminum patio door with single pane
	Reinstall	Ea.		135.20		135.20	tempered safety glass, anodized
	Clean	Ea.	2.05	15.10		17.15	frame, screen, weatherstripped.
	Minimum Charge	Job		169		169	
8' Single Pane							
	Demolish	Ea.		47		47	Cost includes material and labor to
	Install	Ea.	490	225		715	install 8' x 6' 8" sliding aluminum
	Demolish and Install	Ea.	490	272		762	patio door with single pane tempered
	Reinstall	Ea.		180.27		180.27	safety glass, anodized frame, screen,
	Clean	Ea.	2.05	15.10		17.15	weatherstripped.
	Minimum Charge	Job		169		169	
10' Single Pane							
	Demolish	Ea.		47		47	Cost includes material and labor to
	Install	Ea.	640	340		980	install 10' x 6' 8" sliding aluminum
	Demolish and Install	Ea.	640	387		1027	patio door with single pane tempered
	Reinstall	Ea.		270.40		270.40	safety glass, anodized frame, screen,
	Clean	Ea.	2.05	15.10		17.15	weatherstripped.
	Minimum Charge	Job		169		169	
6' Insulated							
	Demolish	Ea.		47		47	Cost includes material and labor to
	Install	Ea.	565	169		734	install 6' x 6' 8" sliding aluminum
	Demolish and Install	Ea.	565	216		781	patio door with double insulated
	Reinstall	Ea.		135.20		135.20	tempered glass, baked-on enamel
	Clean	Ea.	2.05	15.10		17.15	finish, screen, weatherstripped.
	Minimum Charge	Job		169		169	
8' Insulated							
	Demolish	Ea.		47		47	Cost includes material and labor to
	Install	Ea.	670	225		895	install 8' x 6' 8" sliding aluminum
	Demolish and Install	Ea.	670	272		942	patio door with insulated tempered
	Reinstall	Ea.		180.27		180.27	glass, baked-on enamel finish, screen,
	Clean	Ea.	2.05	15.10		17.15	weatherstripped.
	Minimum Charge	Job		169		169	
10' Insulated							
	Demolish	Ea.		47		47	Cost includes material and labor to
	Install	Ea.	755	340		1095	install 10' x 6' 8" sliding aluminum
	Demolish and Install	Ea.	755	387		1142	patio door with double insulated
	Reinstall	Ea.		270.40		270.40	tempered glass, baked-on enamel
	Clean	Ea.	2.05	15.10		17.15	finish, screen, weatherstripped.
	Minimum Charge	Job		169		169	

Sliding Patio Door	Unit	Material	Labor	Equip.	Total	Specification
Vinyl						
6' Single Pane						
Demolish	Ea.		47		47	Cost includes material and labor to
Install	Ea.	405	169		574	install 6' x 6' 8" sliding vinyl patio
Demolish and Install	Ea.	405	216		621	door with single pane tempered safety
Reinstall	Ea.		135.20		135.20	glass, anodized frame, screen,
Clean	Ea.	2.05	15.10		17.15	weatherstripped.
Minimum Charge	Job		169		169	
8' Single Pane						
Demolish	Ea.		47		47	Cost includes material and labor to
Install	Ea.	470	225		695	install 8' x 6' 8" sliding vinyl patio
Demolish and Install	Ea.	470	272		742	door with single pane tempered safety
Reinstall	Ea.		180.27		180.27	glass, anodized frame, screen,
Clean	Ea.	2.05	15.10		17.15	weatherstripped.
Minimum Charge	Job		169		169	
10' Single Pane						
Demolish	Ea.		47		47	Cost includes material and labor to
Install	Ea.	605	340		945	install 10' x 6' 8" sliding vinyl patio
Demolish and Install	Ea.	605	387		992	door with single pane tempered safety
Reinstall	Ea.		270.40		270.40	glass, anodized frame, screen,
Clean	Ea.	2.05	15.10		17.15	weatherstripped.
Minimum Charge	Job		169		169	
6' Insulated						
Demolish	Ea.		47		47	Cost includes material and labor to
Install	Ea.	665	169		834	install 6' x 6' 8" sliding vinyl patio
Demolish and Install	Ea.	665	216		881	door with double insulated tempered
Reinstall	Ea.		135.20		135.20	glass, anodized frame, screen,
Clean	Ea.	2.05	15.10		17.15	weatherstripped.
Minimum Charge	Job		169		169	
8' Insulated						
Demolish	Ea.		47		47	Cost includes material and labor to
Install	Ea.	915	225		1140	install 8' x 6' 8" sliding vinyl patio
Demolish and Install	Ea.	915	272		1187	door with double insulated tempered
Reinstall	Ea.		180.27		180.27	glass, anodized frame, screen,
Clean	Ea.	2.05	15.10		17.15	weatherstripped.
Minimum Charge	Job		169		169	
10' Insulated						
Demolish	Ea.		47		47	Cost includes material and labor to
Install	Ea.	940	340		1280	install 10' x 6' 8" sliding vinyl patio
Demolish and Install	Ea.	940	387		1327	door with double insulated tempered
Reinstall	Ea.		270.40		270.40	glass, anodized frame, screen,
Clean	Ea.	2.05	15.10		17.15	weatherstripped.
Minimum Charge	Job		169		169	
Wood Framed						
6' Premium Insulated						
Demolish	Ea.		47		47	Cost includes material and labor to
Install	Ea.	1275	169		1444	install 6' x 6' 8" sliding wood patio
Demolish and Install	Ea.	1275	216		1491	door with pine frame, double insulated
Reinstall	Ea.		135.20		135.20	glass, screen, lock and dead bolt,
Clean	Ea.	2.05	15.10		17.15	weatherstripped. Add for casing and
Minimum Charge	Job		169		169	finishing.

Exterior Doors and Windows

Sliding Patio Door	Unit	Material	Labor	Equip.	Total	Specification
8' Premium Insulated						
Demolish	Ea.		47		47	Cost includes material and labor to
Install	Ea.	1475	225		1700	install 8' x 6' 8" sliding wood patio
Demolish and Install	Ea.	1475	272		1747	door with pine frame, double insulated
Reinstall	Ea.		180.27		180.27	glass, screen, lock and dead bolt,
Clean	Ea.	2.05	15.10		17.15	weatherstripped. Add for casing and
Minimum Charge	Job		169		169	finishing.
9' Premium Insulated						
Demolish	Ea.		47		47	Cost includes material and labor to
Install	Ea.	1475	225		1700	install 9' x 6' 8" sliding wood patio
Demolish and Install	Ea.	1475	272		1747	door with pine frame, double insulated
Reinstall	Ea.		180.27		180.27	glass, screen, lock and dead bolt,
Clean	Ea.	2.05	15.10		17.15	weatherstripped. Add for casing and
Minimum Charge	Job		169		169	finishing.
Stain						
Install	Ea.	6.10	33		39.10	Includes labor and material to stain
Minimum Charge	Job		148		148	exterior type door and trim on both
						sides.
Casing Trim						
Demolish	Ea.		7.10		7.10	Cost includes material and labor to
Install	Opng.	28	67.50		95.50	install 11/16" x 2-1/2" pine ranch
Demolish and Install	Opng.	28	74.60		102.60	style door casing for one side of a
Reinstall	Opng.		54.08		54.08	double door opening.
Clean	Opng.	.55	4.83		5.38	
Paint	Ea.	3.15	6.75		9.90	
Minimum Charge	Job		169		169	
Sliding Screen						
Demolish	Ea.		6.70		6.70	Includes material and labor to install
Install	Ea.	118	16.90		134.90	sliding screen door, 6' 8" x 3'.
Demolish and Install	Ea.	118	23.60		141.60	
Reinstall	Ea.		13.52		13.52	
Clean	Ea.	.68	5.05		5.73	
Minimum Charge	Job		169		169	

Screen Door	Unit	Material	Labor	Equip.	Total	Specification
Aluminum						
Demolish	Ea.		31.50		31.50	Includes material and labor to install
Install	Ea.	107	48.50		155.50	aluminum frame screen door with
Demolish and Install	Ea.	107	80		187	fiberglass screen cloth, closer, hinges,
Reinstall	Ea.		38.63		38.63	latch.
Clean	Ea.	3.28	10.05		13.33	
Paint	Ea.	3.03	11.85		14.88	
Minimum Charge	Job		169		169	
Wood						
Demolish	Ea.		31.50		31.50	Includes material and labor to install
Install	Ea.	138	56.50		194.50	wood screen door with aluminum cloth
Demolish and Install	Ea.	138	88		226	screen. Frame and hardware not
Reinstall	Ea.		45.07		45.07	included.
Clean	Ea.	3.28	10.05		13.33	
Paint	Ea.	3.03	11.85		14.88	
Minimum Charge	Job		169		169	
Shave & Refit						
Install	Ea.		28		28	Includes labor to shave and rework
Minimum Charge	Job		169		169	door to fit opening at the job site.

Exterior Doors and Windows

Storm Door		Unit	Material	Labor	Equip.	Total	Specification
Good Grade							
	Demolish	Ea.		31.50		31.50	Includes material and labor to install
	Install	Ea.	164	48.50		212.50	aluminum frame storm door with
	Demolish and Install	Ea.	164	80		244	fiberglass screen cloth, upper glass
	Reinstall	Ea.		38.63		38.63	panel, closer, hinges, latch.
	Clean	Ea.	3.28	10.05		13.33	
	Paint	Ea.	3.03	11.85		14.88	
	Minimum Charge	Job		169		169	
Better Grade							
	Demolish	Ea.		31.50		31.50	Includes material and labor to install
	Install	Ea.	216	48.50		264.50	aluminum frame storm door with
	Demolish and Install	Ea.	216	80		296	fiberglass screen cloth, upper and
	Reinstall	Ea.		38.63		38.63	lower glass panel, closer, latch.
	Clean	Ea.	3.28	10.05		13.33	
	Paint	Ea.	3.03	11.85		14.88	
	Minimum Charge	Job		169		169	
Premium Grade							
	Demolish	Ea.		31.50		31.50	Includes material and labor to install
	Install	Ea.	272	48.50		320.50	aluminum frame storm door with
	Demolish and Install	Ea.	272	80		352	fiberglass screen cloth, upper and
	Reinstall	Ea.		38.63		38.63	lower glass panel, closer, latch.
	Clean	Ea.	3.28	10.05		13.33	
	Paint	Ea.	3.03	11.85		14.88	
	Minimum Charge	Job		169		169	
Insulated							
	Demolish	Ea.		31.50		31.50	Includes material and labor to install
	Install	Ea.	251	24		275	aluminum frame storm door with
	Demolish and Install	Ea.	251	55.50		306.50	fiberglass screen cloth, upper and
	Reinstall	Ea.		19.31		19.31	lower glass panel, closer, latch.
	Clean	Ea.	3.28	10.05		13.33	
	Paint	Ea.	3.03	11.85		14.88	
	Minimum Charge	Job		169		169	

Garage Door		Unit	Material	Labor	Equip.	Total	Specification
9' One Piece Metal							
	Demolish	Ea.		79		79	Includes material and labor to install
	Install	Ea.	380	84.50		464.50	one piece steel garage door 9' x 7'
	Demolish and Install	Ea.	380	163.50		543.50	primed including hardware.
	Reinstall	Ea.		67.60		67.60	
	Clean	Ea.	4.10	30		34.10	
	Paint	Ea.	6.20	74		80.20	
	Minimum Charge	Job		169		169	
16' One Piece Metal							
	Demolish	Ea.		94		94	Includes material and labor to install
	Install	Ea.	595	113		708	one piece steel garage door 16' x 7'
	Demolish and Install	Ea.	595	207		802	primed including hardware.
	Reinstall	Ea.		90.13		90.13	
	Clean	Ea.	8.20	40.50		48.70	
	Paint	Ea.	12.40	84.50		96.90	
	Minimum Charge	Job		169		169	

Exterior Doors and Windows

Garage Door		Unit	Material	Labor	Equip.	Total	Specification
9' Metal Sectional							
	Demolish	Ea.		79		79	Includes material and labor to install
	Install	Ea.	525	128		653	sectional metal garage door 9' x 7'
	Demolish and Install	Ea.	525	207		732	primed including hardware.
	Reinstall	Ea.		102.42		102.42	
	Clean	Ea.	4.19	34.50		38.69	
	Paint	Ea.	12.40	84.50		96.90	
	Minimum Charge	Job		169		169	
16' Metal Sectional							
	Demolish	Ea.		94		94	Includes material and labor to install
	Install	Ea.	665	225		890	sectional metal garage door 16' x 7'
	Demolish and Install	Ea.	665	319		984	primed including hardware.
	Reinstall	Ea.		180.27		180.27	
	Clean	Ea.	8.20	40.50		48.70	
	Paint	Ea.	12.40	84.50		96.90	
	Minimum Charge	Job		169		169	
Metal / Steel Roll Up							
8' x 8' Steel							
	Demolish	Ea.		107		107	Cost includes material and labor to
	Install	Ea.	890	520		1410	install 8' x 8' steel overhead roll-up,
	Demolish and Install	Ea.	890	627		1517	chain hoist operated service door.
	Reinstall	Ea.		414.80		414.80	
	Clean	Ea.	4.19	34.50		38.69	
	Paint	Ea.	13.20	99		112.20	
	Minimum Charge	Job		830		830	
10' x 10' Steel							
	Demolish	Ea.		125		125	Cost includes material and labor to
	Install	Ea.	1175	595		1770	install 10' x 10' steel overhead roll-up,
	Demolish and Install	Ea.	1175	720		1895	chain hoist operated service door.
	Reinstall	Ea.		474.06		474.06	
	Clean	Ea.	6.55	48.50		55.05	
	Paint	Ea.	20.50	99		119.50	
	Minimum Charge	Job		830		830	
12' x 12' Steel							
	Demolish	Ea.		150		150	Cost includes material and labor to
	Install	Ea.	1525	690		2215	install 12' x 12' steel overhead roll-up,
	Demolish and Install	Ea.	1525	840		2365	chain hoist operated service door.
	Reinstall	Ea.		553.07		553.07	
	Clean	Ea.	9.45	53.50		62.95	
	Paint	Ea.	29.50	119		148.50	
	Minimum Charge	Job		830		830	
14' x 14' Steel							
	Demolish	Ea.		188		188	Cost includes material and labor to
	Install	Ea.	1975	1025		3000	install 14' x 14' steel overhead roll-up,
	Demolish and Install	Ea.	1975	1213		3188	chain hoist operated service door.
	Reinstall	Ea.		829.60		829.60	
	Clean	Ea.	12.85	69		81.85	
	Paint	Ea.	34.50	148		182.50	
	Minimum Charge	Job		830		830	
18' x 18' Steel							
	Demolish	Ea.		209		209	Cost includes material and labor to
	Install	Opng.	2725	1375		4100	install 18' x 18' steel overhead roll-up,
	Demolish and Install	Opng.	2725	1584		4309	chain hoist operated service door.
	Reinstall	Opng.		1106.13		1106.13	
	Clean	Ea.	12.85	40.50		53.35	
	Paint	Ea.	67	296		363	
	Minimum Charge	Job		830		830	
Automatic Motor							
	Install	Ea.	405	34		439	Includes labor and material to install
	Minimum Charge	Job		169		169	electric opener motor, 1/3 HP.

Exterior Doors and Windows

Garage Door		Unit	Material	Labor	Equip.	Total	Specification
9' One Piece Wood							
	Demolish	Ea.		79		79	Includes material and labor to install
	Install	Ea.	420	84.50		504.50	one piece wood garage door 9' x 7'
	Demolish and Install	Ea.	420	163.50		583.50	primed including hardware.
	Reinstall	Ea.		67.60		67.60	
	Clean	Ea.	4.10	30		34.10	
	Paint	Ea.	6.20	74		80.20	
	Minimum Charge	Job		169		169	
9' Wood Sectional							
	Demolish	Ea.		79		79	Includes material and labor to install
	Install	Ea.	550	84.50		634.50	sectional wood garage door 9' x 7'
	Demolish and Install	Ea.	550	163.50		713.50	unfinished including hardware.
	Reinstall	Ea.		67.60		67.60	
	Clean	Ea.	4.10	30		34.10	
	Paint	Ea.	6.20	74		80.20	
	Minimum Charge	Job		169		169	
16' Wood Sectional							
	Demolish	Ea.		94		94	Includes material and labor to install
	Install	Ea.	1200	113		1313	sectional fiberglass garage door 16' x
	Demolish and Install	Ea.	1200	207		1407	7' unfinished including hardware.
	Reinstall	Ea.		90.13		90.13	
	Clean	Ea.	8.20	40.50		48.70	
	Paint	Ea.	12.40	84.50		96.90	
	Minimum Charge	Job		169		169	
Electric Opener							
	Demolish	Ea.		53.50		53.50	Includes material and labor to install
	Install	Ea.	277	63.50		340.50	electric opener, 1/3 HP economy.
	Demolish and Install	Ea.	277	117		394	
	Reinstall	Ea.		50.73		50.73	
	Clean	Ea.	.17	13.40		13.57	
	Minimum Charge	Job		169		169	
Springs							
	Install	Pr.	41.50	84.50		126	Includes material and labor to replace
	Minimum Charge	Job		169		169	springs for garage door.

Commercial Metal Door		Unit	Material	Labor	Equip.	Total	Specification
Pre-hung Steel Door							
2' 8" x 6' 8"							
	Demolish	Ea.		47		47	Cost includes material and labor to
	Install	Ea.	530	21		551	install 2' 8" x 6' 8" pre-hung 1-3/4"
	Demolish and Install	Ea.	530	68		598	18 gauge steel door with steel frame,
	Reinstall	Ea.		21.13		21.13	hinges, aluminum sill,
	Clean	Ea.	3.28	10.05		13.33	weather-stripped.
	Paint	Ea.	3.87	24.50		28.37	
	Minimum Charge	Job		169		169	
3' x 6' 8"							
	Demolish	Ea.		47		47	Cost includes material and labor to
	Install	Ea.	515	22.50		537.50	install 3' x 6' 8" pre-hung 1-3/4" 18
	Demolish and Install	Ea.	515	69.50		584.50	gauge steel door with steel frame,
	Reinstall	Ea.		22.53		22.53	hinges, aluminum sill,
	Clean	Ea.	3.28	10.05		13.33	weather-stripped.
	Paint	Ea.	3.87	24.50		28.37	
	Minimum Charge	Job		169		169	

Exterior Doors and Windows

Commercial Metal Door		Unit	Material	Labor	Equip.	Total	Specification
3′ 6″ x 6′ 8″							Cost includes material and labor to
	Demolish	Ea.		47		47	install 3′ 6″ x 6′ 8″ pre-hung 1-3/4″
	Install	Ea.	600	26		626	18 gauge steel door with steel frame,
	Demolish and Install	Ea.	600	73		673	hinges, aluminum sill,
	Reinstall	Ea.		26		26	weather-stripped.
	Clean	Ea.	3.28	10.05		13.33	
	Paint	Ea.	3.87	24.50		28.37	
	Minimum Charge	Job		169		169	
4′ x 6′ 8″							Cost includes material and labor to
	Demolish	Ea.		47		47	install 4′ x 6′ 8″ pre-hung 1-3/4″ 18
	Install	Ea.	655	34		689	gauge steel door with steel frame,
	Demolish and Install	Ea.	655	81		736	hinges, aluminum sill,
	Reinstall	Ea.		33.80		33.80	weather-stripped.
	Clean	Ea.	3.28	10.05		13.33	
	Paint	Ea.	3.87	24.50		28.37	
	Minimum Charge	Job		169		169	
Metal Jamb							
6′ 8″ High							Cost includes material and labor to
	Demolish	Ea.		18.80		18.80	install 18 gauge hollow metal door
	Install	Ea.	127	21		148	frame to fit a 4-1/2″ jamb 6′ 8″ high
	Demolish and Install	Ea.	127	39.80		166.80	and up to 3′ 6″ opening.
	Reinstall	Ea.		21.13		21.13	
	Clean	Ea.	.68	5.05		5.73	
	Paint	Ea.	1.07	5.95		7.02	
	Minimum Charge	Job		169		169	
7′ High							Cost includes material and labor to
	Demolish	Ea.		18.80		18.80	install 18 gauge hollow metal door
	Install	Ea.	129	22.50		151.50	frame to fit a 4-1/2″ jamb 7′ high and
	Demolish and Install	Ea.	129	41.30		170.30	up to 3′ 6″ opening.
	Reinstall	Ea.		22.53		22.53	
	Clean	Ea.	.68	5.05		5.73	
	Paint	Ea.	1.07	5.95		7.02	
	Minimum Charge	Job		169		169	
8′ High							Cost includes material and labor to
	Demolish	Ea.		18.80		18.80	install 18 gauge hollow metal door
	Install	Ea.	156	28		184	frame to fit a 4-1/2″ jamb 8′ high and
	Demolish and Install	Ea.	156	46.80		202.80	up to 3′ 6″ opening.
	Reinstall	Ea.		28.17		28.17	
	Clean	Ea.	.97	6.05		7.02	
	Paint	Ea.	1.20	6.60		7.80	
	Minimum Charge	Job		169		169	
9′ High							Cost includes material and labor to
	Demolish	Ea.		18.80		18.80	install 18 gauge hollow metal door
	Install	Ea.	179	34		213	frame to fit a 4-1/2″ jamb 9′ high and
	Demolish and Install	Ea.	179	52.80		231.80	up to 3′ 6″ opening.
	Reinstall	Ea.		33.80		33.80	
	Clean	Ea.	.97	6.05		7.02	
	Paint	Ea.	1.20	6.60		7.80	
	Minimum Charge	Job		169		169	
Hardware							
Doorknob w / Lock							Includes material and labor to install
	Demolish	Ea.		12.50		12.50	keyed entry lock set, bored type
	Install	Ea.	109	28		137	including knob trim, strike and strike
	Demolish and Install	Ea.	109	40.50		149.50	box for standard commercial
	Reinstall	Ea.		22.53		22.53	application.
	Clean	Ea.	.33	7.55		7.88	
	Minimum Charge	Job		169		169	

Exterior Doors and Windows

Commercial Metal Door		Unit	Material	Labor	Equip.	Total	Specification
Lever Handle							
	Demolish	Ea.		12.50		12.50	Includes material and labor to install
	Install	Ea.	58	34		92	passage lockset, keyless bored type
	Demolish and Install	Ea.	58	46.50		104.50	with lever handle, non-locking
	Reinstall	Ea.		27.04		27.04	standard commercial latchset.
	Clean	Ea.	.33	7.55		7.88	
	Minimum Charge	Job		169		169	
Deadbolt							
	Demolish	Ea.		11.75		11.75	Includes material and labor to install a
	Install	Ea.	38.50	24		62.50	deadbolt.
	Demolish and Install	Ea.	38.50	35.75		74.25	
	Reinstall	Ea.		19.31		19.31	
	Clean	Ea.	.33	7.55		7.88	
	Minimum Charge	Job		169		169	
Panic Device							
	Demolish	Ea.		27		27	Includes material and labor to install
	Install	Ea.	365	56.50		421.50	touch bar, low profile, exit only.
	Demolish and Install	Ea.	365	83.50		448.50	
	Reinstall	Ea.		45.07		45.07	
	Clean	Ea.	.17	13.40		13.57	
	Minimum Charge	Job		169		169	
Closer							
	Demolish	Ea.		7.05		7.05	Includes material and labor to install
	Install	Ea.	186	56.50		242.50	pneumatic heavy duty model for
	Demolish and Install	Ea.	186	63.55		249.55	exterior type doors.
	Reinstall	Ea.		45.07		45.07	
	Clean	Ea.	.17	10.05		10.22	
	Minimum Charge	Job		169		169	
Peep-Hole							
	Demolish	Ea.		5.95		5.95	Includes material and labor to install
	Install	Ea.	15.50	10.55		26.05	peep-hole for door.
	Demolish and Install	Ea.	15.50	16.50		32	
	Reinstall	Ea.		8.45		8.45	
	Clean	Ea.	.02	2.67		2.69	
	Minimum Charge	Job		169		169	
Exit Sign							
	Demolish	Ea.		18.55		18.55	Includes material and labor to install a
	Install	Ea.	50.50	46		96.50	wall mounted interior electric exit sign.
	Demolish and Install	Ea.	50.50	64.55		115.05	
	Reinstall	Ea.		36.72		36.72	
	Clean	Ea.	.02	5.45		5.47	
	Minimum Charge	Job		169		169	
Kickplate							
	Demolish	Ea.		10.45		10.45	Includes material and labor to install
	Install	Ea.	49	22.50		71.50	stainless steel kickplate, 10" x 36".
	Demolish and Install	Ea.	49	32.95		81.95	
	Reinstall	Ea.		18.03		18.03	
	Clean	Ea.	.33	7.55		7.88	
	Minimum Charge	Job		169		169	

Aluminum Window		Unit	Material	Labor	Equip.	Total	Specification
Single Hung Two Light							
2' x 2'							
	Demolish	Ea.		16.80		16.80	Cost includes material and labor to
	Install	Ea.	64	75.50		139.50	install 2' x 2' single hung, two light,
	Demolish and Install	Ea.	64	92.30		156.30	double glazed aluminum window
	Reinstall	Ea.		60.33		60.33	including screen.
	Clean	Ea.	.03	6.20		6.23	
	Minimum Charge	Job		207		207	

Exterior Doors and Windows

Aluminum Window		Unit	Material	Labor	Equip.	Total	Specification
2' x 2' 6"							
	Demolish	Ea.		16.80		16.80	Cost includes material and labor to
	Install	Ea.	67	75.50		142.50	install 2' x 2' 6" single hung, two
	Demolish and Install	Ea.	67	92.30		159.30	light, double glazed aluminum
	Reinstall	Ea.		60.33		60.33	window including screen.
	Clean	Ea.	.03	6.20		6.23	
	Minimum Charge	Job		207		207	
3' x 1' 6"							
	Demolish	Ea.		16.80		16.80	Cost includes material and labor to
	Install	Ea.	63.50	83		146.50	install 3' x 1' 6" single hung, two
	Demolish and Install	Ea.	63.50	99.80		163.30	light, double glazed aluminum
	Reinstall	Ea.		66.37		66.37	window including screen.
	Clean	Ea.	.02	5.45		5.47	
	Minimum Charge	Job		207		207	
3' x 2'							
	Demolish	Ea.		16.80		16.80	Cost includes material and labor to
	Install	Ea.	76.50	83		159.50	install 3' x 2' single hung, two light,
	Demolish and Install	Ea.	76.50	99.80		176.30	double glazed aluminum window
	Reinstall	Ea.		66.37		66.37	including screen.
	Clean	Ea.	.02	5.45		5.47	
	Minimum Charge	Job		207		207	
3' x 2' 6"							
	Demolish	Ea.		16.80		16.80	Cost includes material and labor to
	Install	Ea.	80.50	83		163.50	install 3' x 2' 6" single hung, two
	Demolish and Install	Ea.	80.50	99.80		180.30	light, double glazed aluminum
	Reinstall	Ea.		66.37		66.37	window including screen.
	Clean	Ea.	.03	6.20		6.23	
	Minimum Charge	Job		207		207	
3' x 3'							
	Demolish	Ea.		16.80		16.80	Cost includes material and labor to
	Install	Ea.	89	83		172	install 3' x 3' single hung, two light,
	Demolish and Install	Ea.	89	99.80		188.80	double glazed aluminum window
	Reinstall	Ea.		66.37		66.37	including screen.
	Clean	Ea.	.03	6.20		6.23	
	Minimum Charge	Job		207		207	
3' x 3' 6"							
	Demolish	Ea.		16.80		16.80	Cost includes material and labor to
	Install	Ea.	96.50	83		179.50	install 3' x 3' 6" single hung, two
	Demolish and Install	Ea.	96.50	99.80		196.30	light, double glazed aluminum
	Reinstall	Ea.		66.37		66.37	window including screen.
	Clean	Ea.	.04	7.25		7.29	
	Minimum Charge	Job		207		207	
3' x 4'							
	Demolish	Ea.		16.80		16.80	Cost includes material and labor to
	Install	Ea.	105	83		188	install 3' x 4' single hung, two light,
	Demolish and Install	Ea.	105	99.80		204.80	double glazed aluminum window
	Reinstall	Ea.		66.37		66.37	including screen.
	Clean	Ea.	.04	7.25		7.29	
	Minimum Charge	Job		207		207	
3' x 5'							
	Demolish	Ea.		24.50		24.50	Cost includes material and labor to
	Install	Ea.	113	83		196	install 3' x 5' single hung, two light,
	Demolish and Install	Ea.	113	107.50		220.50	double glazed aluminum window
	Reinstall	Ea.		66.37		66.37	including screen.
	Clean	Ea.	.07	8.70		8.77	
	Minimum Charge	Job		207		207	

Exterior Doors and Windows

Aluminum Window	Unit	Material	Labor	Equip.	Total	Specification
3' x 6'						
Demolish	Ea.		24.50		24.50	Cost includes material and labor to
Install	Ea.	125	92		217	install 3' x 6' single hung, two light,
Demolish and Install	Ea.	125	116.50		241.50	double glazed aluminum window
Reinstall	Ea.		73.74		73.74	including screen.
Clean	Ea.	.08	10.20		10.28	
Minimum Charge	Job		207		207	
Double Hung Two Light						
2' x 2'						
Demolish	Ea.		16.80		16.80	Cost includes material and labor to
Install	Ea.	107	75.50		182.50	install 2' x 2' double hung, two light,
Demolish and Install	Ea.	107	92.30		199.30	double glazed aluminum window
Reinstall	Ea.		60.33		60.33	including screen.
Clean	Ea.	.02	5.45		5.47	
Minimum Charge	Job		207		207	
2' x 2' 6"						
Demolish	Ea.		16.80		16.80	Cost includes material and labor to
Install	Ea.	117	75.50		192.50	install 2' x 2' 6" double hung, two
Demolish and Install	Ea.	117	92.30		209.30	light, double glazed aluminum
Reinstall	Ea.		60.33		60.33	window including screen.
Clean	Ea.	.02	5.45		5.47	
Minimum Charge	Job		207		207	
3' x 1' 6"						
Demolish	Ea.		16.80		16.80	Cost includes material and labor to
Install	Ea.	276	83		359	install 3' x 1' 6" double hung, two
Demolish and Install	Ea.	276	99.80		375.80	light, double glazed aluminum
Reinstall	Ea.		66.37		66.37	window including screen.
Clean	Ea.	.02	5.45		5.47	
Minimum Charge	Job		207		207	
3' x 2'						
Demolish	Ea.		16.80		16.80	Cost includes material and labor to
Install	Ea.	310	83		393	install 3' x 2' double hung, two light,
Demolish and Install	Ea.	310	99.80		409.80	double glazed aluminum window
Reinstall	Ea.		66.37		66.37	including screen.
Clean	Ea.	.02	5.45		5.47	
Minimum Charge	Job		207		207	
3' x 2' 6"						
Demolish	Ea.		16.80		16.80	Cost includes material and labor to
Install	Ea.	299	83		382	install 3' x 2' 6" double hung, two
Demolish and Install	Ea.	299	99.80		398.80	light, double glazed aluminum
Reinstall	Ea.		66.37		66.37	window including screen.
Clean	Ea.	.03	6.20		6.23	
Minimum Charge	Job		207		207	
3' x 3'						
Demolish	Ea.		16.80		16.80	Cost includes material and labor to
Install	Ea.	325	83		408	install 3' x 3' double hung, two light,
Demolish and Install	Ea.	325	99.80		424.80	double glazed aluminum window
Reinstall	Ea.		66.37		66.37	including screen.
Clean	Ea.	.03	6.20		6.23	
Minimum Charge	Job		207		207	
3' x 3' 6"						
Demolish	Ea.		16.80		16.80	Cost includes material and labor to
Install	Ea.	360	83		443	install 3' x 3' 6" double hung, two
Demolish and Install	Ea.	360	99.80		459.80	light, double glazed aluminum
Reinstall	Ea.		66.37		66.37	window including screen.
Clean	Ea.	.04	7.25		7.29	
Minimum Charge	Job		207		207	

Exterior Doors and Windows

Aluminum Window		Unit	Material	Labor	Equip.	Total	Specification
3' x 4'							
	Demolish	Ea.		16.80		16.80	Cost includes material and labor to
	Install	Ea.	370	83		453	install 3' x 4' double hung, two light,
	Demolish and Install	Ea.	370	99.80		469.80	double glazed aluminum window
	Reinstall	Ea.		66.37		66.37	including screen.
	Clean	Ea.	.04	7.25		7.29	
	Minimum Charge	Job		207		207	
3' x 5'							
	Demolish	Ea.		24.50		24.50	Cost includes material and labor to
	Install	Ea.	415	83		498	install 3' x 5' double hung, two light,
	Demolish and Install	Ea.	415	107.50		522.50	double glazed aluminum window
	Reinstall	Ea.		66.37		66.37	including screen.
	Clean	Ea.	.07	8.70		8.77	
	Minimum Charge	Job		207		207	
3' x 6'							
	Demolish	Ea.		24.50		24.50	Cost includes material and labor to
	Install	Ea.	550	92		642	install 3' x 6' double hung, two light,
	Demolish and Install	Ea.	550	116.50		666.50	double glazed aluminum window
	Reinstall	Ea.		73.74		73.74	including screen.
	Clean	Ea.	.08	10.20		10.28	
	Minimum Charge	Job		207		207	
Sliding Sash							
1' 6" x 3'							
	Demolish	Ea.		16.80		16.80	Cost includes material and labor to
	Install	Ea.	61	75.50		136.50	install 1' 6" x 3' sliding sash, two
	Demolish and Install	Ea.	61	92.30		153.30	light, double glazed aluminum
	Reinstall	Ea.		60.33		60.33	window including screen.
	Clean	Ea.	.02	5.45		5.47	
	Minimum Charge	Job		207		207	
2' x 2'							
	Demolish	Ea.		16.80		16.80	Cost includes material and labor to
	Install	Ea.	54.50	83		137.50	install 2' x 2' sliding sash, two light,
	Demolish and Install	Ea.	54.50	99.80		154.30	double glazed aluminum window
	Reinstall	Ea.		66.37		66.37	including screen.
	Clean	Ea.	.02	5.45		5.47	
	Minimum Charge	Job		207		207	
3' x 2'							
	Demolish	Ea.		16.80		16.80	Cost includes material and labor to
	Install	Ea.	71	83		154	install 3' x 2' sliding sash, two light,
	Demolish and Install	Ea.	71	99.80		170.80	double glazed aluminum window
	Reinstall	Ea.		66.37		66.37	including screen.
	Clean	Ea.	.02	5.45		5.47	
	Minimum Charge	Job		207		207	
3' x 2' 6"							
	Demolish	Ea.		16.80		16.80	Cost includes material and labor to
	Install	Ea.	73.50	83		156.50	install 3' x 2' 6" sliding sash, two
	Demolish and Install	Ea.	73.50	99.80		173.30	light, double glazed aluminum
	Reinstall	Ea.		66.37		66.37	window including screen.
	Clean	Ea.	.03	6.20		6.23	
	Minimum Charge	Job		207		207	
3' x 3'							
	Demolish	Ea.		16.80		16.80	Cost includes material and labor to
	Install	Ea.	75.50	83		158.50	install 3' x 3' sliding sash, two light,
	Demolish and Install	Ea.	75.50	99.80		175.30	double glazed aluminum window
	Reinstall	Ea.		66.37		66.37	including screen.
	Clean	Ea.	.03	6.20		6.23	
	Minimum Charge	Job		207		207	

Exterior Doors and Windows

Aluminum Window

Aluminum Window	Unit	Material	Labor	Equip.	Total	Specification
3' x 3' 6"						
Demolish	Ea.		16.80		16.80	Cost includes material and labor to
Install	Ea.	88	83		171	install 3' x 3' 6" sliding sash, two
Demolish and Install	Ea.	88	99.80		187.80	light, double glazed aluminum
Reinstall	Ea.		66.37		66.37	window including screen.
Clean	Ea.	.04	7.25		7.29	
Minimum Charge	Job		207		207	
3' x 4'						
Demolish	Ea.		16.80		16.80	Cost includes material and labor to
Install	Ea.	95.50	92		187.50	install 3' x 4' sliding sash, two light,
Demolish and Install	Ea.	95.50	108.80		204.30	double glazed aluminum window
Reinstall	Ea.		73.74		73.74	including screen.
Clean	Ea.	.04	7.25		7.29	
Minimum Charge	Job		207		207	
3' x 5'						
Demolish	Ea.		24.50		24.50	Cost includes material and labor to
Install	Ea.	121	92		213	install 3' x 5' sliding sash, two light,
Demolish and Install	Ea.	121	116.50		237.50	double glazed aluminum window
Reinstall	Ea.		73.74		73.74	including screen.
Clean	Ea.	.07	8.70		8.77	
Minimum Charge	Job		207		207	
4' x 5'						
Demolish	Ea.		24.50		24.50	Cost includes material and labor to
Install	Ea.	123	119		242	install 4' x 5' sliding sash, two light,
Demolish and Install	Ea.	123	143.50		266.50	double glazed aluminum window
Reinstall	Ea.		94.81		94.81	including screen.
Clean	Ea.	.10	12.40		12.50	
Minimum Charge	Job		207		207	
5' x 5'						
Demolish	Ea.		24.50		24.50	Cost includes material and labor to
Install	Ea.	172	119		291	install 5' x 5' sliding sash, two light,
Demolish and Install	Ea.	172	143.50		315.50	double glazed aluminum window
Reinstall	Ea.		94.81		94.81	including screen.
Clean	Ea.	.10	12.40		12.50	
Minimum Charge	Job		207		207	

Awning Sash

Awning Sash	Unit	Material	Labor	Equip.	Total	Specification
26" x 26", 2 Light						
Demolish	Ea.		16.80		16.80	Cost includes material and labor to
Install	Ea.	276	92		368	install 26" x 26" two light awning,
Demolish and Install	Ea.	276	108.80		384.80	double glazed aluminum window
Reinstall	Ea.		73.74		73.74	including screen.
Clean	Ea.	.03	6.20		6.23	
Minimum Charge	Job		207		207	
26" x 37", 2 Light						
Demolish	Ea.		16.80		16.80	Cost includes material and labor to
Install	Ea.	315	92		407	install 26" x 37" two light awning,
Demolish and Install	Ea.	315	108.80		423.80	double glazed aluminum window
Reinstall	Ea.		73.74		73.74	including screen.
Clean	Ea.	.03	6.20		6.23	
Minimum Charge	Job		207		207	
38" x 26", 3 Light						
Demolish	Ea.		16.80		16.80	Cost includes material and labor to
Install	Ea.	350	92		442	install 38" x 26" three light awning,
Demolish and Install	Ea.	350	108.80		458.80	double glazed aluminum window
Reinstall	Ea.		73.74		73.74	including screen.
Clean	Ea.	.04	7.25		7.29	
Minimum Charge	Job		207		207	

Aluminum Window		Unit	Material	Labor	Equip.	Total	Specification
38" x 37", 3 Light							
	Demolish	Ea.		16.80		16.80	Cost includes material and labor to
	Install	Ea.	360	104		464	install 38" x 37" three light awning,
	Demolish and Install	Ea.	360	120.80		480.80	double glazed aluminum window
	Reinstall	Ea.		82.96		82.96	including screen.
	Clean	Ea.	.04	7.25		7.29	
	Minimum Charge	Job		207		207	
38" x 53", 3 Light							
	Demolish	Ea.		24.50		24.50	Cost includes material and labor to
	Install	Ea.	595	104		699	install 38" x 53" three light awning,
	Demolish and Install	Ea.	595	128.50		723.50	double glazed aluminum window
	Reinstall	Ea.		82.96		82.96	including screen.
	Clean	Ea.	.07	8.70		8.77	
	Minimum Charge	Job		207		207	
50" x 37", 4 Light							
	Demolish	Ea.		24.50		24.50	Cost includes material and labor to
	Install	Ea.	595	104		699	install 50" x 37" four light awning,
	Demolish and Install	Ea.	595	128.50		723.50	double glazed aluminum window
	Reinstall	Ea.		82.96		82.96	including screen.
	Clean	Ea.	.07	8.70		8.77	
	Minimum Charge	Job		207		207	
50" x 53", 4 Light							
	Demolish	Ea.		24.50		24.50	Cost includes material and labor to
	Install	Ea.	985	119		1104	install 50" x 53" four light awning,
	Demolish and Install	Ea.	985	143.50		1128.50	double glazed aluminum window
	Reinstall	Ea.		94.81		94.81	including screen.
	Clean	Ea.	.08	10.20		10.28	
	Minimum Charge	Job		207		207	
63" x 37", 5 Light							
	Demolish	Ea.		24.50		24.50	Cost includes material and labor to
	Install	Ea.	770	119		889	install 63" x 37" five light awning,
	Demolish and Install	Ea.	770	143.50		913.50	double glazed aluminum window
	Reinstall	Ea.		94.81		94.81	including screen.
	Clean	Ea.	.08	10.20		10.28	
	Minimum Charge	Job		207		207	
63" x 53", 5 Light							
	Demolish	Ea.		24.50		24.50	Cost includes material and labor to
	Install	Ea.	1075	119		1194	install 63" x 53" five light awning,
	Demolish and Install	Ea.	1075	143.50		1218.50	double glazed aluminum window
	Reinstall	Ea.		94.81		94.81	including screen.
	Clean	Ea.	.10	12.40		12.50	
	Minimum Charge	Job		207		207	
Mullion Bars							
	Install	Ea.	10.10	33		43.10	Includes labor and material to install
	Minimum Charge	Job		207		207	mullions for sliding aluminum window.
Glass Louvered							
2' x 2'							
	Demolish	Ea.		16.80		16.80	Cost includes material and labor to
	Install	Ea.	50	75.50		125.50	install 2' x 2' aluminum window
	Demolish and Install	Ea.	50	92.30		142.30	including 4" glass louvers, hardware,
	Reinstall	Ea.		60.33		60.33	screen.
	Clean	Ea.	.02	5.45		5.47	
	Minimum Charge	Job		207		207	

Exterior Doors and Windows

Aluminum Window	Unit	Material	Labor	Equip.	Total	Specification
2' x 2' 6"						
Demolish	Ea.		16.80		16.80	Cost includes material and labor to
Install	Ea.	61	75.50		136.50	install 2' x 2' 6" aluminum window
Demolish and Install	Ea.	61	92.30		153.30	including 4" glass louvers, hardware,
Reinstall	Ea.		60.33		60.33	screen.
Clean	Ea.	.02	5.45		5.47	
Minimum Charge	Job		207		207	
3' x 1' 6"						
Demolish	Ea.		16.80		16.80	Cost includes material and labor to
Install	Ea.	64	83		147	install 3' x 1' 6" aluminum window
Demolish and Install	Ea.	64	99.80		163.80	including 4" glass louvers, hardware,
Reinstall	Ea.		66.37		66.37	screen.
Clean	Ea.	.02	5.45		5.47	
Minimum Charge	Job		207		207	
3' x 2'						
Demolish	Ea.		16.80		16.80	Cost includes material and labor to
Install	Ea.	73.50	83		156.50	install 3' x 2' aluminum window
Demolish and Install	Ea.	73.50	99.80		173.30	including 4" glass louvers, hardware,
Reinstall	Ea.		66.37		66.37	screen.
Clean	Ea.	.02	5.45		5.47	
Minimum Charge	Job		207		207	
3' x 2' 6"						
Demolish	Ea.		16.80		16.80	Cost includes material and labor to
Install	Ea.	80.50	83		163.50	install 3' x 2' 6" aluminum window
Demolish and Install	Ea.	80.50	99.80		180.30	including 4" glass louvers, hardware,
Reinstall	Ea.		66.37		66.37	screen.
Clean	Ea.	.03	6.20		6.23	
Minimum Charge	Job		207		207	
3' x 3'						
Demolish	Ea.		16.80		16.80	Cost includes material and labor to
Install	Ea.	93.50	92		185.50	install 3' x 3' aluminum window
Demolish and Install	Ea.	93.50	108.80		202.30	including 4" glass louvers, hardware,
Reinstall	Ea.		73.74		73.74	screen.
Clean	Ea.	.03	6.20		6.23	
Minimum Charge	Job		207		207	
3' x 3' 6"						
Demolish	Ea.		16.80		16.80	Cost includes material and labor to
Install	Ea.	106	92		198	install 3' x 3' 6" aluminum window
Demolish and Install	Ea.	106	108.80		214.80	including 4" glass louvers, hardware,
Reinstall	Ea.		73.74		73.74	screen.
Clean	Ea.	.04	7.25		7.29	
Minimum Charge	Job		207		207	
3' x 4'						
Demolish	Ea.		16.80		16.80	Cost includes material and labor to
Install	Ea.	119	92		211	install 3' x 4' aluminum window
Demolish and Install	Ea.	119	108.80		227.80	including 4" glass louvers, hardware,
Reinstall	Ea.		73.74		73.74	screen.
Clean	Ea.	.04	7.25		7.29	
Minimum Charge	Job		207		207	
3' x 5'						
Demolish	Ea.		24.50		24.50	Cost includes material and labor to
Install	Ea.	151	104		255	install 3' x 5' aluminum window with
Demolish and Install	Ea.	151	128.50		279.50	4" glass louvers, hardware, screen.
Reinstall	Ea.		82.96		82.96	
Clean	Ea.	.07	8.70		8.77	
Minimum Charge	Job		207		207	

Exterior Doors and Windows

Aluminum Window		Unit	Material	Labor	Equip.	Total	Specification
3' x 6'							
	Demolish	Ea.		24.50		24.50	Cost includes material and labor to
	Install	Ea.	172	104		276	install 3' x 6' aluminum window with
	Demolish and Install	Ea.	172	128.50		300.50	4" glass louvers, hardware, screen.
	Reinstall	Ea.		82.96		82.96	
	Clean	Ea.	.08	10.20		10.28	
	Minimum Charge	Job		207		207	
Bay							
4' 8" x 6' 4"							
	Demolish	Ea.		17.90		17.90	Cost includes material and labor to
	Install	Ea.	1250	84.50		1334.50	install 4' 8" x 6' 4" angle bay double
	Demolish and Install	Ea.	1250	102.40		1352.40	hung unit, pine frame with aluminum
	Reinstall	Ea.		67.60		67.60	cladding, picture center sash,
	Clean	Ea.	.14	14.45		14.59	hardware, screens.
	Minimum Charge	Job		207		207	
4' 8" x 7' 2"							
	Demolish	Ea.		17.90		17.90	Cost includes material and labor to
	Install	Ea.	1275	84.50		1359.50	install 4' 8" x 7' 2" angle bay double
	Demolish and Install	Ea.	1275	102.40		1377.40	hung unit, pine frame with aluminum
	Reinstall	Ea.		67.60		67.60	cladding, picture center sash,
	Clean	Ea.	.14	14.45		14.59	hardware, screens.
	Minimum Charge	Job		207		207	
4' 8" x 7' 10"							
	Demolish	Ea.		17.90		17.90	Cost includes material and labor to
	Install	Ea.	1300	84.50		1384.50	install 4' 8" x 7' 10" angle bay
	Demolish and Install	Ea.	1300	102.40		1402.40	double hung unit, pine frame with
	Reinstall	Ea.		67.60		67.60	aluminum cladding, picture center
	Clean	Ea.	.14	14.45		14.59	sash, hardware, screens.
	Minimum Charge	Job		207		207	
4' 8" x 8' 6"							
	Demolish	Ea.		20.50		20.50	Cost includes material and labor to
	Install	Ea.	1350	96.50		1446.50	install 4' 8" x 8' 6" angle bay double
	Demolish and Install	Ea.	1350	117		1467	hung unit, pine frame with aluminum
	Reinstall	Ea.		77.26		77.26	cladding, picture center sash,
	Clean	Ea.	.02	19.30		19.32	hardware, screens.
	Minimum Charge	Job		207		207	
4' 8" x 8' 10"							
	Demolish	Ea.		20.50		20.50	Cost includes material and labor to
	Install	Ea.	1400	96.50		1496.50	install 4' 8" x 8' 10" angle bay
	Demolish and Install	Ea.	1400	117		1517	double hung unit, pine frame with
	Reinstall	Ea.		77.26		77.26	aluminum cladding, picture center
	Clean	Ea.	.02	19.30		19.32	sash, hardware, screens.
	Minimum Charge	Job		207		207	
4' 8" x 9' 2"							
	Demolish	Ea.		20.50		20.50	Cost includes material and labor to
	Install	Ea.	1450	169		1619	install 4' 8" x 9' 2" angle bay double
	Demolish and Install	Ea.	1450	189.50		1639.50	hung unit, pine frame with aluminum
	Reinstall	Ea.		135.20		135.20	cladding, picture center sash,
	Clean	Ea.	.02	19.30		19.32	hardware, screens.
	Minimum Charge	Job		207		207	
4' 8" x 9' 10"							
	Demolish	Ea.		20.50		20.50	Cost includes material and labor to
	Install	Ea.	1550	169		1719	install 4' 8" x 9' 10" angle bay
	Demolish and Install	Ea.	1550	189.50		1739.50	double hung unit, pine frame with
	Reinstall	Ea.		135.20		135.20	aluminum cladding, picture center
	Clean	Ea.	.02	19.30		19.32	sash, hardware, screens.
	Minimum Charge	Job		207		207	

Exterior Doors and Windows

Aluminum Window		Unit	Material	Labor	Equip.	Total	Specification
Casing							
	Demolish	Ea.		7.80		7.80	Cost includes material and labor to
	Install	Opng.	33	26		59	install 11/16" x 2-1/2" pine ranch
	Demolish and Install	Opng.	33	33.80		66.80	style window casing.
	Reinstall	Opng.		20.80		20.80	
	Clean	Opng.	.46	2.42		2.88	
	Paint	Ea.	3.15	5.95		9.10	
	Minimum Charge	Job		207		207	
Storm Window							
2' x 3' 6"							
	Demolish	Ea.		9.95		9.95	Includes material and labor to install
	Install	Ea.	76.50	22.50		99	residential aluminum storm window
	Demolish and Install	Ea.	76.50	32.45		108.95	with mill finish, 2' x 3' 5" high.
	Reinstall	Ea.		18.03		18.03	
	Clean	Ea.	.02	5.45		5.47	
	Minimum Charge	Job		207		207	
2' 6" x 5'							
	Demolish	Ea.		9.95		9.95	Includes material and labor to install
	Install	Ea.	85.50	24		109.50	residential aluminum storm window
	Demolish and Install	Ea.	85.50	33.95		119.45	with mill finish, 2' 6" x 5' high.
	Reinstall	Ea.		19.31		19.31	
	Clean	Ea.	.04	7.25		7.29	
	Minimum Charge	Job		207		207	
4' x 6'							
	Demolish	Ea.		19.15		19.15	Includes material and labor to install
	Install	Ea.	102	27		129	residential aluminum storm window, 4'
	Demolish and Install	Ea.	102	46.15		148.15	x 6' high.
	Reinstall	Ea.		21.63		21.63	
	Clean	Ea.	.10	10.85		10.95	
	Minimum Charge	Job		207		207	
Hardware							
Weatherstripping							
	Install	L.F.	.28	2.82		3.10	Includes material and labor to install
	Minimum Charge	Job		113		113	vinyl V strip weatherstripping for
							double-hung window.
Trim Set							
	Demolish	Ea.		7.80		7.80	Cost includes material and labor to
	Install	Opng.	33	26		59	install 11/16" x 2-1/2" pine ranch
	Demolish and Install	Opng.	33	33.80		66.80	style window casing.
	Reinstall	Opng.		20.80		20.80	
	Clean	Opng.	.46	2.42		2.88	
	Paint	Ea.	3.15	5.95		9.10	
	Minimum Charge	Job		169		169	

Vinyl Window		Unit	Material	Labor	Equip.	Total	Specification
Double Hung							
2' x 2' 6"							
	Demolish	Ea.		12.20		12.20	Cost includes material and labor to
	Install	Ea.	180	45		225	install 2' x 2' 6" double hung, two
	Demolish and Install	Ea.	180	57.20		237.20	light, vinyl window with screen.
	Reinstall	Ea.		36.05		36.05	
	Clean	Ea.	.02	5.45		5.47	
	Minimum Charge	Job		340		340	

Exterior Doors and Windows

Vinyl Window		Unit	Material	Labor	Equip.	Total	Specification
2' x 3' 6"							
	Demolish	Ea.		12.20		12.20	Cost includes material and labor to
	Install	Ea.	188	48.50		236.50	install 2' x 3' 6" double hung, two
	Demolish and Install	Ea.	188	60.70		248.70	light, vinyl window with screen.
	Reinstall	Ea.		38.63		38.63	
	Clean	Ea.	.03	6.20		6.23	
	Minimum Charge	Job		340		340	
2' 6" x 4' 6"							
	Demolish	Ea.		12.20		12.20	Cost includes material and labor to
	Install	Ea.	227	52		279	install 2' 6" x 4' 6" double hung, two
	Demolish and Install	Ea.	227	64.20		291.20	light, vinyl window with screen.
	Reinstall	Ea.		41.60		41.60	
	Clean	Ea.	.04	7.25		7.29	
	Minimum Charge	Job		340		340	
3' x 4'							
	Demolish	Ea.		12.20		12.20	Includes material and labor to install
	Install	Ea.	230	67.50		297.50	medium size double hung, 3' x 4',
	Demolish and Install	Ea.	230	79.70		309.70	two light, vinyl window with screen.
	Reinstall	Ea.		54.08		54.08	
	Clean	Ea.	.04	7.25		7.29	
	Minimum Charge	Job		340		340	
3' x 4' 6"							
	Demolish	Ea.		14.90		14.90	Cost includes material and labor to
	Install	Ea.	254	75		329	install 3' x 4' 6" double hung, two
	Demolish and Install	Ea.	254	89.90		343.90	light, vinyl window with screen.
	Reinstall	Ea.		60.09		60.09	
	Clean	Ea.	.07	8.70		8.77	
	Minimum Charge	Job		340		340	
4' x 4' 6"							
	Demolish	Ea.		14.90		14.90	Cost includes material and labor to
	Install	Ea.	281	84.50		365.50	install 4' x 4' 6" double hung, two
	Demolish and Install	Ea.	281	99.40		380.40	light, vinyl window with screen.
	Reinstall	Ea.		67.60		67.60	
	Clean	Ea.	.08	10.20		10.28	
	Minimum Charge	Job		340		340	
4' x 6'							
	Demolish	Ea.		14.90		14.90	Includes material and labor to install
	Install	Ea.	325	96.50		421.50	large double hung, 4' x 6', two light,
	Demolish and Install	Ea.	325	111.40		436.40	vinyl window with screen.
	Reinstall	Ea.		77.26		77.26	
	Clean	Ea.	.10	12.40		12.50	
	Minimum Charge	Job		340		340	
Trim Set							
	Demolish	Ea.		7.80		7.80	Cost includes material and labor to
	Install	Opng.	33	26		59	install 11/16" x 2-1/2" pine ranch
	Demolish and Install	Opng.	33	33.80		66.80	style window casing.
	Reinstall	Opng.		20.80		20.80	
	Clean	Opng.	.46	2.42		2.88	
	Paint	Ea.	3.15	5.95		9.10	
	Minimum Charge	Job		207		207	

Wood Window		Unit	Material	Labor	Equip.	Total	Specification
2' 2" x 3' 4"							
	Demolish	Ea.		12.20		12.20	Cost includes material and labor to
	Install	Ea.	197	45		242	install 2' 2" x 3' 4" double hung, two
	Demolish and Install	Ea.	197	57.20		254.20	light, double glazed wood window
	Reinstall	Ea.		36.05		36.05	with screen.
	Clean	Ea.	.03	6.20		6.23	
	Paint	Ea.	1.97	12.35		14.32	
	Minimum Charge	Job		169		169	

Wood Window		Unit	Material	Labor	Equip.	Total	Specification
2' 2" x 4' 4"							
	Demolish	Ea.		12.20		12.20	Cost includes material and labor to
	Install	Ea.	221	48.50		269.50	install 2' 2" x 4' 4" double hung, two
	Demolish and Install	Ea.	221	60.70		281.70	light, double glazed wood window
	Reinstall	Ea.		38.63		38.63	with screen.
	Clean	Ea.	.03	6.20		6.23	
	Paint	Ea.	1.97	12.35		14.32	
	Minimum Charge	Job		169		169	
2' 6" x 3' 4"							
	Demolish	Ea.		12.20		12.20	Cost includes material and labor to
	Install	Ea.	207	52		259	install 2' 6" x 3' 4" double hung, two
	Demolish and Install	Ea.	207	64.20		271.20	light, double glazed wood window
	Reinstall	Ea.		41.60		41.60	with screen.
	Clean	Ea.	.04	7.25		7.29	
	Paint	Ea.	2.52	14.80		17.32	
	Minimum Charge	Job		169		169	
2' 6" x 4'							
	Demolish	Ea.		12.20		12.20	Cost includes material and labor to
	Install	Ea.	223	56.50		279.50	install 2' 6" x 4' double hung, two
	Demolish and Install	Ea.	223	68.70		291.70	light, double glazed wood window
	Reinstall	Ea.		45.07		45.07	with screen.
	Clean	Ea.	.04	7.25		7.29	
	Paint	Ea.	2.52	14.80		17.32	
	Minimum Charge	Job		169		169	
2' 6" x 4' 8"							
	Demolish	Ea.		12.20		12.20	Cost includes material and labor to
	Install	Ea.	246	56.50		302.50	install 2' 6" x 4' 8" double hung, two
	Demolish and Install	Ea.	246	68.70		314.70	light, double glazed wood window
	Reinstall	Ea.		45.07		45.07	with screen.
	Clean	Ea.	.04	7.25		7.29	
	Paint	Ea.	2.52	14.80		17.32	
	Minimum Charge	Job		169		169	
2' 10" x 3' 4"							
	Demolish	Ea.		12.20		12.20	Cost includes material and labor to
	Install	Ea.	218	67.50		285.50	install 2' 10" x 3' 4" double hung, two
	Demolish and Install	Ea.	218	79.70		297.70	light, double glazed wood window
	Reinstall	Ea.		54.08		54.08	with screen.
	Clean	Ea.	.04	7.25		7.29	
	Paint	Ea.	2.52	14.80		17.32	
	Minimum Charge	Job		169		169	
2' 10" x 4'							
	Demolish	Ea.		12.20		12.20	Cost includes material and labor to
	Install	Ea.	240	67.50		307.50	install 2' 10" x 4' double hung, two
	Demolish and Install	Ea.	240	79.70		319.70	light, double glazed wood window
	Reinstall	Ea.		54.08		54.08	with screen.
	Clean	Ea.	.04	7.25		7.29	
	Paint	Ea.	2.52	14.80		17.32	
	Minimum Charge	Job		169		169	
3' 7" x 3' 4"							
	Demolish	Ea.		12.20		12.20	Cost includes material and labor to
	Install	Ea.	246	75		321	install 3' 7" x 3' 4" double hung, two
	Demolish and Install	Ea.	246	87.20		333.20	light, double glazed wood window
	Reinstall	Ea.		60.09		60.09	with screen.
	Clean	Ea.	.04	7.25		7.29	
	Paint	Ea.	2.52	14.80		17.32	
	Minimum Charge	Job		169		169	

Exterior Doors and Windows

Wood Window		Unit	Material	Labor	Equip.	Total	Specification
3′ 7″ x 5′ 4″							
	Demolish	Ea.		14.90		14.90	Cost includes material and labor to
	Install	Ea.	315	75		390	install 3′ 7″ x 5′ 4″ double hung, two
	Demolish and Install	Ea.	315	89.90		404.90	light, double glazed wood window
	Reinstall	Ea.		60.09		60.09	with screen.
	Clean	Ea.	.10	12.40		12.50	
	Paint	Ea.	4.20	24.50		28.70	
	Minimum Charge	Job		169		169	
4′ 3″ x 5′ 4″							
	Demolish	Ea.		14.90		14.90	Cost includes material and labor to
	Install	Ea.	495	84.50		579.50	install 4′ 3″ x 5′ 4″ double hung, two
	Demolish and Install	Ea.	495	99.40		594.40	light, double glazed wood window
	Reinstall	Ea.		67.60		67.60	with screen.
	Clean	Ea.	.10	12.40		12.50	
	Paint	Ea.	2.70	29.50		32.20	
	Minimum Charge	Job		169		169	
Casement							
2′ x 3′ 4″							
	Demolish	Ea.		12.20		12.20	Cost includes material and labor to
	Install	Ea.	206	34		240	install 2′ x 3′ 4″ casement window
	Demolish and Install	Ea.	206	46.20		252.20	with 3/4″ insulated glass, screens,
	Reinstall	Ea.		27.04		27.04	weatherstripping, hardware.
	Clean	Ea.	.03	6.20		6.23	
	Paint	Ea.	1.97	12.35		14.32	
	Minimum Charge	Job		340		340	
2′ x 4′							
	Demolish	Ea.		12.20		12.20	Cost includes material and labor to
	Install	Ea.	231	37.50		268.50	install 2′ x 4′ casement window with
	Demolish and Install	Ea.	231	49.70		280.70	3/4″ insulated glass, screens,
	Reinstall	Ea.		30.04		30.04	weatherstripping, hardware.
	Clean	Ea.	.03	6.20		6.23	
	Paint	Ea.	1.97	12.35		14.32	
	Minimum Charge	Job		340		340	
2′ x 5′							
	Demolish	Ea.		12.20		12.20	Cost includes material and labor to
	Install	Ea.	260	40		300	install 2′ x 5′ casement window with
	Demolish and Install	Ea.	260	52.20		312.20	3/4″ insulated glass, screens,
	Reinstall	Ea.		31.81		31.81	weatherstripping, hardware.
	Clean	Ea.	.04	7.25		7.29	
	Paint	Ea.	2.52	14.80		17.32	
	Minimum Charge	Job		340		340	
2′ x 6′							
	Demolish	Ea.		12.20		12.20	Cost includes material and labor to
	Install	Ea.	293	42.50		335.50	install 2′ x 6′ casement window with
	Demolish and Install	Ea.	293	54.70		347.70	3/4″ insulated glass, screens,
	Reinstall	Ea.		33.80		33.80	weatherstripping, hardware.
	Clean	Ea.	.04	7.25		7.29	
	Paint	Ea.	2.52	14.80		17.32	
	Minimum Charge	Job		340		340	
4′ x 3′ 4″							
	Demolish	Ea.		14.90		14.90	Cost includes material and labor to
	Install	Ea.	595	45		640	install 4′ x 3′ 4″ casement windows
	Demolish and Install	Ea.	595	59.90		654.90	with 3/4″ insulated glass, screens,
	Reinstall	Ea.		36.05		36.05	weatherstripping, hardware.
	Clean	Ea.	.04	7.25		7.29	
	Paint	Ea.	1.34	14.80		16.14	
	Minimum Charge	Job		340		340	

Wood Window		Unit	Material	Labor	Equip.	Total	Specification
4' x 4'							
	Demolish	Ea.		14.90		14.90	Cost includes material and labor to
	Install	Ea.	655	45		700	install 4' x 4' casement windows with
	Demolish and Install	Ea.	655	59.90		714.90	3/4" insulated glass, screens,
	Reinstall	Ea.		36.05		36.05	weatherstripping, hardware.
	Clean	Ea.	.08	10.20		10.28	
	Paint	Ea.	1.60	18.55		20.15	
	Minimum Charge	Job		340		340	
4' x 5'							
	Demolish	Ea.		14.90		14.90	Cost includes material and labor to
	Install	Ea.	745	48.50		793.50	install 4' x 5' casement windows with
	Demolish and Install	Ea.	745	63.40		808.40	3/4" insulated glass, screens,
	Reinstall	Ea.		38.63		38.63	weatherstripping, hardware.
	Clean	Ea.	.10	12.40		12.50	
	Paint	Ea.	2.08	24.50		26.58	
	Minimum Charge	Job		340		340	
4' x 6'							
	Demolish	Ea.		14.90		14.90	Cost includes material and labor to
	Install	Ea.	835	56.50		891.50	install 4' x 6' casement windows with
	Demolish and Install	Ea.	835	71.40		906.40	3/4" insulated glass, screens,
	Reinstall	Ea.		45.07		45.07	weatherstripping, hardware.
	Clean	Ea.	.10	12.40		12.50	
	Paint	Ea.	2.45	29.50		31.95	
	Minimum Charge	Job		340		340	
Casement Bow							
4' 8" x 8' 1"							
	Demolish	Ea.		20.50		20.50	Includes material and labor to install
	Install	Ea.	1600	127		1727	casement bow window unit with 4
	Demolish and Install	Ea.	1600	147.50		1747.50	lights, pine frame, 1/2" insulated
	Reinstall	Ea.		101.40		101.40	glass, weatherstripping, hardware,
	Clean	Ea.	.14	14.45		14.59	screens.
	Paint	Ea.	4.53	59.50		64.03	
	Minimum Charge	Job		340		340	
4' 8" x 10'							
	Demolish	Ea.		20.50		20.50	Includes material and labor to install
	Install	Ea.	1675	169		1844	casement bow window unit with 5
	Demolish and Install	Ea.	1675	189.50		1864.50	lights, pine frame, 1/2" insulated
	Reinstall	Ea.		135.20		135.20	glass, weatherstripping, hardware,
	Clean	Ea.	.02	19.30		19.32	screens.
	Paint	Ea.	5.60	74		79.60	
	Minimum Charge	Job		340		340	
5' 4" x 8' 1"							
	Demolish	Ea.		20.50		20.50	Includes material and labor to install
	Install	Ea.	1725	127		1852	casement bow window unit with 4
	Demolish and Install	Ea.	1725	147.50		1872.50	lights, pine frame, 1/2" insulated
	Reinstall	Ea.		101.40		101.40	glass, weatherstripping, hardware,
	Clean	Ea.	.02	19.30		19.32	screens.
	Paint	Ea.	5.15	59.50		64.65	
	Minimum Charge	Job		340		340	
5' 4" x 10'							
	Demolish	Ea.		20.50		20.50	Includes material and labor to install
	Install	Ea.	1925	169		2094	casement bow window unit with 5
	Demolish and Install	Ea.	1925	189.50		2114.50	lights, pine frame, 1/2" insulated
	Reinstall	Ea.		135.20		135.20	glass, weatherstripping, hardware,
	Clean	Ea.	.02	19.30		19.32	screens.
	Paint	Ea.	6.50	74		80.50	
	Minimum Charge	Job		340		340	

Exterior Doors and Windows

Wood Window		Unit	Material	Labor	Equip.	Total	Specification
6' x 8' 1"							
	Demolish	Ea.		20.50		20.50	Includes material and labor to install
	Install	Ea.	1675	169		1844	casement bow window unit with 5
	Demolish and Install	Ea.	1675	189.50		1864.50	lights, pine frame, 1/2" insulated
	Reinstall	Ea.		135.20		135.20	glass, weatherstripping, hardware,
	Clean	Ea.	.02	19.30		19.32	screens.
	Paint	Ea.	5.90	74		79.90	
	Minimum Charge	Job		340		340	
6' x 10'							
	Demolish	Ea.		20.50		20.50	Includes material and labor to install
	Install	Ea.	2000	169		2169	casement bow window unit with 5
	Demolish and Install	Ea.	2000	189.50		2189.50	lights, pine frame, 1/2" insulated
	Reinstall	Ea.		135.20		135.20	glass, weatherstripping, hardware,
	Clean	Ea.	.02	19.30		19.32	screens.
	Paint	Ea.	7.35	99		106.35	
	Minimum Charge	Job		340		340	
Casing							
	Demolish	Ea.		7.80		7.80	Cost includes material and labor to
	Install	Opng.	33	26		59	install 11/16" x 2-1/2" pine ranch
	Demolish and Install	Opng.	33	33.80		66.80	style window casing.
	Reinstall	Opng.		20.80		20.80	
	Clean	Opng.	.46	2.42		2.88	
	Paint	Ea.	3.15	5.95		9.10	
	Minimum Charge	Job		207		207	
Trim Set							
	Demolish	Ea.		5.95		5.95	Cost includes material and labor to
	Install	Opng.	33	26		59	install 11/16" x 2-1/2" pine ranch
	Demolish and Install	Opng.	33	31.95		64.95	style window casing.
	Reinstall	Opng.		20.80		20.80	
	Clean	Opng.	.46	2.42		2.88	
	Paint	Ea.	3.15	5.95		9.10	
	Minimum Charge	Job		169		169	
Wood Picture Window							
3' 6" x 4'							
	Demolish	Ea.		14.90		14.90	Includes material and labor to install a
	Install	Ea.	340	56.50		396.50	wood framed picture window
	Demolish and Install	Ea.	340	71.40		411.40	including exterior trim.
	Clean	Ea.	.10	12.40		12.50	
	Paint	Ea.	1.72	23		24.72	
	Minimum Charge	Job		340		340	
4' x 4' 6"							
	Demolish	Ea.		14.90		14.90	Includes material and labor to install a
	Install	Ea.	380	61.50		441.50	wood framed picture window
	Demolish and Install	Ea.	380	76.40		456.40	including exterior trim.
	Clean	Ea.	.10	12.40		12.50	
	Paint	Ea.	1.72	23		24.72	
	Minimum Charge	Job		340		340	
5' x 4'							
	Demolish	Ea.		14.90		14.90	Includes material and labor to install a
	Install	Ea.	480	61.50		541.50	wood framed picture window
	Demolish and Install	Ea.	480	76.40		556.40	including exterior trim.
	Clean	Ea.	.10	12.40		12.50	
	Paint	Ea.	3.67	37		40.67	
	Minimum Charge	Job		340		340	

Exterior Doors and Windows

Wood Window

	Unit	Material	Labor	Equip.	Total	Specification
6' x 4' 6"						
Demolish	Ea.		14.90		14.90	Includes material and labor to install a
Install	Ea.	615	67.50		682.50	wood framed picture window
Demolish and Install	Ea.	615	82.40		697.40	including exterior trim.
Clean	Ea.	.10	12.40		12.50	
Paint	Ea.	3.67	37		40.67	
Minimum Charge	Job		340		340	

Window Screens

	Unit	Material	Labor	Equip.	Total	Specification
Small						
Demolish	Ea.		2.80		2.80	Includes material and labor to install
Install	Ea.	17	10.25		27.25	small window screen, 4 S.F.
Demolish and Install	Ea.	17	13.05		30.05	
Reinstall	Ea.		8.19		8.19	
Clean	Ea.	.33	5.05		5.38	
Minimum Charge	Job		84.50		84.50	
Standard Size						
Demolish	Ea.		2.80		2.80	Includes material and labor to install
Install	Ea.	41.50	10.25		51.75	medium window screen, 12 S.F.
Demolish and Install	Ea.	41.50	13.05		54.55	
Reinstall	Ea.		8.19		8.19	
Clean	Ea.	.33	5.05		5.38	
Minimum Charge	Job		84.50		84.50	
Large						
Demolish	Ea.		2.80		2.80	Includes material and labor to install
Install	Ea.	55	10.25		65.25	large window screen, 16 S.F.
Demolish and Install	Ea.	55	13.05		68.05	
Reinstall	Ea.		8.19		8.19	
Clean	Ea.	.33	5.05		5.38	
Minimum Charge	Job		84.50		84.50	

Window Glass

	Unit	Material	Labor	Equip.	Total	Specification
Plate, Clear, 1/4"						
Demolish	S.F.		1.34		1.34	Includes material and labor to install
Install	S.F.	7.50	5.35		12.85	clear glass.
Demolish and Install	S.F.	7.50	6.69		14.19	
Reinstall	S.F.		4.28		4.28	
Clean	S.F.	.02	.14		.16	
Minimum Charge	Job		161		161	
Insulated						
Demolish	S.F.		1.34		1.34	Includes material and labor to install
Install	S.F.	12.50	9.20		21.70	insulated glass.
Demolish and Install	S.F.	12.50	10.54		23.04	
Reinstall	S.F.		7.34		7.34	
Clean	S.F.	.02	.14		.16	
Minimum Charge	Job		161		161	
Tinted						
Demolish	S.F.		1.34		1.34	Includes material and labor to install
Install	S.F.	7.65	4.94		12.59	tinted glass.
Demolish and Install	S.F.	7.65	6.28		13.93	
Reinstall	S.F.		3.95		3.95	
Clean	S.F.	.02	.14		.16	
Minimum Charge	Job		161		161	

Window Glass	Unit	Material	Labor	Equip.	Total	Specification
Obscure / Florentine						
Demolish	S.F.		1.34		1.34	Includes material and labor to install
Install	S.F.	8.45	4.59		13.04	obscure glass.
Demolish and Install	S.F.	8.45	5.93		14.38	
Reinstall	S.F.		3.67		3.67	
Clean	S.F.	.02	.14		.16	
Minimum Charge	Job		161		161	
Laminated Safety						
Demolish	S.F.		1.34		1.34	Includes material and labor to install
Install	S.F.	10.40	7.15		17.55	clear laminated glass.
Demolish and Install	S.F.	10.40	8.49		18.89	
Reinstall	S.F.		5.71		5.71	
Clean	S.F.	.02	.14		.16	
Minimum Charge	Job		161		161	
Wired						
Demolish	S.F.		1.34		1.34	Includes material and labor to install
Install	S.F.	13.70	4.76		18.46	rough obscure wire glass.
Demolish and Install	S.F.	13.70	6.10		19.80	
Reinstall	S.F.		3.81		3.81	
Clean	S.F.	.02	.14		.16	
Minimum Charge	Job		161		161	
Tempered						
Demolish	S.F.		1.34		1.34	Includes material and labor to install
Install	S.F.	7.50	5.35		12.85	clear glass.
Demolish and Install	S.F.	7.50	6.69		14.19	
Reinstall	S.F.		4.28		4.28	
Clean	S.F.	.02	.14		.16	
Minimum Charge	Job		161		161	
Stained						
Demolish	S.F.		1.34		1.34	Includes material and labor to install
Install	S.F.	56	5.35		61.35	stained glass.
Demolish and Install	S.F.	56	6.69		62.69	
Reinstall	S.F.		4.28		4.28	
Clean	S.F.	.02	.14		.16	
Minimum Charge	Job		161		161	
Etched						
Demolish	S.F.		1.34		1.34	Includes material and labor to install
Install	S.F.	46	5.35		51.35	etched glass.
Demolish and Install	S.F.	46	6.69		52.69	
Reinstall	S.F.		4.28		4.28	
Clean	S.F.	.02	.14		.16	
Minimum Charge	Job		161		161	
Wall Mirror						
Demolish	S.F.		1.34		1.34	Includes material and labor to install
Install	S.F.	7.35	5.15		12.50	distortion-free float glass mirror, 1/4"
Demolish and Install	S.F.	7.35	6.49		13.84	thick, cut and polished edges.
Reinstall	S.F.		4.11		4.11	
Clean	S.F.	.02	.14		.16	
Minimum Charge	Job		161		161	
Bullet Resistant						
Demolish	S.F.		1.34		1.34	Includes material and labor to install
Install	S.F.	100	64		164	bullet resistant clear laminated glass.
Demolish and Install	S.F.	100	65.34		165.34	
Reinstall	S.F.		51.39		51.39	
Clean	S.F.	.02	.14		.16	
Minimum Charge	Job		161		161	

Storefront System	Unit	Material	Labor	Equip.	Total	Specification
Clear Glass						
Stub Wall to 8'						
Demolish	S.F.		2.10		2.10	Includes material and labor to install
Install	S.F.	17.40	8.85		26.25	commercial grade 4-1/2" section,
Demolish and Install	S.F.	17.40	10.95		28.35	anodized aluminum with clear glass in
Reinstall	S.F.		7.07		7.07	upper section and safety glass in
Clean	S.F.		.29		.29	lower section.
Minimum Charge	Job		207		207	
Floor to 10'						
Demolish	S.F.		2.10		2.10	Includes material and labor to install
Install	S.F.	19.05	8.85		27.90	commercial grade 4-1/2" section,
Demolish and Install	S.F.	19.05	10.95		30	anodized aluminum with clear glass in
Reinstall	S.F.		7.07		7.07	upper section and safety glass in
Clean	S.F.		.29		.29	lower section.
Minimum Charge	Job		207		207	
Polished Plate Glass						
Stub Wall to 9'						
Demolish	S.F.		2.10		2.10	Cost includes material and labor to
Install	S.F.	13.20	8.85		22.05	install 4' x 9' polished plate glass
Demolish and Install	S.F.	13.20	10.95		24.15	window built on 8" stub wall.
Reinstall	S.F.		7.07		7.07	
Clean	S.F.		.29		.29	
Minimum Charge	Job		207		207	
Stub Wall to 13'						
Demolish	S.F.		2.10		2.10	Cost includes material and labor to
Install	S.F.	13.75	8.85		22.60	install 4' x 13' polished plate glass
Demolish and Install	S.F.	13.75	10.95		24.70	window built on 8" stub wall.
Reinstall	S.F.		7.07		7.07	
Clean	S.F.		.29		.29	
Minimum Charge	Job		207		207	
Floor to 9'						
Demolish	S.F.		2.10		2.10	Cost includes material and labor to
Install	S.F.	14.30	8.85		23.15	install 4' x 9' polished plate glass
Demolish and Install	S.F.	14.30	10.95		25.25	window built on floor.
Reinstall	S.F.		7.07		7.07	
Clean	S.F.		.29		.29	
Minimum Charge	Job		207		207	
Floor to 13'						
Demolish	S.F.		2.10		2.10	Cost includes material and labor to
Install	S.F.	15.40	8.85		24.25	install 4' x 13' polished plate glass
Demolish and Install	S.F.	15.40	10.95		26.35	window built on floor.
Reinstall	S.F.		7.07		7.07	
Clean	S.F.		.29		.29	
Minimum Charge	Job		207		207	
Clear Plate Glass						
Floor to 9'						
Demolish	S.F.		2.10		2.10	Cost includes material and labor to
Install	S.F.	20.50	8.85		29.35	install 4' x 9' clear plate glass window
Demolish and Install	S.F.	20.50	10.95		31.45	built on floor.
Reinstall	S.F.		7.07		7.07	
Clean	S.F.		.29		.29	
Minimum Charge	Job		207		207	

Exterior Doors and Windows

Storefront System		Unit	Material	Labor	Equip.	Total	Specification
Floor to 11'							
	Demolish	S.F.		2.10		2.10	Cost includes material and labor to
	Install	S.F.	20	8.85		28.85	install 4' x 11' clear plate glass
	Demolish and Install	S.F.	20	10.95		30.95	window built on floor.
	Reinstall	S.F.		7.07		7.07	
	Clean	S.F.		.29		.29	
	Minimum Charge	Job		207		207	
Floor to 13'							
	Demolish	S.F.		2.10		2.10	Cost includes material and labor to
	Install	S.F.	20	8.85		28.85	install 4' x 13' clear plate glass
	Demolish and Install	S.F.	20	10.95		30.95	window built on floor.
	Reinstall	S.F.		7.07		7.07	
	Clean	S.F.		.29		.29	
	Minimum Charge	Job		207		207	
Decorative Type							
	Demolish	S.F.		2.10		2.10	Includes material and labor to install
	Install	S.F.	15.75	6.40		22.15	decorative storefront system.
	Demolish and Install	S.F.	15.75	8.50		24.25	
	Reinstall	S.F.		5.14		5.14	
	Clean	S.F.		.29		.29	
	Minimum Charge	Job		161		161	
Decorative Curved							
	Demolish	S.F.		2.10		2.10	Includes material and labor to install
	Install	S.F.	29.50	12.85		42.35	decorative curved storefront system.
	Demolish and Install	S.F.	29.50	14.95		44.45	
	Reinstall	S.F.		10.28		10.28	
	Clean	S.F.		.29		.29	
	Minimum Charge	Job		161		161	
Entrance							
Narrow Stile 3' x 7"							
	Demolish	Ea.		18.80		18.80	Cost includes material and labor to
	Install	Ea.	775	225		1000	install 3' x 7' narrow stile door, center
	Demolish and Install	Ea.	775	243.80		1018.80	pivot concealed closer.
	Reinstall	Ea.		180.27		180.27	
	Clean	Ea.	4.10	30		34.10	
	Minimum Charge	Job		207		207	
Tempered Glass							
	Demolish	Ea.		18.80		18.80	Cost includes material and labor to
	Install	Ea.	2700	450		3150	install 3' x 7' tempered glass door,
	Demolish and Install	Ea.	2700	468.80		3168.80	center pivot concealed closer,.
	Reinstall	Ea.		358.75		358.75	
	Clean	Ea.	4.10	30		34.10	
	Minimum Charge	Job		207		207	
Heavy Section							
	Demolish	Ea.		18.80		18.80	Includes material and labor to install
	Install	Ea.	815	415		1230	wide stile door with 3' 6" x 7'
	Demolish and Install	Ea.	815	433.80		1248.80	opening.
	Reinstall	Ea.		331.84		331.84	
	Clean	Ea.	.83	10.05		10.88	
	Minimum Charge	Job		640		640	

Exterior Doors and Windows

Curtain Wall System		Unit	Material	Labor	Equip.	Total	Specification
Regular Weight							
Bronze Anodized							
	Demolish	S.F.		2.10		2.10	Includes material and labor to install
	Install	S.F.	32.50	8.85		41.35	structural curtain wall system with
	Demolish and Install	S.F.	32.50	10.95		43.45	bronze anodized frame of heavy
	Clean	S.F.		.29		.29	weight, float glass.
	Minimum Charge	Job		169		169	
Black Anodized							
	Demolish	S.F.		2.10		2.10	Includes material and labor to install
	Install	S.F.	34.50	8.85		43.35	structural curtain wall system with
	Demolish and Install	S.F.	34.50	10.95		45.45	black anodized frame of regular
	Clean	S.F.		.29		.29	weight, float glass.
	Minimum Charge	Job		169		169	
Heavy Weight							
Bronze Anodized							
	Demolish	S.F.		2.10		2.10	Includes material and labor to install
	Install	S.F.	32.50	8.85		41.35	structural curtain wall system with
	Demolish and Install	S.F.	32.50	10.95		43.45	bronze anodized frame of heavy
	Clean	S.F.		.29		.29	weight, float glass.
	Minimum Charge	Job		169		169	
Black Anodized							
	Demolish	S.F.		2.10		2.10	Includes material and labor to install
	Install	S.F.	33	8.85		41.85	structural curtain wall system with
	Demolish and Install	S.F.	33	10.95		43.95	black anodized frame of heavy
	Clean	S.F.		.29		.29	weight, float glass.
	Minimum Charge	Job		169		169	
Glass							
1″ Insulated							
	Install	S.F.	14.80	8.55		23.35	Cost includes material and labor to
	Minimum Charge	Job		161		161	install 1″ insulated glass.
Heat Absorbing							
	Install	S.F.	8.65	8.55		17.20	Includes material and labor to install
	Minimum Charge	Job		161		161	heat absorbing glass.
Low Transmission							
	Install	S.F.	5.65	8.55		14.20	Includes material and labor to install
	Minimum Charge	Job		161		161	low transmission glass.

Skylight		Unit	Material	Labor	Equip.	Total	Specification
Dome							
22″ x 22″							
	Demolish	Ea.		65		65	Cost includes material and labor to
	Install	Ea.	49	103		152	install 22″ x 22″ fixed dome skylight.
	Demolish and Install	Ea.	49	168		217	
	Reinstall	Ea.		82.09		82.09	
	Clean	Ea.	.37	8.65		9.02	
	Minimum Charge	Job		615		615	
22″ x 46″							
	Demolish	Ea.		65		65	Cost includes material and labor to
	Install	Ea.	110	123		233	install 22″ x 46″ fixed dome skylight.
	Demolish and Install	Ea.	110	188		298	
	Reinstall	Ea.		98.51		98.51	
	Clean	Ea.		13.40		13.40	
	Minimum Charge	Job		615		615	

Skylight	Unit	Material	Labor	Equip.	Total	Specification
30″ x 30″						
Demolish	Ea.		65		65	Cost includes material and labor to
Install	Ea.	92	103		195	install 30″ x 30″ fixed dome skylight.
Demolish and Install	Ea.	92	168		260	
Reinstall	Ea.		82.09		82.09	
Clean	Ea.		13.40		13.40	
Minimum Charge	Job		615		615	
30″ x 46″						
Demolish	Ea.		65		65	Cost includes material and labor to
Install	Ea.	160	123		283	install 30″ x 46″ fixed dome skylight.
Demolish and Install	Ea.	160	188		348	
Reinstall	Ea.		98.51		98.51	
Clean	Ea.		13.40		13.40	
Minimum Charge	Job		615		615	
Fixed						
22″ x 27″						
Demolish	Ea.		65		65	Includes material and labor to install
Install	Ea.	234	103		337	double glazed skylight, 22″ x 27″,
Demolish and Install	Ea.	234	168		402	fixed.
Reinstall	Ea.		82.09		82.09	
Clean	Ea.	.37	8.65		9.02	
Minimum Charge	Job		615		615	
22″ x 46″						
Demolish	Ea.		65		65	Includes material and labor to install
Install	Ea.	295	123		418	double glazed skylight, 22″ x 46″,
Demolish and Install	Ea.	295	188		483	fixed.
Reinstall	Ea.		98.51		98.51	
Clean	Ea.		13.40		13.40	
Minimum Charge	Job		615		615	
44″ x 46″						
Demolish	Ea.		65		65	Includes material and labor to install
Install	Ea.	420	123		543	double glazed skylight, 44″ x 46″,
Demolish and Install	Ea.	420	188		608	fixed.
Reinstall	Ea.		98.51		98.51	
Clean	Ea.		15.10		15.10	
Minimum Charge	Job		615		615	
Operable						
22″ x 27″						
Demolish	Ea.		65		65	Includes material and labor to install
Install	Ea.	345	103		448	double glazed skylight, 22″ x 27″,
Demolish and Install	Ea.	345	168		513	operable.
Reinstall	Ea.		82.09		82.09	
Clean	Ea.	.37	8.65		9.02	
Minimum Charge	Job		615		615	
22″ x 46″						
Demolish	Ea.		65		65	Includes material and labor to install
Install	Ea.	405	123		528	double glazed skylight, 22″ x 46″,
Demolish and Install	Ea.	405	188		593	operable.
Reinstall	Ea.		98.51		98.51	
Clean	Ea.		13.40		13.40	
Minimum Charge	Job		615		615	
44″ x 46″						
Demolish	Ea.		65		65	Includes material and labor to install
Install	Ea.	605	123		728	double glazed skylight, 44″ x 46″,
Demolish and Install	Ea.	605	188		793	operable.
Reinstall	Ea.		98.51		98.51	
Clean	Ea.		15.10		15.10	
Minimum Charge	Job		615		615	

Exterior Doors and Windows

Industrial Steel Window	Unit	Material	Labor	Equip.	Total	Specification
100% Fixed						
Demolish	S.F.		1.72		1.72	Includes material and labor to install
Install	S.F.	28.50	4.15		32.65	commercial grade steel windows,
Demolish and Install	S.F.	28.50	5.87		34.37	including glazing.
Reinstall	S.F.		3.32		3.32	
Clean	Ea.	.14	.90		1.04	
Paint	S.F.	.11	.66		.77	
Minimum Charge	Job		207		207	
50% Vented						
Demolish	S.F.		1.72		1.72	Includes material and labor to install
Install	S.F.	42	4.15		46.15	commercial grade steel windows,
Demolish and Install	S.F.	42	5.87		47.87	including glazing.
Reinstall	S.F.		3.32		3.32	
Clean	Ea.	.14	.90		1.04	
Paint	S.F.	.11	.66		.77	
Minimum Charge	Job		207		207	

Brick Veneer　　　　　**Concrete Masonry Units**

Masonry Wall	Unit	Material	Labor	Equip.	Total	Specification
Unreinforced CMU						
4" x 8" x 16"						
Demolish	S.F.		1.49		1.49	Includes material and labor to install
Install	S.F.	1.33	3.35		4.68	normal weight 4" x 8" x 16" block.
Demolish and Install	S.F.	1.33	4.84		6.17	
Clean	S.F.	.04	.37		.41	
Paint	S.F.	.14	.67		.81	
Minimum Charge	Job		170		170	
6" x 8" x 16"						
Demolish	S.F.		1.58		1.58	Includes material and labor to install
Install	S.F.	1.94	3.50		5.44	normal weight 6" x 8" x 16" block.
Demolish and Install	S.F.	1.94	5.08		7.02	
Clean	S.F.	.04	.37		.41	
Paint	S.F.	.14	.67		.81	
Minimum Charge	Job		170		170	
8" x 8" x 16"						
Demolish	S.F.		1.79		1.79	Includes material and labor to install
Install	S.F.	2.12	3.85		5.97	normal weight 8" x 8" x 16" block.
Demolish and Install	S.F.	2.12	5.64		7.76	
Clean	S.F.	.04	.37		.41	
Paint	S.F.	.14	.67		.81	
Minimum Charge	Job		170		170	
12" x 8" x 16"						
Demolish	S.F.		1.79		1.79	Includes material and labor to install
Install	S.F.	3.06	5.80		8.86	normal weight 12" x 8" x 16" block.
Demolish and Install	S.F.	3.06	7.59		10.65	
Clean	S.F.	.04	.37		.41	
Paint	S.F.	.14	.67		.81	
Minimum Charge	Job		170		170	
Reinforced CMU						
4" x 8" x 16"						
Demolish	S.F.		1.49	.73	2.22	Includes material and labor to install
Install	S.F.	1.42	3.42		4.84	normal weight 4" x 8" x 16" block,
Demolish and Install	S.F.	1.42	4.91	.73	7.06	mortar and 1/2" re-bar installed
Clean	S.F.	.04	.37		.41	vertically every 24" O.C.
Paint	S.F.	.14	.67		.81	
Minimum Charge	Job		300		300	
6" x 8" x 16"						
Demolish	S.F.		1.57	.35	1.92	Includes material and labor to install
Install	S.F.	2.04	3.58		5.62	normal weight 6" x 8" x 16" block,
Demolish and Install	S.F.	2.04	5.15	.35	7.54	mortar and 1/2" re-bar installed
Clean	S.F.	.04	.37		.41	vertically every 24" O.C.
Paint	S.F.	.14	.67		.81	
Minimum Charge	Job		300		300	

Masonry

Masonry Wall		Unit	Material	Labor	Equip.	Total	Specification
8″ x 8″ x 16″							
	Demolish	S.F.		1.66	.37	2.03	Includes material and labor to install
	Install	S.F.	2.22	3.90		6.12	normal weight 8″ x 8″ x 16″ block,
	Demolish and Install	S.F.	2.22	5.56	.37	8.15	mortar and 1/2″ re-bar installed
	Clean	S.F.	.04	.37		.41	vertically every 24″ O.C.
	Paint	S.F.	.14	.67		.81	
	Minimum Charge	Job		300		300	
12″ x 8″ x 16″							
	Demolish	S.F.		1.71	.38	2.09	Includes material and labor to install
	Install	S.F.	3.17	6		9.17	normal weight 12″ x 8″ x 16″ block,
	Demolish and Install	S.F.	3.17	7.71	.38	11.26	mortar and 1/2″ re-bar installed
	Clean	S.F.	.04	.37		.41	vertically every 24″ O.C.
	Paint	S.F.	.14	.67		.81	
	Minimum Charge	Job		300		300	
Structural Brick							
4″ Single Row							
	Demolish	S.F.		2.15		2.15	Cost includes material and labor to
	Install	S.F.	3.55	7.15		10.70	install 4″ red brick wall, running bond
	Demolish and Install	S.F.	3.55	9.30		12.85	with 3/8″ concave joints, mortar and
	Clean	S.F.	.04	.32		.36	ties.
	Paint	S.F.	.14	.67		.81	
	Minimum Charge	Job		300		300	
8″ Double Row							
	Demolish	S.F.		2.15		2.15	Cost includes material and labor to
	Install	S.F.	6.05	11.40		17.45	install 8″ red brick wall, running bond,
	Demolish and Install	S.F.	6.05	13.55		19.60	with 3/8″ concave joints, mortar and
	Clean	S.F.	.04	.32		.36	wall ties.
	Paint	S.F.	.14	.67		.81	
	Minimum Charge	Job		300		300	
12″ Triple Row							
	Demolish	S.F.		2.98		2.98	Cost includes material and labor to
	Install	S.F.	9.10	16.20		25.30	install 12″ red brick wall, running
	Demolish and Install	S.F.	9.10	19.18		28.28	bond, with 3/8″ concave joints,
	Clean	S.F.	.04	.32		.36	mortar and wall ties.
	Paint	S.F.	.14	.67		.81	
	Minimum Charge	Job		300		300	
Brick Veneer							
4″ Red Common							
	Demolish	S.F.		2.15		2.15	Cost includes material and labor to
	Install	S.F.	2.84	6.70		9.54	install 4″ common red face brick,
	Demolish and Install	S.F.	2.84	8.85		11.69	running bond with 3/8″ concave
	Clean	S.F.	.04	.32		.36	joints, mortar and wall ties.
	Paint	S.F.	.14	.67		.81	
	Minimum Charge	Job		170		170	
4″ Used							
	Demolish	S.F.		2.15		2.15	Cost includes material and labor to
	Install	S.F.	3.55	7.15		10.70	install 4″ red brick wall, running bond
	Demolish and Install	S.F.	3.55	9.30		12.85	with 3/8″ concave joints, mortar and
	Clean	S.F.	.04	.32		.36	ties.
	Paint	S.F.	.14	.67		.81	
	Minimum Charge	Job		170		170	

Masonry Wall		Unit	Material	Labor	Equip.	Total	Specification
6" Jumbo							
	Demolish	S.F.		2.15		2.15	Cost includes material and labor to
	Install	S.F.	4.43	3.54		7.97	install 6" red jumbo brick veneer,
	Demolish and Install	S.F.	4.43	5.69		10.12	running bond with 3/8" concave
	Clean	S.F.	.04	.32		.36	joints, mortar and ties.
	Paint	S.F.	.14	.67		.81	
	Minimum Charge	Job		170		170	
Norman (11-1/2")							
	Demolish	S.F.		2.15		2.15	Cost includes material and labor to
	Install	S.F.	4.87	4.81		9.68	install 4" norman red brick wall,
	Demolish and Install	S.F.	4.87	6.96		11.83	running bond with 3/8" concave
	Clean	S.F.	.04	.32		.36	joints, mortar and ties.
	Paint	S.F.	.14	.67		.81	
	Minimum Charge	Job		170		170	
Glazed							
	Demolish	S.F.		2.15		2.15	Includes material, labor and
	Install	S.F.	8.60	7.35		15.95	equipment to install glazed brick.
	Demolish and Install	S.F.	8.60	9.50		18.10	
	Clean	S.F.	.04	.32		.36	
	Paint	S.F.	.14	.67		.81	
	Minimum Charge	Job		170		170	
Slumpstone							
6" x 4"							
	Demolish	S.F.		1.49		1.49	Cost includes material and labor to
	Install	S.F.	5.30	3.75		9.05	install 6" x 4" x 16" concrete slump
	Demolish and Install	S.F.	5.30	5.24		10.54	block.
	Clean	S.F.	.04	.37		.41	
	Paint	S.F.	.14	.67		.81	
	Minimum Charge	Job		170		170	
6" x 6"							
	Demolish	S.F.		1.58		1.58	Cost includes material and labor to
	Install	S.F.	4.73	3.87		8.60	install 6" x 6" x 16" concrete slump
	Demolish and Install	S.F.	4.73	5.45		10.18	block.
	Clean	S.F.	.04	.37		.41	
	Paint	S.F.	.14	.67		.81	
	Minimum Charge	Job		170		170	
8" x 6"							
	Demolish	S.F.		1.58		1.58	Cost includes material and labor to
	Install	S.F.	6.40	3.99		10.39	install 8" x 6" x 16" concrete slump
	Demolish and Install	S.F.	6.40	5.57		11.97	block.
	Clean	S.F.	.04	.37		.41	
	Paint	S.F.	.14	.67		.81	
	Minimum Charge	Job		170		170	
12" x 4"							
	Demolish	S.F.		1.49		1.49	Cost includes material and labor to
	Install	S.F.	11.15	4.61		15.76	install 12" x 4" x 16" concrete slump
	Demolish and Install	S.F.	11.15	6.10		17.25	block.
	Clean	S.F.	.04	.37		.41	
	Paint	S.F.	.14	.67		.81	
	Minimum Charge	Job		170		170	
12" x 6"							
	Demolish	S.F.		1.58		1.58	Cost includes material and labor to
	Install	S.F.	10.20	4.99		15.19	install 12" x 6" x 16" concrete slump
	Demolish and Install	S.F.	10.20	6.57		16.77	block.
	Clean	S.F.	.04	.37		.41	
	Paint	S.F.	.14	.67		.81	
	Minimum Charge	Job		170		170	

Masonry Wall		Unit	Material	Labor	Equip.	Total	Specification
Stucco							
Three Coat							
	Demolish	S.Y.		1.85		1.85	Includes material and labor to install
	Install	S.Y.	3.66	7.10	.58	11.34	stucco on exterior masonry walls.
	Demolish and Install	S.Y.	3.66	8.95	.58	13.19	
	Clean	S.Y.		3.36		3.36	
	Paint	S.F.	.12	.48		.60	
	Minimum Charge	Job		150		150	
Three Coat w / Wire Mesh							
	Demolish	S.Y.		3.36		3.36	Includes material and labor to install
	Install	S.Y.	5.90	27	1.84	34.74	stucco on wire mesh over wood
	Demolish and Install	S.Y.	5.90	30.36	1.84	38.10	framing.
	Clean	S.Y.		3.36		3.36	
	Paint	S.F.	.12	.48		.60	
	Minimum Charge	Job		150		150	
Fill Cracks							
	Install	S.F.	.41	2.40		2.81	Includes labor and material to fill-in
	Minimum Charge	Job		150		150	hairline cracks up to 1/8" with filler,
							including sand and fill per S.F.
Furring							
1" x 2"							
	Demolish	L.F.		.11		.11	Cost includes material and labor to
	Install	L.F.	.21	.68		.89	install 1" x 2" furring on masonry wall.
	Demolish and Install	L.F.	.21	.79		1	
	Minimum Charge	Job		169		169	
2" x 2"							
	Demolish	L.F.		.12		.12	Cost includes material and labor to
	Install	L.F.	.29	.75		1.04	install 2" x 2" furring on masonry wall.
	Demolish and Install	L.F.	.29	.87		1.16	
	Minimum Charge	Job		169		169	
2" x 4"							
	Demolish	L.F.		.13		.13	Cost includes material and labor to
	Install	L.F.	.41	.86		1.27	install 2" x 4" furring on masonry wall.
	Demolish and Install	L.F.	.41	.99		1.40	
	Reinstall	L.F.		.68		.68	
	Minimum Charge	Job		169		169	
Maintenance / Cleaning							
Repoint							
	Install	S.F.	.47	4.26		4.73	Includes labor and material to cut and
	Minimum Charge	Hr.		42.50		42.50	repoint brick, hard mortar, running
							bond.
Acid Etch							
	Install	S.F.	.03	.61		.64	Includes labor and material to acid
	Minimum Charge	Job		121	27	148	wash smooth brick.
Pressure Wash							
	Clean	S.F.		.82	.11	.93	Includes labor and material to high
	Minimum Charge	Job		121	27	148	pressure water wash masonry,
							average.
Waterproof							
	Install	S.F.	.20	.62		.82	Includes labor and material to install
	Minimum Charge	Job		155		155	masory waterproofing.

Masonry

Chimney		Unit	Material	Labor	Equip.	Total	Specification
Demo & Haul Exterior							
	Demolish	C.F.		4.39		4.39	Includes minimum labor and
	Minimum Charge	Job		134		134	equipment to remove masonry and haul debris to truck or dumpster.
Demo & Haul Interior							
	Demolish	C.F.		6.05		6.05	Includes minimum labor and
	Minimum Charge	Job		134		134	equipment to remove masonry and haul debris to truck or dumpster.
Install New (w / flue)							
Brick							
	Install	V.L.F.	29.50	37.50		67	Includes material and labor to install
	Minimum Charge	Job		170		170	brick chimney with flue from ground level up. Foundation not included.
Natural Stone							
	Install	L.F.	92.50	60		152.50	Includes labor and material to install
	Minimum Charge	Job		170		170	natural stone chimney.
Install From Roof Up							
Brick							
	Install	V.L.F.	29.50	37.50		67	Includes material and labor to install
	Minimum Charge	Job		170		170	brick chimney with flue from ground level up. Foundation not included.
Natural Stone							
	Install	L.F.	92.50	50		142.50	Includes labor and material to install
	Minimum Charge	Job		170		170	natural stone chimney.
Install Base Pad							
	Install	Ea.	237	184	.98	421.98	Includes labor and material to replace
	Minimum Charge	Job		775	23.50	798.50	spread footing, 5' square x 16" deep.
Metal Cap							
	Install	Ea.	41.50	21.50		63	Includes labor and materials to install
	Minimum Charge	Job		169		169	bird and squirrel screens in a chimney flue.

Fireplace		Unit	Material	Labor	Equip.	Total	Specification
Common Brick							
30"							
	Demolish	Ea.		960		960	Includes material and labor to install a
	Install	Ea.	1475	3000		4475	complete unit which includes common
	Demolish and Install	Ea.	1475	3960		5435	brick, foundation, damper, flue lining
	Minimum Charge	Job		170		170	stack to 15' high and a 30" fire box.
36"							
	Demolish	Ea.		960		960	Includes material and labor to install a
	Install	Ea.	1600	3325		4925	complete unit which includes common
	Demolish and Install	Ea.	1600	4285		5885	brick, foundation, damper, flue lining
	Minimum Charge	Job		170		170	stack to 15' high and a 36" fire box.
42"							
	Demolish	Ea.		960		960	Includes material and labor to install a
	Install	Ea.	1725	3750		5475	complete unit which includes common
	Demolish and Install	Ea.	1725	4710		6435	brick, foundation, damper, flue lining
	Minimum Charge	Job		170		170	stack to 15' high and a 42" fire box.

Fireplace		Unit	Material	Labor	Equip.	Total	Specification
48"							Includes material and labor to install a
	Demolish	Ea.		960		960	complete unit which includes common
	Install	Ea.	1825	4275		6100	brick, foundation, damper, flue lining
	Demolish and Install	Ea.	1825	5235		7060	stack to 15' high and a 48" fire box.
	Minimum Charge	Job		170		170	
Raised Hearth							
Brick Hearth							
	Demolish	S.F.		4.19		4.19	Includes material and labor to install a
	Install	S.F.	13.95	25		38.95	brick fireplace hearth based on 30" x
	Demolish and Install	S.F.	13.95	29.19		43.14	29" opening.
	Clean	S.F.	.04	.32		.36	
	Minimum Charge	Job		170		170	
Marble Hearth / Facing							
	Demolish	Ea.		27		27	Includes material and labor to install
	Install	Ea.	600	101		701	marble fireplace hearth.
	Demolish and Install	Ea.	600	128		728	
	Reinstall	Ea.		81.17		81.17	
	Clean	Ea.	24	4.83		28.83	
	Minimum Charge	Job		170		170	
Breast (face) Brick							
	Demolish	S.F.		4.19		4.19	Includes material and labor to install
	Install	S.F.	3.61	7		10.61	standard red brick, running bond.
	Demolish and Install	S.F.	3.61	11.19		14.80	
	Clean	S.F.		.82	.11	.93	
	Minimum Charge	Hr.		42.50		42.50	
Fire Box							
	Demolish	Ea.		215		215	Includes material and labor to install
	Install	Ea.	150	300		450	fireplace box only (110 brick).
	Demolish and Install	Ea.	150	515		665	
	Clean	Ea.	.12	10.85		10.97	
	Minimum Charge	Hr.		42.50		42.50	
Natural Stone							
	Demolish	S.F.		4.19		4.19	Includes material and labor to install
	Install	S.F.	22	12.80		34.80	high priced stone, split or rock face.
	Demolish and Install	S.F.	22	16.99		38.99	
	Clean	S.F.	.07	.75		.82	
	Minimum Charge	Job		300		300	
Slumpstone Veneer							
	Demolish	S.F.		1.58		1.58	Cost includes material and labor to
	Install	S.F.	6.40	3.99		10.39	install 8" x 6" x 16" concrete slump
	Demolish and Install	S.F.	6.40	5.57		11.97	block.
	Clean	S.F.	.04	.37		.41	
	Paint	S.F.	.14	.67		.81	
	Minimum Charge	Job		170		170	
Prefabricated							
Firebox							
	Demolish	Ea.		38.50		38.50	Includes material and labor to install
	Install	Ea.	1550	305		1855	up to 43" radiant heat, zero
	Demolish and Install	Ea.	1550	343.50		1893.50	clearance, prefabricated fireplace.
	Reinstall	Ea.		245.82		245.82	Flues, doors or blowers not included.
	Clean	Ea.	1.64	15.10		16.74	
	Minimum Charge	Job		169		169	

Masonry

Fireplace

	Unit	Material	Labor	Equip.	Total	Specification
Flue						
Demolish	V.L.F.		4.47		4.47	Includes material and labor to install
Install	V.L.F.	12.05	12.95		25	prefabricated chimney vent, 8"
Demolish and Install	V.L.F.	12.05	17.42		29.47	exposed metal flue.
Reinstall	V.L.F.		10.36		10.36	
Clean	L.F.	.66	3.02		3.68	
Minimum Charge	Job		335		335	
Blower						
Demolish	Ea.		3.58		3.58	Includes material and labor to install
Install	Ea.	117	23.50		140.50	blower for metal fireplace flue.
Demolish and Install	Ea.	117	27.08		144.08	
Reinstall	Ea.		18.70		18.70	
Clean	Ea.	3.28	30		33.28	
Minimum Charge	Job		335		335	

Accessories

	Unit	Material	Labor	Equip.	Total	Specification
Wood Mantle						
Demolish	L.F.		4.47		4.47	Includes material and labor to install
Install	L.F.	6.80	9.65		16.45	wood fireplace mantel.
Demolish and Install	L.F.	6.80	14.12		20.92	
Reinstall	L.F.		9.66		9.66	
Clean	Ea.	.02	.67		.69	
Paint	L.F.	.07	8.05		8.12	
Minimum Charge	Job		169		169	
Colonial Style						
Demolish	Ea.		42		42	Includes material and labor to install
Install	Opng.	2400	169		2569	wood fireplace mantel.
Demolish and Install	Opng.	2400	211		2611	
Reinstall	Opng.		169		169	
Clean	Ea.		.48		.48	
Paint	Ea.	11.90	111		122.90	
Minimum Charge	Job		169		169	
Modern Design						
Demolish	Ea.		42		42	Includes material and labor to install
Install	Opng.	190	67.50		257.50	wood fireplace mantel.
Demolish and Install	Opng.	190	109.50		299.50	
Reinstall	Opng.		67.60		67.60	
Clean	Ea.		.48		.48	
Paint	Ea.	11.90	111		122.90	
Minimum Charge	Job		169		169	
Glass Door Unit						
Demolish	Ea.		5.60		5.60	Includes material and labor to install a
Install	Ea.	128	7.10		135.10	fireplace glass door unit.
Demolish and Install	Ea.	128	12.70		140.70	
Reinstall	Ea.		7.09		7.09	
Clean	Ea.	.40	9.30		9.70	
Minimum Charge	Job		169		169	
Fire Screen						
Demolish	Ea.		2.80		2.80	Includes material and labor to install a
Install	Ea.	39	3.55		42.55	fireplace fire screen.
Demolish and Install	Ea.	39	6.35		45.35	
Reinstall	Ea.		3.55		3.55	
Clean	Ea.	.20	8.05		8.25	
Minimum Charge	Job		169		169	

Siding

Shingles

Board & Batten

Texture 1-11

Exterior Siding	Unit	Material	Labor	Equip.	Total	Specification
Trim						
1″ x 2″						
Demolish	L.F.		.22		.22	Cost includes material and labor to
Install	L.F.	.40	1.02		1.42	install 1″ x 2″ pine trim.
Demolish and Install	L.F.	.40	1.24		1.64	
Clean	L.F.	.01	.16		.17	
Paint	L.F.	.02	.46		.48	
Minimum Charge	Job		169		169	
1″ x 3″						
Demolish	L.F.		.22		.22	Cost includes material and labor to
Install	L.F.	.31	1.13		1.44	install 1″ x 3″ pine trim.
Demolish and Install	L.F.	.31	1.35		1.66	
Clean	L.F.	.01	.16		.17	
Paint	L.F.	.02	.46		.48	
Minimum Charge	Job		169		169	
1″ x 4″						
Demolish	L.F.		.22		.22	Cost includes material and labor to
Install	L.F.	.78	1.35		2.13	install 1″ x 4″ pine trim.
Demolish and Install	L.F.	.78	1.57		2.35	
Clean	L.F.	.01	.16		.17	
Paint	L.F.	.02	.46		.48	
Minimum Charge	Job		169		169	
1″ x 6″						
Demolish	L.F.		.22		.22	Cost includes material and labor to
Install	L.F.	1.36	1.35		2.71	install 1″ x 6″ pine trim.
Demolish and Install	L.F.	1.36	1.57		2.93	
Clean	L.F.	.01	.16		.17	
Paint	L.F.	.07	.46		.53	
Minimum Charge	Job		169		169	
1″ x 8″						
Demolish	L.F.		.24		.24	Cost includes material and labor to
Install	L.F.	1.58	1.69		3.27	install 1″ x 8″ pine trim.
Demolish and Install	L.F.	1.58	1.93		3.51	
Clean	L.F.	.02	.17		.19	
Paint	L.F.	.07	.46		.53	
Minimum Charge	Job		169		169	
Aluminum						
8″ Non-Insulated						
Demolish	S.F.		.60		.60	Includes material and labor to install
Install	S.F.	1.33	1.33		2.66	horizontal, colored clapboard, 8″ to
Demolish and Install	S.F.	1.33	1.93		3.26	10″ wide, plain.
Reinstall	S.F.		1.33		1.33	
Clean	S.F.		.24		.24	
Minimum Charge	Job		169		169	

Siding

Exterior Siding	Unit	Material	Labor	Equip.	Total	Specification
8″ Insulated						
Demolish	S.F.		.60		.60	Includes material and labor to install
Install	S.F.	1.34	1.33		2.67	horizontal, colored insulated
Demolish and Install	S.F.	1.34	1.93		3.27	clapboard, 8″ wide, plain.
Reinstall	S.F.		1.33		1.33	
Clean	S.F.		.24		.24	
Minimum Charge	Job		169		169	
12″ Non-Insulated						
Demolish	S.F.		.60		.60	Includes material and labor to install
Install	S.F.	1.38	1.15		2.53	horizontal, colored clapboard, plain.
Demolish and Install	S.F.	1.38	1.75		3.13	
Reinstall	S.F.		1.15		1.15	
Clean	S.F.		.24		.24	
Minimum Charge	Job		169		169	
12″ Insulated						
Demolish	S.F.		.60		.60	Includes material and labor to install
Install	S.F.	1.60	1.15		2.75	horizontal, colored insulated
Demolish and Install	S.F.	1.60	1.75		3.35	clapboard, plain.
Reinstall	S.F.		1.15		1.15	
Clean	S.F.		.24		.24	
Minimum Charge	Job		169		169	
Clapboard 8″ to 10″						
Demolish	S.F.		.60		.60	Includes material and labor to install
Install	S.F.	1.33	1.33		2.66	horizontal, colored clapboard, 8″ to
Demolish and Install	S.F.	1.33	1.93		3.26	10″ wide, plain.
Reinstall	S.F.		1.33		1.33	
Clean	S.F.		.24		.24	
Minimum Charge	Job		169		169	
Corner Strips						
Demolish	L.F.		.32		.32	Includes material and labor to install
Install	V.L.F.	2.05	1.33		3.38	corners for horizontal lapped
Demolish and Install	V.L.F.	2.05	1.65		3.70	aluminum siding.
Reinstall	V.L.F.		1.33		1.33	
Clean	L.F.		.16		.16	
Minimum Charge	Job		169		169	
Vinyl						
Non-Insulated						
Demolish	S.F.		.47		.47	Cost includes material and labor to
Install	S.F.	.81	1.33		2.14	install 10″ non-insulated solid vinyl
Demolish and Install	S.F.	.81	1.80		2.61	siding.
Reinstall	S.F.		1.06		1.06	
Clean	S.F.		.24		.24	
Minimum Charge	Job		169		169	
Insulated						
Demolish	S.F.		.47		.47	Includes material and labor to install
Install	S.F.	.92	1.33		2.25	insulated solid vinyl siding panels.
Demolish and Install	S.F.	.92	1.80		2.72	
Reinstall	S.F.		1.06		1.06	
Clean	S.F.		.24		.24	
Minimum Charge	Job		169		169	
Corner Strips						
Demolish	L.F.		.30		.30	Includes material and labor to install
Install	L.F.	1.58	1.65		3.23	corner posts for vinyl siding panels.
Demolish and Install	L.F.	1.58	1.95		3.53	
Reinstall	L.F.		1.32		1.32	
Clean	L.F.		.16		.16	
Minimum Charge	Job		169		169	

Siding

Exterior Siding		Unit	Material	Labor	Equip.	Total	Specification
Hardboard							
4' x 8' Panel							
	Demolish	S.F.		.51		.51	Cost includes material and labor to
	Install	S.F.	1.10	.84		1.94	install 4' x 8' masonite siding 7/16"
	Demolish and Install	S.F.	1.10	1.35		2.45	thick.
	Clean	S.F.		.24		.24	
	Paint	S.F.	.25	1		1.25	
	Minimum Charge	Job		169		169	
Lap							
	Demolish	S.F.		.50		.50	Cost includes material and labor to
	Install	S.F.	1.01	.87		1.88	install 8" or 12" masonite lap siding
	Demolish and Install	S.F.	1.01	1.37		2.38	7/16" thick with wood grain finish.
	Clean	S.F.		.24		.24	
	Paint	S.F.	.25	1		1.25	
	Minimum Charge	Job		169		169	
Plywood Panel							
Douglas Fir (T-1-11)							
	Demolish	S.F.		.37		.37	Includes material and labor to install
	Install	S.F.	1.11	1.02		2.13	Douglas fir 4' x 8' plywood exterior
	Demolish and Install	S.F.	1.11	1.39		2.50	siding plain or patterns.
	Reinstall	S.F.		1.02		1.02	
	Clean	S.F.		.24		.24	
	Paint	S.F.	.25	1		1.25	
	Minimum Charge	Job		169		169	
Redwood							
	Demolish	S.F.		.37		.37	Includes material and labor to install
	Install	S.F.	1.98	1.02		3	redwood 4' x 8' plywood exterior
	Demolish and Install	S.F.	1.98	1.39		3.37	siding plain or patterns.
	Reinstall	S.F.		1.02		1.02	
	Clean	S.F.		.24		.24	
	Paint	S.F.	.25	1		1.25	
	Minimum Charge	Job		169		169	
Cedar							
	Demolish	S.F.		.37		.37	Cost includes material and labor to
	Install	S.F.	2.62	1.02		3.64	install 4' x 8' cedar plywood exterior
	Demolish and Install	S.F.	2.62	1.39		4.01	siding plain or patterns.
	Reinstall	S.F.		1.02		1.02	
	Clean	S.F.		.24		.24	
	Paint	S.F.	.25	1		1.25	
	Minimum Charge	Job		169		169	
Tongue and Groove							
Fir							
	Demolish	S.F.		.67		.67	Includes material and labor to install
	Install	S.F.	1.51	.90		2.41	fir tongue and groove exterior siding
	Demolish and Install	S.F.	1.51	1.57		3.08	1" x 8".
	Reinstall	S.F.		.72		.72	
	Clean	S.F.		.24		.24	
	Paint	S.F.	.25	1		1.25	
	Minimum Charge	Job		169		169	
Pine							
	Demolish	S.F.		.67		.67	Includes material and labor to install
	Install	S.F.	.73	1.83		2.56	pine tongue and groove exterior
	Demolish and Install	S.F.	.73	2.50		3.23	siding 1" x 8".
	Reinstall	S.F.		1.83		1.83	
	Clean	S.F.		.24		.24	
	Paint	S.F.	.25	1		1.25	
	Minimum Charge	Job		169		169	

Siding

Exterior Siding		Unit	Material	Labor	Equip.	Total	Specification
Cedar							
	Demolish	S.F.		.67		.67	Includes material and labor to install
	Install	S.F.	3.31	.90		4.21	cedar tongue and groove exterior
	Demolish and Install	S.F.	3.31	1.57		4.88	siding 1" x 8".
	Reinstall	S.F.		.72		.72	
	Clean	S.F.		.24		.24	
	Paint	S.F.	.25	1		1.25	
	Minimum Charge	Job		169		169	
Redwood							
	Demolish	S.F.		.67		.67	Includes material and labor to install
	Install	S.F.	3.19	2.29		5.48	redwood tongue and groove exterior
	Demolish and Install	S.F.	3.19	2.96		6.15	siding.
	Reinstall	S.F.		2.29		2.29	
	Clean	S.F.		.24		.24	
	Paint	S.F.	.25	1		1.25	
	Minimum Charge	Job		169		169	
Board and Batten							
Fir							
	Demolish	S.F.		.67		.67	Cost includes material and labor to
	Install	S.F.	1.49	.85		2.34	install 1" x 8" fir boards with 1" x 3"
	Demolish and Install	S.F.	1.49	1.52		3.01	battens.
	Reinstall	S.F.		.68		.68	
	Clean	S.F.		.24		.24	
	Paint	S.F.	.25	1		1.25	
	Minimum Charge	Job		169		169	
Pine							
	Demolish	S.F.		.67		.67	Cost includes material and labor to
	Install	S.F.	.77	.85		1.62	install 1" x 10" pine boards with 1" x
	Demolish and Install	S.F.	.77	1.52		2.29	3" battens.
	Reinstall	S.F.		.68		.68	
	Clean	S.F.		.24		.24	
	Paint	S.F.	.25	1		1.25	
	Minimum Charge	Job		169		169	
Cedar							
	Demolish	S.F.		.67		.67	Cost includes material and labor to
	Install	S.F.	2.18	.85		3.03	install 1" x 10" cedar boards with 1" x
	Demolish and Install	S.F.	2.18	1.52		3.70	3" battens.
	Reinstall	S.F.		.68		.68	
	Clean	S.F.		.24		.24	
	Paint	S.F.	.25	1		1.25	
	Minimum Charge	Job		169		169	
Redwood							
	Demolish	S.F.		.67		.67	Cost includes material and labor to
	Install	S.F.	3.95	.85		4.80	install 1" x 10" redwood boards with
	Demolish and Install	S.F.	3.95	1.52		5.47	1" x 3" battens.
	Reinstall	S.F.		.68		.68	
	Clean	S.F.		.24		.24	
	Paint	S.F.	.25	1		1.25	
	Minimum Charge	Job		169		169	
Wood Shingle							
	Demolish	Sq.		62.50		62.50	Includes material and labor to install
	Install	Sq.	135	150		285	wood shingles/shakes.
	Demolish and Install	Sq.	135	212.50		347.50	
	Clean	Sq.		25.50		25.50	
	Paint	Sq.	16.10	89		105.10	
	Minimum Charge	Job		169		169	

Siding

Exterior Siding	Unit	Material	Labor	Equip.	Total	Specification
Synthetic Stucco System						
Demolish	S.F.		2.24		2.24	Cost includes material and labor to
Install	S.F.	3.11	6.45	.53	10.09	install 2 coats of adhesive mixture,
Demolish and Install	S.F.	3.11	8.69	.53	12.33	glass fiber mesh, 1″ insulation board,
Clean	S.F.		.22		.22	and textured finish coat.
Paint	S.F.	.12	.48		.60	
Minimum Charge	Job		710	58	768	

Front-end Loader   **Bulldozer**

Grading		Unit	Material	Labor	Equip.	Total	Specification
Hand (fine)							
	Install	S.F.		.48		.48	Includes labor and material to install
	Minimum Charge	Job		121		121	fine grade for slab on grade, hand grade.
Machine (fine)							
	Install	S.Y.		.42	.37	.79	Includes labor and material to install
	Minimum Charge	Job		288	252	540	fine grade for roadway, base or leveling course.

Asphalt Paving		Unit	Material	Labor	Equip.	Total	Specification
Seal Coat							
	Install	S.F.	.07	.09		.16	Includes labor and material to install
	Minimum Charge	Job		121		121	sealcoat for a small area.
Wearing Course							
	Install	S.F.	.08	.26	.18	.52	Cost includes labor and material to
	Minimum Charge	Job		121		121	install 1" thick wearing course.

Masonry Pavers		Unit	Material	Labor	Equip.	Total	Specification
Natural Concrete							
Sand Base							
	Demolish	S.F.		2.68		2.68	Includes material and labor to install
	Install	S.F.	2.21	5.65		7.86	concrete pavers on sand base.
	Demolish and Install	S.F.	2.21	8.33		10.54	
	Reinstall	S.F.		4.54		4.54	
	Clean	S.F.		.97		.97	
	Minimum Charge	Job		170		170	
Mortar Base							
	Demolish	S.F.		4.46	1.38	5.84	Includes material and labor to install
	Install	S.F.	2.89	4.73		7.62	concrete pavers on mortar base.
	Demolish and Install	S.F.	2.89	9.19	1.38	13.46	
	Reinstall	S.F.		3.78		3.78	
	Clean	S.F.		.88		.88	
	Minimum Charge	Job		170		170	

Masonry Pavers	Unit	Material	Labor	Equip.	Total	Specification
Adobe						
Sand Base						
Demolish	S.F.		2.68		2.68	Includes material and labor to install
Install	S.F.	1.86	3.78		5.64	adobe pavers on sand base.
Demolish and Install	S.F.	1.86	6.46		8.32	
Reinstall	S.F.		3.03		3.03	
Clean	S.F.		.97		.97	
Minimum Charge	Job		170		170	
Mortar Base						
Demolish	S.F.		4.46	1.38	5.84	Includes material and labor to install
Install	S.F.	2.57	4.73		7.30	adobe pavers on mortar base.
Demolish and Install	S.F.	2.57	9.19	1.38	13.14	
Reinstall	S.F.		3.78		3.78	
Clean	S.F.		.88		.88	
Minimum Charge	Job		170		170	
Standard Brick						
Sand Base						
Demolish	S.F.		2.68		2.68	Includes material and labor to install
Install	S.F.	2.78	6		8.78	standard brick pavers on sand base.
Demolish and Install	S.F.	2.78	8.68		11.46	
Reinstall	S.F.		4.79		4.79	
Clean	S.F.		.97		.97	
Minimum Charge	Job		170		170	
Mortar Base						
Demolish	S.F.		4.46	1.38	5.84	Includes material and labor to install
Install	S.F.	5.10	8.50		13.60	standard brick pavers on mortar base.
Demolish and Install	S.F.	5.10	12.96	1.38	19.44	
Reinstall	S.F.		6.81		6.81	
Clean	S.F.		.88		.88	
Minimum Charge	Job		170		170	
Deluxe Brick						
Sand Base						
Demolish	S.F.		2.68		2.68	Includes material and labor to install
Install	S.F.	5.10	4.73		9.83	deluxe brick pavers on sand base.
Demolish and Install	S.F.	5.10	7.41		12.51	
Reinstall	S.F.		3.78		3.78	
Clean	S.F.		.97		.97	
Minimum Charge	Job		170		170	
Mortar Base						
Demolish	S.F.		4.46	1.38	5.84	Includes material and labor to install
Install	S.F.	5.50	8.50		14	deluxe brick pavers on mortar base.
Demolish and Install	S.F.	5.50	12.96	1.38	19.84	
Reinstall	S.F.		6.81		6.81	
Clean	S.F.		.88		.88	
Minimum Charge	Job		170		170	

Fencing	Unit	Material	Labor	Equip.	Total	Specification
Block w / Footing						
4"						
Demolish	S.F.		1.49		1.49	Cost includes material and labor to
Install	S.F.	1.25	3.58		4.83	install 4" non-reinforced concrete
Demolish and Install	S.F.	1.25	5.07		6.32	block.
Clean	S.F.	.14	1.03	.14	1.31	
Paint	S.F.	.17	.39		.56	
Minimum Charge	Job		170		170	

Site Work

Fencing		Unit	Material	Labor	Equip.	Total	Specification
4" Reinforced							
	Demolish	S.F.		1.49	.73	2.22	Cost includes material and labor to
	Install	S.F.	1.34	3.62		4.96	install 4" reinforced concrete block.
	Demolish and Install	S.F.	1.34	5.11	.73	7.18	
	Clean	S.F.	.14	1.03	.14	1.31	
	Paint	S.F.	.17	.39		.56	
	Minimum Charge	Job		170		170	
6"							
	Demolish	S.F.		1.58		1.58	Cost includes material and labor to
	Install	S.F.	1.86	3.85		5.71	install 6" non-reinforced concrete
	Demolish and Install	S.F.	1.86	5.43		7.29	block.
	Clean	S.F.	.14	1.03	.14	1.31	
	Paint	S.F.	.17	.39		.56	
	Minimum Charge	Job		170		170	
6" Reinforced							
	Demolish	S.F.		1.57	.35	1.92	Cost includes material and labor to
	Install	S.F.	1.96	3.90		5.86	install 6" reinforced concrete block.
	Demolish and Install	S.F.	1.96	5.47	.35	7.78	
	Clean	S.F.	.14	1.03	.14	1.31	
	Paint	S.F.	.17	.39		.56	
	Minimum Charge	Job		170		170	
Cap for Block Wall							
	Demolish	L.F.		1.68		1.68	Includes labor and equipment to install
	Install	L.F.	16.50	8		24.50	precast concrete coping.
	Demolish and Install	L.F.	16.50	9.68		26.18	
	Clean	L.F.	.01	2.52		2.53	
	Paint	S.F.	.17	.39		.56	
	Minimum Charge	Job		170		170	
Block Pilaster							
	Demolish	Ea.		115		115	Includes material and labor to install
	Install	Ea.	69.50	120		189.50	block pilaster for fence.
	Demolish and Install	Ea.	69.50	235		304.50	
	Reinstall	Ea.		119.84		119.84	
	Paint	Ea.	4.22	13.15		17.37	
	Minimum Charge	Job		170		170	

Block w / o Footing

		Unit	Material	Labor	Equip.	Total	Specification
4"							
	Demolish	S.F.		1.49		1.49	Cost includes material and labor to
	Install	S.F.	1.47	2.72		4.19	install 4" non-reinforced hollow
	Demolish and Install	S.F.	1.47	4.21		5.68	lightweight block.
	Clean	S.F.	.14	1.03	.14	1.31	
	Paint	S.F.	.17	.39		.56	
	Minimum Charge	Job		170		170	
4" Reinforced							
	Demolish	S.F.		1.49	.73	2.22	Cost includes material and labor to
	Install	S.F.	2.16	3.54		5.70	install 4" reinforced hollow lightweight
	Demolish and Install	S.F.	2.16	5.03	.73	7.92	block.
	Clean	S.F.	.14	1.03	.14	1.31	
	Paint	S.F.	.17	.39		.56	
	Minimum Charge	Job		170		170	
6"							
	Demolish	S.F.		1.58		1.58	Cost includes material and labor to
	Install	S.F.	2.02	3		5.02	install 6" non-reinforced hollow
	Demolish and Install	S.F.	2.02	4.58		6.60	lightweight block.
	Clean	S.F.	.14	1.03	.14	1.31	
	Paint	S.F.	.17	.39		.56	
	Minimum Charge	Job		170		170	

139

Fencing	Unit	Material	Labor	Equip.	Total	Specification
6" Reinforced						
Demolish	S.F.		1.57	.35	1.92	Cost includes material and labor to
Install	S.F.	2.11	3.80		5.91	install 6" reinforced hollow lightweight
Demolish and Install	S.F.	2.11	5.37	.35	7.83	block.
Clean	S.F.	.14	1.03	.14	1.31	
Paint	S.F.	.17	.39		.56	
Minimum Charge	Job		170		170	
Cap for Block Wall						
Demolish	L.F.		1.68		1.68	Includes labor and equipment to install
Install	L.F.	16.50	8		24.50	precast concrete coping.
Demolish and Install	L.F.	16.50	9.68		26.18	
Clean	L.F.	.01	2.52		2.53	
Paint	S.F.	.17	.39		.56	
Minimum Charge	Job		170		170	
Block Pilaster						
Demolish	Ea.		115		115	Includes material and labor to install
Install	Ea.	69.50	120		189.50	block pilaster for fence.
Demolish and Install	Ea.	69.50	235		304.50	
Reinstall	Ea.		119.84		119.84	
Paint	Ea.	4.22	13.15		17.37	
Minimum Charge	Job		170		170	
Chain Link						
3' High w / o Top Rail						
Demolish	L.F.		2.10	.65	2.75	Includes material and labor to install
Install	L.F.	3.44	1.42	.28	5.14	residential galvanized fence 3' high
Demolish and Install	L.F.	3.44	3.52	.93	7.89	with 1-3/8" steel post 10' O.C. set
Reinstall	L.F.		1.13	.22	1.35	without concrete.
Clean	L.F.	.10	.88		.98	
Minimum Charge	Job		375		375	
4' High w / o Top Rail						
Demolish	L.F.		2.10	.65	2.75	Includes material and labor to install
Install	L.F.	3.65	1.75	.35	5.75	residential galvanized fence 4' high
Demolish and Install	L.F.	3.65	3.85	1	8.50	with 1-3/8" steel post 10' O.C. set
Reinstall	L.F.		1.40	.28	1.68	without concrete.
Clean	L.F.	.13	.97		1.10	
Minimum Charge	Job		375		375	
5' High w / o Top Rail						
Demolish	L.F.		2.10	.65	2.75	Includes material and labor to install
Install	L.F.	4.29	2.29	.45	7.03	residential galvanized fence 5' high
Demolish and Install	L.F.	4.29	4.39	1.10	9.78	with 1-3/8" steel post 10' O.C. set
Reinstall	L.F.		1.83	.36	2.19	without concrete.
Clean	L.F.	.17	1.07		1.24	
Minimum Charge	Job		375		375	
3' High w / Top Rail						
Demolish	L.F.		2.23	.69	2.92	Includes material and labor to install
Install	L.F.	4.25	1.49	.30	6.04	residential galvanized fence 3' high
Demolish and Install	L.F.	4.25	3.72	.99	8.96	with 1-3/8" steel post 10' O.C. set
Reinstall	L.F.		1.19	.24	1.43	without concrete.
Clean	L.F.	.10	.89		.99	
Minimum Charge	Job		375		375	
4' High w / Top Rail						
Demolish	L.F.		2.23	.69	2.92	Includes material and labor to install
Install	L.F.	4.82	1.86	.37	7.05	residential galvanized fence 4' high
Demolish and Install	L.F.	4.82	4.09	1.06	9.97	with 1-5/8" steel post 10' O.C. set
Reinstall	L.F.		1.49	.30	1.79	without concrete and 1-5/8" top rail
Clean	L.F.	.13	.99		1.12	with sleeves.
Minimum Charge	Job		375		375	

Fencing	Unit	Material	Labor	Equip.	Total	Specification
5' High w / Top Rail						
Demolish	L.F.		2.23	.69	2.92	Includes material and labor to install
Install	L.F.	5.45	2.48	.49	8.42	residential galvanized fence 5' high
Demolish and Install	L.F.	5.45	4.71	1.18	11.34	with 1-5/8" steel post 10' O.C. set
Reinstall	L.F.		1.99	.39	2.38	without concrete and 1-5/8" top rail
Clean	L.F.	.17	1.10		1.27	with sleeves.
Minimum Charge	Job		375		375	
6' High w / Top Rail						
Demolish	L.F.		2.23	.69	2.92	Includes material and labor to install
Install	L.F.	5.85	3.72	.74	10.31	residential galvanized fence 6' high
Demolish and Install	L.F.	5.85	5.95	1.43	13.23	with 2" steel post 10' O.C. set without
Reinstall	L.F.		2.98	.59	3.57	concrete and 1-3/8" top rail with
Clean	L.F.	.20	1.24		1.44	sleeves.
Minimum Charge	Job		375		375	
Barbed Wire						
Demolish	L.F.		1.25		1.25	Cost includes material and labor to
Install	L.F.	.36	.48	.13	.97	install 3 strand galvanized barbed
Demolish and Install	L.F.	.36	1.73	.13	2.22	wire fence.
Reinstall	L.F.		.39	.10	.49	
Minimum Charge	Job		375	293	668	
6' Redwood						
Demolish	L.F.		8.35		8.35	Cost includes material and labor to
Install	L.F.	10.85	5.95	1.18	17.98	install 6' high redwood fencing.
Demolish and Install	L.F.	10.85	14.30	1.18	26.33	
Reinstall	L.F.		4.76	.94	5.70	
Clean	L.F.	.20	1.12		1.32	
Paint	L.F.	.66	3.12		3.78	
Minimum Charge	Job		375		375	
6' Red Cedar						
Demolish	L.F.		8.35		8.35	Cost includes material and labor to
Install	L.F.	16.55	5.95	1.18	23.68	install 6' cedar fence.
Demolish and Install	L.F.	16.55	14.30	1.18	32.03	
Reinstall	L.F.		4.76	.94	5.70	
Clean	L.F.	.20	1.12		1.32	
Paint	L.F.	.66	3.12		3.78	
Minimum Charge	Job		375		375	
6' Shadowbox						
Demolish	L.F.		8.35		8.35	Cost includes material and labor to
Install	L.F.	12.65	4.96	.98	18.59	install 6' pressure treated pine fence.
Demolish and Install	L.F.	12.65	13.31	.98	26.94	
Reinstall	L.F.		3.97	.79	4.76	
Clean	L.F.	.20	1.12		1.32	
Paint	L.F.	.66	3.12		3.78	
Minimum Charge	Job		375		375	
3' Picket						
Demolish	L.F.		8.35		8.35	Includes material, labor and
Install	L.F.	4.81	5.30	1.05	11.16	equipment to install pressure treated
Demolish and Install	L.F.	4.81	13.65	1.05	19.51	pine picket fence.
Reinstall	L.F.		4.25	.84	5.09	
Clean	L.F.	.10	.97		1.07	
Paint	L.F.	.33	1.69		2.02	
Minimum Charge	Job		375		375	

Fencing	Unit	Material	Labor	Equip.	Total	Specification
Wrought Iron						
Demolish	L.F.		17.90		17.90	Cost includes material and labor to
Install	L.F.	24	10.05		34.05	install 5' high wrought iron fence with
Demolish and Install	L.F.	24	27.95		51.95	pickets at 4-1/2" O.C.
Reinstall	L.F.		8.02		8.02	
Clean	L.F.		.97		.97	
Paint	L.F.	.58	3.49		4.07	
Minimum Charge	Job		375		375	
Post						
Dirt Based Wood						
Demolish	Ea.		21		21	Includes material and labor to install
Install	Ea.	11.70	94		105.70	wood post set in earth.
Demolish and Install	Ea.	11.70	115		126.70	
Reinstall	Ea.		75.19		75.19	
Clean	Ea.		3.02		3.02	
Paint	Ea.	.73	1.19		1.92	
Minimum Charge	Job		375		375	
Concrete Based Wood						
Demolish	Ea.		30	9.20	39.20	Includes material and labor to install
Install	Ea.	11.70	125		136.70	wood post set in concrete.
Demolish and Install	Ea.	11.70	155	9.20	175.90	
Clean	Ea.		3.02		3.02	
Paint	Ea.	.73	1.19		1.92	
Minimum Charge	Job		375		375	
Block Pilaster						
Demolish	Ea.		115		115	Includes material and labor to install
Install	Ea.	69.50	120		189.50	block pilaster for fence.
Demolish and Install	Ea.	69.50	235		304.50	
Reinstall	Ea.		119.84		119.84	
Paint	Ea.	4.22	13.15		17.37	
Minimum Charge	Job		170		170	
Gate						
Chain Link						
Demolish	Ea.		28		28	Cost includes material and labor to
Install	Ea.	107	42		149	install 3' wide chain link pass gate
Demolish and Install	Ea.	107	70		177	and hardware.
Reinstall	Ea.		33.42		33.42	
Clean	Ea.		3.36		3.36	
Minimum Charge	Job		375		375	
Redwood						
Demolish	Ea.		24		24	Includes material, labor and
Install	Ea.	139	42		181	equipment to install redwood pass
Demolish and Install	Ea.	139	66		205	gate and hardware.
Reinstall	Ea.		33.42		33.42	
Clean	Ea.		3.36		3.36	
Paint	Ea.	1.09	5.95		7.04	
Minimum Charge	Job		375		375	
Red Cedar						
Demolish	Ea.		24		24	Cost includes material and labor to
Install	Ea.	73	42		115	install 3' wide red cedar pass gate
Demolish and Install	Ea.	73	66		139	and hardware.
Reinstall	Ea.		33.42		33.42	
Clean	Ea.		3.36		3.36	
Paint	Ea.	1.09	5.95		7.04	
Minimum Charge	Job		375		375	

Site Work

Fencing

Fencing	Unit	Material	Labor	Equip.	Total	Specification
Wrought Iron						
Demolish	Ea.		17.90		17.90	Includes material, labor and
Install	Ea.	128	94		222	equipment to install wrought iron gate
Demolish and Install	Ea.	128	111.90		239.90	with hardware.
Reinstall	Ea.		75.19		75.19	
Clean	Ea.		3.36		3.36	
Paint	Ea.	1.17	5.95		7.12	
Minimum Charge	Job		375		375	

Rough Mechanical

Fire-extinguishing System

Space Heater

Hot-air Furnace

Plumbing	Unit	Material	Labor	Equip.	Total	Specification
Components						
3/4" Supply						
Demolish	L.F.		2.06		2.06	Includes material and labor to install
Install	L.F.	2.86	4.88		7.74	type L copper tubing including
Demolish and Install	L.F.	2.86	6.94		9.80	couplings and hangers at 10' O.C.
Reinstall	L.F.		4.88		4.88	
Minimum Charge	Job		186		186	
1/2" Fixture Supply						
Install	Ea.	29.50	124		153.50	Includes labor and materials to install supply pipe only to a plumbing fixture.
2" Single Vent Stack						
Demolish	Ea.		34.50		34.50	Includes material and labor to install
Install	Ea.	14.80	186		200.80	PVC vent stack pipe with couplings at
Demolish and Install	Ea.	14.80	220.50		235.30	10' O.C.
Reinstall	Ea.		185.60		185.60	
Minimum Charge	Job		186		186	
4" Drain						
Demolish	L.F.		2.75		2.75	Includes material and labor to install
Install	L.F.	8.55	6.05		14.60	DVW type PVC pipe with couplings at
Demolish and Install	L.F.	8.55	8.80		17.35	10' O.C. and 3 strap hangers per 10'
Reinstall	L.F.		6.07		6.07	run.
Minimum Charge	Job		186		186	
1-1/2" Drain						
Demolish	L.F.		2.06		2.06	Includes material and labor to install
Install	L.F.	.76	3.95		4.71	DVW type PVC pipe with couplings at
Demolish and Install	L.F.	.76	6.01		6.77	10' O.C. and 3 strap hangers per 10'
Reinstall	L.F.		3.95		3.95	run.
Minimum Charge	Job		186		186	
Sewer / Septic						
Install	L.F.	2.48	3.42	1.06	6.96	Includes labor and material to install sewerage and drainage piping, 6" diameter.
Maintenance						
Flush Out / Unclog						
Clean	Job		186		186	Minimum labor and equipment charge
Minimum Charge	Job		186		186	to install water cooler.

Plumbing

Plumbing	Unit	Material	Labor	Equip.	Total	Specification
Vent Cap						
Large						
Demolish	Ea.		9.35		9.35	Includes labor and material to install a
Install	Ea.	34.50	18.55		53.05	PVC plumbing pipe vent cap.
Demolish and Install	Ea.	34.50	27.90		62.40	
Clean	Ea.	.45	5.50		5.95	
Paint	Ea.	.15	6.75		6.90	
Minimum Charge	Job		93		93	
Medium						
Demolish	Ea.		9.35		9.35	Includes material and labor to install a
Install	Ea.	28.50	16.85		45.35	PVC plumbing pipe vent cap.
Demolish and Install	Ea.	28.50	26.20		54.70	
Clean	Ea.	.45	5.50		5.95	
Paint	Ea.	.15	6.75		6.90	
Minimum Charge	Job		93		93	
Small						
Demolish	Ea.		9.35		9.35	Includes material and labor to install a
Install	Ea.	22	15.45		37.45	PVC plumbing pipe vent cap.
Demolish and Install	Ea.	22	24.80		46.80	
Clean	Ea.	.45	5.50		5.95	
Paint	Ea.	.15	6.75		6.90	
Minimum Charge	Job		93		93	
Sump Pump						
1/4 HP						
Demolish	Ea.		37		37	Includes material and labor to install a
Install	Ea.	125	62		187	submersible sump pump.
Demolish and Install	Ea.	125	99		224	
Reinstall	Ea.		49.49		49.49	
Clean	Ea.	1.14	15.10		16.24	
Minimum Charge	Job		186		186	
1/3 HP						
Demolish	Ea.		37		37	Includes material and labor to install a
Install	Ea.	152	74		226	submersible sump pump.
Demolish and Install	Ea.	152	111		263	
Reinstall	Ea.		59.39		59.39	
Clean	Ea.	1.14	15.10		16.24	
Minimum Charge	Job		186		186	
1/2 HP						
Demolish	Ea.		37		37	Includes material and labor to install a
Install	Ea.	185	74		259	submersible sump pump.
Demolish and Install	Ea.	185	111		296	
Reinstall	Ea.		59.39		59.39	
Clean	Ea.	1.14	15.10		16.24	
Minimum Charge	Job		186		186	
Clean and Service						
Clean	Ea.		23		23	Includes labor to clean and service a
Minimum Charge	Job		186		186	sump pump.

Rough Mechanical

Water Heater		Unit	Material	Labor	Equip.	Total	Specification
30 Gallon							
	Demolish	Ea.		116		116	Includes material and labor to install a
	Install	Ea.	320	169		489	30 gallon electric water heater.
	Demolish and Install	Ea.	320	285		605	
	Reinstall	Ea.		134.98		134.98	
	Clean	Ea.	1.14	15.10		16.24	
	Minimum Charge	Job		186		186	
40 Gallon							
	Demolish	Ea.		116		116	Includes material and labor to install a
	Install	Ea.	340	186		526	40 gallon electric water heater.
	Demolish and Install	Ea.	340	302		642	
	Reinstall	Ea.		148.48		148.48	
	Clean	Ea.	1.14	15.10		16.24	
	Minimum Charge	Job		186		186	
52 Gallon							
	Demolish	Ea.		116		116	Includes material and labor to install a
	Install	Ea.	405	186		591	52 gallon electric water heater.
	Demolish and Install	Ea.	405	302		707	
	Reinstall	Ea.		148.48		148.48	
	Clean	Ea.	1.14	15.10		16.24	
	Minimum Charge	Job		186		186	
82 Gallon							
	Demolish	Ea.		116		116	Includes material and labor to install
	Install	Ea.	610	232		842	an 82 gallon electric water heater.
	Demolish and Install	Ea.	610	348		958	
	Reinstall	Ea.		185.60		185.60	
	Clean	Ea.	1.14	15.10		16.24	
	Minimum Charge	Job		186		186	
Gas-fired							
30 Gallon Gas-fired							
	Demolish	Ea.		143		143	Includes material and labor to install a
	Install	Ea.	695	186		881	30 gallon gas water heater.
	Demolish and Install	Ea.	695	329		1024	
	Reinstall	Ea.		148.48		148.48	
	Clean	Ea.	1.14	15.10		16.24	
	Minimum Charge	Job		186		186	
30 Gallon Gas-fired, Flue							
	Install	Ea.	35	53		88	Includes labor and materials to install
	Minimum Charge	Job		186		186	a flue for a water heater (up to 50 gallon size).
40 Gallon Gas-fired							
	Demolish	Ea.		143		143	Includes material and labor to install a
	Install	Ea.	700	195		895	40 gallon gas water heater.
	Demolish and Install	Ea.	700	338		1038	
	Reinstall	Ea.		156.29		156.29	
	Clean	Ea.	1.14	15.10		16.24	
	Minimum Charge	Job		186		186	
40 Gallon Gas-fired, Flue							
	Install	Ea.	35	53		88	Includes labor and materials to install
	Minimum Charge	Job		186		186	a flue for a water heater (up to 50 gallon size).
50 Gallon Gas-fired							
	Demolish	Ea.		143		143	Includes material and labor to install a
	Install	Ea.	730	206		936	50 gallon gas water heater.
	Demolish and Install	Ea.	730	349		1079	
	Reinstall	Ea.		164.98		164.98	
	Clean	Ea.	1.14	15.10		16.24	
	Minimum Charge	Job		186		186	

Rough Mechanical

Water Heater

Water Heater	Unit	Material	Labor	Equip.	Total	Specification
50 Gallon Gas-fired, Flue						
Install	Ea.	35	53		88	Includes labor and materials to install
Minimum Charge	Job		186		186	a flue for a water heater (up to 50 gallon size).
Solar Hot Water System						
Water Heater						
Demolish	Ea.		93		93	Includes material and labor to install a
Install	Ea.	865	247		1112	solar hot water heat exchanger
Demolish and Install	Ea.	865	340		1205	storage tank, 100 gallon.
Reinstall	Ea.		197.97		197.97	
Minimum Charge	Job		187		187	
Solar Panel						
Demolish	S.F.		1.36		1.36	Includes material and labor to install
Install	S.F.	21.50	2.44		23.94	solar panel.
Demolish and Install	S.F.	21.50	3.80		25.30	
Reinstall	S.F.		1.95		1.95	
Minimum Charge	Job		187		187	
Repair Plumbing						
Install	Job		187		187	Includes 4 hour minimum labor charge
Minimum Charge	Job		187		187	for a steamfitter or pipefitter.
Insulating Wrap						
Demolish	Ea.		4.30		4.30	Includes material and labor to install
Install	Ea.	71.50	11.60		83.10	insulating wrap around a 100 gallon
Demolish and Install	Ea.	71.50	15.90		87.40	solar hot water heat exchanger
Reinstall	Ea.		9.28		9.28	storage tank.
Minimum Charge	Job		186		186	
Water Softener						
Demolish	Ea.		68.50		68.50	Includes material and labor to install a
Install	Ea.	1350	167		1517	14 GPM water softener.
Demolish and Install	Ea.	1350	235.50		1585.50	
Reinstall	Ea.		133.60		133.60	
Clean	Ea.	1.14	15.10		16.24	
Minimum Charge	Ea.		186		186	

Ductwork

Ductwork	Unit	Material	Labor	Equip.	Total	Specification
Galvanized, Square or Rectangular						
6"						
Demolish	L.F.		1.73		1.73	Includes material and labor to install
Install	L.F.	2.42	15.60		18.02	rectangular ductwork with a greatest
Demolish and Install	L.F.	2.42	17.33		19.75	dimension of 6".
Reinstall	L.F.		12.47		12.47	
Clean	L.F.		1.99		1.99	
Minimum Charge	Job		187		187	
8"						
Demolish	L.F.		1.73		1.73	Includes material and labor to install
Install	L.F.	3.21	15.60		18.81	rectangular ductwork with a greatest
Demolish and Install	L.F.	3.21	17.33		20.54	dimension of 8".
Reinstall	L.F.		12.47		12.47	
Clean	L.F.		2.66		2.66	
Minimum Charge	Job		187		187	

Rough Mechanical

Ductwork

Ductwork	Unit	Material	Labor	Equip.	Total	Specification
10″						
Demolish	L.F.		1.73		1.73	Includes material and labor to install
Install	L.F.	4.03	15.60		19.63	rectangular ductwork with a greatest
Demolish and Install	L.F.	4.03	17.33		21.36	dimension of 10″.
Reinstall	L.F.		12.47		12.47	
Clean	L.F.		3.32		3.32	
Minimum Charge	Job		187		187	
Galvanized, Spiral						
4″						
Demolish	L.F.		2.08		2.08	Cost includes material and labor to
Install	L.F.	1.33	1.87		3.20	install 4″ diameter spiral pre-formed
Demolish and Install	L.F.	1.33	3.95		5.28	steel galvanized ductwork.
Reinstall	L.F.		1.50		1.50	
Clean	L.F.		1.04		1.04	
Minimum Charge	Job		187		187	
6″						
Demolish	L.F.		2.08		2.08	Cost includes material and labor to
Install	L.F.	1.94	2.40		4.34	install 6″ diameter spiral pre-formed
Demolish and Install	L.F.	1.94	4.48		6.42	steel galvanized ductwork.
Reinstall	L.F.		1.92		1.92	
Clean	L.F.		1.57		1.57	
Minimum Charge	Job		187		187	
8″						
Demolish	L.F.		2.08		2.08	Cost includes material and labor to
Install	L.F.	2.54	3.37		5.91	install 8″ diameter spiral pre-formed
Demolish and Install	L.F.	2.54	5.45		7.99	steel galvanized ductwork.
Reinstall	L.F.		2.69		2.69	
Clean	L.F.		2.09		2.09	
Minimum Charge	Job		187		187	
10″						
Demolish	L.F.		2.60		2.60	Cost includes material and labor to
Install	L.F.	3.15	4.21		7.36	install 10″ diameter spiral pre-formed
Demolish and Install	L.F.	3.15	6.81		9.96	steel galvanized ductwork.
Reinstall	L.F.		3.37		3.37	
Clean	L.F.		2.61		2.61	
Minimum Charge	Job		187		187	
12″						
Demolish	L.F.		3.46		3.46	Cost includes material and labor to
Install	L.F.	3.87	5.60		9.47	install 12″ diameter spiral pre-formed
Demolish and Install	L.F.	3.87	9.06		12.93	steel galvanized ductwork.
Reinstall	L.F.		4.49		4.49	
Clean	L.F.		3.13		3.13	
Minimum Charge	Job		187		187	
16″						
Demolish	L.F.		6.95		6.95	Cost includes material and labor to
Install	L.F.	6.65	11.20		17.85	install 16″ diameter spiral pre-formed
Demolish and Install	L.F.	6.65	18.15		24.80	steel galvanized ductwork.
Reinstall	L.F.		8.98		8.98	
Clean	L.F.		4.18		4.18	
Minimum Charge	Job		187		187	
Flexible Fiberglass						
6″ Non-insulated						
Demolish	L.F.		1.60		1.60	Cost includes material and labor to
Install	L.F.	1.76	2.40		4.16	install 6″ diameter non-insulated
Demolish and Install	L.F.	1.76	4		5.76	flexible fiberglass fabric on corrosion
Reinstall	L.F.		1.92		1.92	resistant metal ductwork.
Clean	L.F.		1.57		1.57	
Minimum Charge	Job		187		187	

149

Ductwork	Unit	Material	Labor	Equip.	Total	Specification
8″ Non-insulated						
Demolish	L.F.		2.31		2.31	Cost includes material and labor to
Install	L.F.	2.42	3.37		5.79	install 8″ diameter non-insulated
Demolish and Install	L.F.	2.42	5.68		8.10	flexible fiberglass fabric on corrosion
Reinstall	L.F.		2.69		2.69	resistant metal ductwork.
Clean	L.F.		2.09		2.09	
Minimum Charge	Job		187		187	
10″ Non-insulated						
Demolish	L.F.		2.97		2.97	Cost includes material and labor to
Install	L.F.	3.04	4.21		7.25	install 10″ diameter non-insulated
Demolish and Install	L.F.	3.04	7.18		10.22	flexible fiberglass fabric on corrosion
Reinstall	L.F.		3.37		3.37	resistant metal ductwork.
Clean	L.F.		2.61		2.61	
Minimum Charge	Job		187		187	
12″ Non-insulated						
Demolish	L.F.		4.16		4.16	Cost includes material and labor to
Install	L.F.	3.67	5.60		9.27	install 12″ diameter non-insulated
Demolish and Install	L.F.	3.67	9.76		13.43	flexible fiberglass fabric on corrosion
Reinstall	L.F.		4.49		4.49	resistant metal ductwork.
Clean	L.F.		3.13		3.13	
Minimum Charge	Job		187		187	
4″ Insulated						
Demolish	L.F.		1.22		1.22	Cost includes material and labor to
Install	L.F.	2.44	1.98		4.42	install 4″ diameter insulated flexible
Demolish and Install	L.F.	2.44	3.20		5.64	fiberglass fabric on corrosion resistant
Reinstall	L.F.		1.58		1.58	metal ductwork.
Clean	L.F.		1.04		1.04	
Minimum Charge	Job		187		187	
6″ Insulated						
Demolish	L.F.		1.60		1.60	Cost includes material and labor to
Install	L.F.	2.79	2.59		5.38	install 6″ diameter insulated flexible
Demolish and Install	L.F.	2.79	4.19		6.98	fiberglass fabric on corrosion resistant
Reinstall	L.F.		2.07		2.07	metal ductwork.
Clean	L.F.		1.57		1.57	
Minimum Charge	Job		187		187	
8″ Insulated						
Demolish	L.F.		2.31		2.31	Cost includes material and labor to
Install	L.F.	3.47	3.74		7.21	install 8″ diameter insulated flexible
Demolish and Install	L.F.	3.47	6.05		9.52	fiberglass fabric on corrosion resistant
Reinstall	L.F.		2.99		2.99	metal ductwork.
Clean	L.F.		2.09		2.09	
Minimum Charge	Job		187		187	
10″ Insulated						
Demolish	L.F.		2.97		2.97	Cost includes material and labor to
Install	L.F.	4.08	4.81		8.89	install 10″ diameter insulated flexible
Demolish and Install	L.F.	4.08	7.78		11.86	fiberglass fabric on corrosion resistant
Reinstall	L.F.		3.85		3.85	metal ductwork.
Clean	L.F.		2.61		2.61	
Minimum Charge	Job		187		187	
12″ Insulated						
Demolish	L.F.		4.16		4.16	Cost includes material and labor to
Install	L.F.	5.10	6.75		11.85	install 12″ diameter insulated flexible
Demolish and Install	L.F.	5.10	10.91		16.01	fiberglass fabric on corrosion resistant
Reinstall	L.F.		5.39		5.39	metal ductwork.
Clean	L.F.		3.13		3.13	
Minimum Charge	Job		187		187	

Rough Mechanical

Heating		Unit	Material	Labor	Equip.	Total	Specification
Hot Water Baseboard							
	Demolish	L.F.		6.40		6.40	Includes material and labor to install
	Install	L.F.	6.90	11.20		18.10	baseboard radiation.
	Demolish and Install	L.F.	6.90	17.60		24.50	
	Reinstall	L.F.		8.97		8.97	
	Clean	L.F.	1.14	3.78		4.92	
	Minimum Charge	Job		186		186	
3'							
	Demolish	Ea.		20.50		20.50	Includes material and labor to install
	Install	Ea.	22	37.50		59.50	fin tube baseboard radiation in 3'
	Demolish and Install	Ea.	22	58		80	sections.
	Reinstall	Ea.		29.89		29.89	
	Clean	Ea.	.33	11.50		11.83	
	Minimum Charge	Job		186		186	
4'							
	Demolish	Ea.		27.50		27.50	Includes material and labor to install
	Install	Ea.	29	42		71	fin tube baseboard radiation in 4'
	Demolish and Install	Ea.	29	69.50		98.50	sections.
	Reinstall	Ea.		33.62		33.62	
	Clean	Ea.	.66	15.10		15.76	
	Minimum Charge	Job		186		186	
5'							
	Demolish	Ea.		34.50		34.50	Includes material and labor to install
	Install	Ea.	37	48		85	fin tube baseboard radiation in 5'
	Demolish and Install	Ea.	37	82.50		119.50	sections.
	Reinstall	Ea.		38.43		38.43	
	Clean	Ea.	.66	18.60		19.26	
	Minimum Charge	Job		186		186	
6'							
	Demolish	Ea.		41		41	Includes material and labor to install
	Install	Ea.	44	56		100	fin tube baseboard radiation in 6'
	Demolish and Install	Ea.	44	97		141	sections.
	Reinstall	Ea.		44.83		44.83	
	Clean	Ea.	.66	22		22.66	
	Minimum Charge	Job		186		186	
7'							
	Demolish	Ea.		46		46	Includes material and labor to install
	Install	Ea.	51.50	74		125.50	fin tube baseboard radiation in 7'
	Demolish and Install	Ea.	51.50	120		171.50	sections.
	Reinstall	Ea.		59.39		59.39	
	Clean	Ea.	.83	27		27.83	
	Minimum Charge	Job		186		186	
9'							
	Demolish	Ea.		68.50		68.50	Includes material and labor to install
	Install	Ea.	66	106		172	fin tube baseboard radiation in 9'
	Demolish and Install	Ea.	66	174.50		240.50	sections.
	Reinstall	Ea.		84.85		84.85	
	Clean	Ea.	1.31	34.50		35.81	
	Minimum Charge	Job		186		186	
50 MBH							
	Demolish	Ea.		33.50		33.50	Includes material and labor to install a
	Install	Ea.	650	96		746	cabinet space heater with an output of
	Demolish and Install	Ea.	650	129.50		779.50	50 MBH.
	Reinstall	Ea.		76.85		76.85	
	Clean	Ea.	.83	30		30.83	
	Minimum Charge	Job		187		187	

Rough Mechanical

Heating		Unit	Material	Labor	Equip.	Total	Specification
75 MBH							
	Demolish	Ea.		33.50		33.50	Includes material and labor to install a
	Install	Ea.	745	112		857	cabinet space heater with an output of
	Demolish and Install	Ea.	745	145.50		890.50	75 MBH.
	Reinstall	Ea.		89.66		89.66	
	Clean	Ea.	.83	30		30.83	
	Minimum Charge	Job		187		187	
125 MBH							
	Demolish	Ea.		40.50		40.50	Includes material and labor to install a
	Install	Ea.	965	135		1100	cabinet space heater with an output of
	Demolish and Install	Ea.	965	175.50		1140.50	125 MBH.
	Reinstall	Ea.		107.60		107.60	
	Clean	Ea.	.83	30		30.83	
	Minimum Charge	Job		187		187	
175 MBH							
	Demolish	Ea.		50.50		50.50	Includes material and labor to install a
	Install	Ea.	1250	192		1442	cabinet space heater with an output of
	Demolish and Install	Ea.	1250	242.50		1492.50	175 MBH.
	Reinstall	Ea.		153.71		153.71	
	Clean	Ea.	.83	30		30.83	
	Minimum Charge	Job		187		187	

Furnace Unit		Unit	Material	Labor	Equip.	Total	Specification
Wall (gas-fired)							
25 MBH w / Thermostat							
	Demolish	Ea.		84		84	Includes material and labor to install a
	Install	Ea.	415	135		550	gas-fired hot air furnace rated at 25
	Demolish and Install	Ea.	415	219		634	MBH.
	Reinstall	Ea.		107.72		107.72	
	Clean	Ea.	1.64	67		68.64	
	Minimum Charge	Job		187		187	
35 MBH w / Thermostat							
	Demolish	Ea.		84		84	Includes material and labor to install a
	Install	Ea.	400	150		550	gas-fired hot air furnace rated at 35
	Demolish and Install	Ea.	400	234		634	MBH.
	Reinstall	Ea.		119.69		119.69	
	Clean	Ea.	1.64	67		68.64	
	Minimum Charge	Job		187		187	
50 MBH w / Thermostat							
	Demolish	Ea.		84		84	Includes material and labor to install a
	Install	Ea.	815	168		983	gas-fired hot air furnace rated at 50
	Demolish and Install	Ea.	815	252		1067	MBH.
	Reinstall	Ea.		134.66		134.66	
	Clean	Ea.	1.64	67		68.64	
	Minimum Charge	Job		187		187	
60 MBH w / Thermostat							
	Demolish	Ea.		84		84	Includes material and labor to install a
	Install	Ea.	890	187		1077	gas-fired hot air furnace rated at 60
	Demolish and Install	Ea.	890	271		1161	MBH.
	Reinstall	Ea.		149.62		149.62	
	Clean	Ea.	1.64	67		68.64	
	Minimum Charge	Job		187		187	

Rough Mechanical

Furnace Unit	Unit	Material	Labor	Equip.	Total	Specification
Floor (gas-fired)						
30 MBH w / Thermostat						
Demolish	Ea.		84		84	Includes material and labor to install a
Install	Ea.	695	135		830	gas-fired hot-air floor furnace rated at
Demolish and Install	Ea.	695	219		914	30 MBH.
Reinstall	Ea.		107.72		107.72	
Clean	Ea.	1.64	67		68.64	
Minimum Charge	Job		187		187	
50 MBH w / Thermostat						
Demolish	Ea.		84		84	Includes material and labor to install a
Install	Ea.	760	168		928	gas-fired hot-air floor furnace rated at
Demolish and Install	Ea.	760	252		1012	50 MBH.
Reinstall	Ea.		134.66		134.66	
Clean	Ea.	1.64	67		68.64	
Minimum Charge	Job		187		187	
Gas Forced Air						
50 MBH						
Demolish	Ea.		84		84	Includes material and labor to install
Install	Ea.	615	168		783	an up-flow gas-fired hot-air furnace
Demolish and Install	Ea.	615	252		867	rated at 50 MBH.
Reinstall	Ea.		134.66		134.66	
Clean	Ea.	1.64	67		68.64	
Minimum Charge	Job		187		187	
80 MBH						
Demolish	Ea.		84		84	Includes material and labor to install
Install	Ea.	675	187		862	an up-flow gas-fired hot-air furnace
Demolish and Install	Ea.	675	271		946	rated at 80 MBH.
Reinstall	Ea.		149.62		149.62	
Clean	Ea.	1.64	67		68.64	
Minimum Charge	Job		187		187	
100 MBH						
Demolish	Ea.		84		84	Includes material and labor to install
Install	Ea.	730	210		940	an up-flow gas-fired hot-air furnace
Demolish and Install	Ea.	730	294		1024	rated at 100 MBH.
Reinstall	Ea.		168.32		168.32	
Clean	Ea.	1.64	67		68.64	
Minimum Charge	Job		187		187	
120 MBH						
Demolish	Ea.		84		84	Includes material and labor to install
Install	Ea.	815	224		1039	an up-flow gas-fired hot-air furnace
Demolish and Install	Ea.	815	308		1123	rated at 120 MBH.
Reinstall	Ea.		179.54		179.54	
Clean	Ea.	1.64	67		68.64	
Minimum Charge	Job		187		187	
Electric Forced Air						
31.4 MBH						
Demolish	Ea.		84		84	Includes material and labor to install
Install	Ea.	1000	190		1190	an electric hot-air furnace rated at
Demolish and Install	Ea.	1000	274		1274	31.4 MBH.
Reinstall	Ea.		152.32		152.32	
Clean	Ea.	1.64	67		68.64	
Minimum Charge	Job		187		187	

Rough Mechanical

Furnace Unit		Unit	Material	Labor	Equip.	Total	Specification
47.1 MBH							
	Demolish	Ea.		84		84	Includes material and labor to install
	Install	Ea.	1175	199		1374	an electric hot-air furnace rated at
	Demolish and Install	Ea.	1175	283		1458	47.1 MBH.
	Reinstall	Ea.		159.40		159.40	
	Clean	Ea.	1.64	67		68.64	
	Minimum Charge	Job		187		187	
62.9 MBH							
	Demolish	Ea.		84		84	Includes material and labor to install
	Install	Ea.	1400	214		1614	an electric hot-air furnace rated at
	Demolish and Install	Ea.	1400	298		1698	62.9 MBH.
	Reinstall	Ea.		171.36		171.36	
	Clean	Ea.	1.64	67		68.64	
	Minimum Charge	Job		187		187	
78.5 MBH							
	Demolish	Ea.		84		84	Includes material and labor to install
	Install	Ea.	1575	220		1795	an electric hot-air furnace rated at
	Demolish and Install	Ea.	1575	304		1879	78.5 MBH.
	Reinstall	Ea.		175.75		175.75	
	Clean	Ea.	1.64	67		68.64	
	Minimum Charge	Job		187		187	
110 MBH							
	Demolish	Ea.		84		84	Includes material and labor to install
	Install	Ea.	2175	232		2407	an electric hot-air furnace rated at 110
	Demolish and Install	Ea.	2175	316		2491	MBH.
	Reinstall	Ea.		185.25		185.25	
	Clean	Ea.	1.64	67		68.64	
	Minimum Charge	Job		187		187	
A/C Junction Box							
	Install	Ea.	50.50	92		142.50	Includes labor and materials to install
	Minimum Charge	Job		187		187	a heating system emergency shut-off switch. Cost includes up to 40 feet of non-metallic cable.

Boiler		Unit	Material	Labor	Equip.	Total	Specification
Gas-Fired							
60 MBH							
	Demolish	Ea.		680		680	Includes material and labor to install a
	Install	Ea.	2025	1225		3250	gas-fired cast iron boiler rated at 60
	Demolish and Install	Ea.	2025	1905		3930	MBH.
	Reinstall	Ea.		978.15		978.15	
	Clean	Ea.	3.28	40.50		43.78	
	Minimum Charge	Job		187		187	
Gas-Fired							
125 MBH							
	Demolish	Ea.		1875		1875	Includes material and labor to install a
	Install	Ea.	2925	1500		4425	gas-fired cast iron boiler rated at 125
	Demolish and Install	Ea.	2925	3375		6300	MBH.
	Reinstall	Ea.		1195.52		1195.52	
	Clean	Ea.	3.28	40.50		43.78	
	Minimum Charge	Job		187		187	

Rough Mechanical

Boiler

Boiler	Unit	Material	Labor	Equip.	Total	Specification
Gas-Fired						
400 MBH						Includes material and labor to install a gas-fired cast iron boiler rated at 400 MBH.
Demolish	Ea.		1875		1875	
Install	Ea.	4075	2100		6175	
Demolish and Install	Ea.	4075	3975		8050	
Reinstall	Ea.		2101.50		2101.50	
Clean	Ea.	3.28	40.50		43.78	
Minimum Charge	Job		187		187	

Heat Pump

Heat Pump	Unit	Material	Labor	Equip.	Total	Specification
2 Ton						
Demolish	Ea.		88		88	Includes material and labor to install a 2 ton heat pump.
Install	Ea.	1800	560		2360	
Demolish and Install	Ea.	1800	648		2448	
Reinstall	Ea.		448.32		448.32	
Clean	Ea.	3.28	30		33.28	
Minimum Charge	Job		187		187	
3 Ton						
Demolish	Ea.		167		167	Includes material and labor to install a 3 ton heat pump.
Install	Ea.	2175	840		3015	
Demolish and Install	Ea.	2175	1007		3182	
Reinstall	Ea.		672.48		672.48	
Clean	Ea.	3.28	30		33.28	
Minimum Charge	Job		187		187	
4 Ton						
Demolish	Ea.		305		305	Includes material and labor to install a 4 ton heat pump.
Install	Ea.	2725	1125		3850	
Demolish and Install	Ea.	2725	1430		4155	
Reinstall	Ea.		896.64		896.64	
Clean	Ea.	3.28	30		33.28	
Minimum Charge	Job		187		187	
5 Ton						
Demolish	Ea.		445		445	Includes material and labor to install a 5 ton heat pump.
Install	Ea.	3100	1350		4450	
Demolish and Install	Ea.	3100	1795		4895	
Reinstall	Ea.		1075.97		1075.97	
Clean	Ea.	3.28	30		33.28	
Minimum Charge	Job		187		187	
Thermostat						
Demolish	Ea.		10.40		10.40	Includes material and labor to install an electric thermostat, 2 set back.
Install	Ea.	196	47		243	
Demolish and Install	Ea.	196	57.40		253.40	
Reinstall	Ea.		37.40		37.40	
Clean	Ea.	.07	7.55		7.62	
Minimum Charge	Job		187		187	
A/C Junction Box						
Install	Ea.	470	282		752	Includes labor and materials to install a heat pump circuit including up to 40 feet of non-metallic cable.
Minimum Charge	Job		184		184	

Heating		Unit	Material	Labor	Equip.	Total	Specification
Electric Baseboard							
	Demolish	L.F.		5.10		5.10	Includes material and labor to install
	Install	L.F.	12.05	12.25		24.30	electric baseboard heating elements.
	Demolish and Install	L.F.	12.05	17.35		29.40	
	Reinstall	L.F.		9.79		9.79	
	Clean	L.F.	1.14	3.78		4.92	
	Minimum Charge	Job		184		184	
2' 6"							
	Demolish	Ea.		17		17	Includes material and labor to install a
	Install	Ea.	36.50	46		82.50	2' 6" long section of electric
	Demolish and Install	Ea.	36.50	63		99.50	baseboard heating elements.
	Reinstall	Ea.		36.72		36.72	
	Clean	Ea.	.33	9.30		9.63	
	Minimum Charge	Job		184		184	
3'							
	Demolish	Ea.		18.55		18.55	Includes material and labor to install a
	Install	Ea.	43	46		89	3' long section of electric baseboard
	Demolish and Install	Ea.	43	64.55		107.55	heating elements.
	Reinstall	Ea.		36.72		36.72	
	Clean	Ea.	.33	11.50		11.83	
	Minimum Charge	Job		184		184	
4'							
	Demolish	Ea.		20.50		20.50	Includes material and labor to install a
	Install	Ea.	50.50	55		105.50	4' long section of electric baseboard
	Demolish and Install	Ea.	50.50	75.50		126	heating elements.
	Reinstall	Ea.		43.84		43.84	
	Clean	Ea.	.13	2.52		2.65	
	Minimum Charge	Job		184		184	
5'							
	Demolish	Ea.		22.50		22.50	Includes material and labor to install a
	Install	Ea.	60.50	64.50		125	5' long section of electric baseboard
	Demolish and Install	Ea.	60.50	87		147.50	heating elements.
	Reinstall	Ea.		51.54		51.54	
	Clean	Ea.	.66	18.60		19.26	
	Minimum Charge	Job		184		184	
6'							
	Demolish	Ea.		25.50		25.50	Includes material and labor to install a
	Install	Ea.	67	73.50		140.50	6' long section of electric baseboard
	Demolish and Install	Ea.	67	99		166	heating elements.
	Reinstall	Ea.		58.75		58.75	
	Clean	Ea.	.13	3.02		3.15	
	Minimum Charge	Job		184		184	
8'							
	Demolish	Ea.		34		34	Includes material and labor to install
	Install	Ea.	84	92		176	an 8' long section of electric
	Demolish and Install	Ea.	84	126		210	baseboard heating elements.
	Reinstall	Ea.		73.44		73.44	
	Clean	Ea.	.13	3.78		3.91	
	Minimum Charge	Job		184		184	
10'							
	Demolish	Ea.		41		41	Includes material and labor to install a
	Install	Ea.	105	111		216	10' long section of electric baseboard
	Demolish and Install	Ea.	105	152		257	heating elements.
	Reinstall	Ea.		89.02		89.02	
	Clean	Ea.	1.31	38		39.31	
	Minimum Charge	Job		184		184	

Rough Mechanical

Heating	Unit	Material	Labor	Equip.	Total	Specification
Electric Radiant Wall Panel						
Demolish	Ea.		27		27	Includes material and labor to install
Install	Ea.	405	73.50		478.50	an 18" x 23" electric radiant heat wall
Demolish and Install	Ea.	405	100.50		505.50	panel.
Reinstall	Ea.		58.75		58.75	
Clean	Ea.	.17	7.55		7.72	
Minimum Charge	Job		184		184	

Heating Control	Unit	Material	Labor	Equip.	Total	Specification
Thermostat						
Demolish	Ea.		10.40		10.40	Includes material and labor to install
Install	Ea.	196	47		243	an electric thermostat, 2 set back.
Demolish and Install	Ea.	196	57.40		253.40	
Reinstall	Ea.		37.40		37.40	
Clean	Ea.	.07	7.55		7.62	
Minimum Charge	Job		187		187	

Air Conditioning	Unit	Material	Labor	Equip.	Total	Specification
2 Ton						
Demolish	Ea.		345		345	Includes material and labor to install a
Install	Ea.	1700	580		2280	rooftop air conditioning unit, 2 ton.
Demolish and Install	Ea.	1700	925		2625	No hoisting is included.
Reinstall	Ea.		580.22		580.22	
Clean	Ea.	3.28	30		33.28	
Minimum Charge	Job		187		187	
3 Ton						
Demolish	Ea.		345		345	Includes material and labor to install a
Install	Ea.	1800	765		2565	rooftop air conditioning unit, 3 ton.
Demolish and Install	Ea.	1800	1110		2910	No hoisting is included.
Reinstall	Ea.		765.05		765.05	
Clean	Ea.	3.28	30		33.28	
Minimum Charge	Job		187		187	
4 Ton						
Demolish	Ea.		345		345	Includes material and labor to install a
Install	Ea.	2125	885		3010	rooftop air conditioning unit, 4 ton.
Demolish and Install	Ea.	2125	1230		3355	No hoisting is included.
Reinstall	Ea.		887.18		887.18	
Clean	Ea.	3.28	30		33.28	
Minimum Charge	Job		187		187	
5 Ton						
Demolish	Ea.		345		345	Includes material and labor to install a
Install	Ea.	2250	960		3210	rooftop air conditioning unit, 5 ton.
Demolish and Install	Ea.	2250	1305		3555	No hoisting is included.
Reinstall	Ea.		957.95		957.95	
Clean	Ea.	3.28	30		33.28	
Minimum Charge	Job		187		187	
Compressor						
2 Ton						
Demolish	Ea.		310		310	Includes material and labor to install a
Install	Ea.	565	187		752	refrigeration compressor.
Demolish and Install	Ea.	565	497		1062	
Reinstall	Ea.		149.44		149.44	
Clean	Ea.	3.28	30		33.28	
Minimum Charge	Job		187		187	

Rough Mechanical

Air Conditioning		Unit	Material	Labor	Equip.	Total	Specification
2.5 Ton							
	Demolish	Ea.		310		310	Includes material and labor to install a
	Install	Ea.	405	202		607	refrigeration compressor.
	Demolish and Install	Ea.	405	512		917	
	Reinstall	Ea.		161.91		161.91	
	Clean	Ea.	3.28	30		33.28	
	Minimum Charge	Job		187		187	
3 Ton							
	Demolish	Ea.		310		310	Includes material and labor to install a
	Install	Ea.	500	211		711	refrigeration compressor.
	Demolish and Install	Ea.	500	521		1021	
	Reinstall	Ea.		168.95		168.95	
	Clean	Ea.	3.28	30		33.28	
	Minimum Charge	Job		187		187	
4 Ton							
	Demolish	Ea.		310		310	Includes material and labor to install a
	Install	Ea.	710	170		880	refrigeration compressor.
	Demolish and Install	Ea.	710	480		1190	
	Reinstall	Ea.		135.85		135.85	
	Clean	Ea.	3.28	30		33.28	
	Minimum Charge	Job		187		187	
5 Ton							
	Demolish	Ea.		310		310	Includes material and labor to install a
	Install	Ea.	875	231		1106	refrigeration compressor.
	Demolish and Install	Ea.	875	541		1416	
	Reinstall	Ea.		185.07		185.07	
	Clean	Ea.	3.28	30		33.28	
	Minimum Charge	Job		187		187	
Condensing Unit							
2 Ton Air-cooled							
	Demolish	Ea.		89		89	Includes material and labor to install a
	Install	Ea.	735	320		1055	2 ton condensing unit.
	Demolish and Install	Ea.	735	409		1144	
	Reinstall	Ea.		256.18		256.18	
	Clean	Ea.	3.28	30		33.28	
	Minimum Charge	Job		187		187	
2.5 Ton Air-cooled							
	Demolish	Ea.		104		104	Includes material and labor to install a
	Install	Ea.	845	395		1240	2.5 ton condensing unit.
	Demolish and Install	Ea.	845	499		1344	
	Reinstall	Ea.		316.46		316.46	
	Clean	Ea.	3.28	30		33.28	
	Minimum Charge	Job		187		187	
3 Ton Air-cooled							
	Demolish	Ea.		125		125	Includes material and labor to install a
	Install	Ea.	1000	515		1515	3 ton condensing unit.
	Demolish and Install	Ea.	1000	640		1640	
	Reinstall	Ea.		413.83		413.83	
	Clean	Ea.	3.28	30		33.28	
	Minimum Charge	Job		187		187	
4 Ton Air-cooled							
	Demolish	Ea.		187		187	Includes material and labor to install a
	Install	Ea.	1300	745		2045	4 ton condensing unit.
	Demolish and Install	Ea.	1300	932		2232	
	Reinstall	Ea.		597.76		597.76	
	Clean	Ea.	3.28	30		33.28	
	Minimum Charge	Job		187		187	

Rough Mechanical

Air Conditioning		Unit	Material	Labor	Equip.	Total	Specification
5 Ton Air-cooled							
	Demolish	Ea.		280		280	Includes material and labor to install a
	Install	Ea.	1575	1125		2700	5 ton condensing unit.
	Demolish and Install	Ea.	1575	1405		2980	
	Reinstall	Ea.		896.64		896.64	
	Clean	Ea.	3.28	30		33.28	
	Minimum Charge	Job		187		187	
Air Handler							
1,200 CFM							
	Demolish	Ea.		224		224	Includes material and labor to install a
	Install	Ea.	2150	515		2665	1,200 CFM capacity air handling
	Demolish and Install	Ea.	2150	739		2889	unit.
	Reinstall	Ea.		413.83		413.83	
	Clean	Ea.	3.28	30		33.28	
	Minimum Charge	Job		187		187	
2,000 CFM							
	Demolish	Ea.		224		224	Includes material and labor to install a
	Install	Ea.	3575	610		4185	2,000 CFM capacity air handling
	Demolish and Install	Ea.	3575	834		4409	unit.
	Reinstall	Ea.		489.08		489.08	
	Clean	Ea.	3.28	30		33.28	
	Minimum Charge	Job		187		187	
3,000 CFM							
	Demolish	Ea.		249		249	Includes material and labor to install a
	Install	Ea.	4450	670		5120	3,000 CFM capacity air handling
	Demolish and Install	Ea.	4450	919		5369	unit.
	Reinstall	Ea.		537.98		537.98	
	Clean	Ea.	3.28	30		33.28	
	Minimum Charge	Job		187		187	
Evaporative Cooler							
4,300 CFM, 1/3 HP							
	Demolish	Ea.		91.50		91.50	Includes material and labor to install
	Install	Ea.	705	210		915	an evaporative cooler.
	Demolish and Install	Ea.	705	301.50		1006.50	
	Reinstall	Ea.		168.32		168.32	
	Clean	Ea.	1.31	30		31.31	
	Minimum Charge	Job		187		187	
4,700 CFM, 1/2 HP							
	Demolish	Ea.		91.50		91.50	Includes material and labor to install
	Install	Ea.	675	217		892	an evaporative cooler.
	Demolish and Install	Ea.	675	308.50		983.50	
	Reinstall	Ea.		173.75		173.75	
	Clean	Ea.	1.31	30		31.31	
	Minimum Charge	Job		187		187	
5,600 CFM, 3/4 HP							
	Demolish	Ea.		91.50		91.50	Includes material and labor to install
	Install	Ea.	1150	224		1374	an evaporative cooler.
	Demolish and Install	Ea.	1150	315.50		1465.50	
	Reinstall	Ea.		179.54		179.54	
	Clean	Ea.	1.31	30		31.31	
	Minimum Charge	Job		187		187	
Master Cool Brand							
	Demolish	Ea.		91.50		91.50	Includes material and labor to install
	Install	Ea.	705	232		937	an evaporative cooler.
	Demolish and Install	Ea.	705	323.50		1028.50	
	Reinstall	Ea.		185.73		185.73	
	Clean	Ea.	1.31	30		31.31	
	Minimum Charge	Job		187		187	

Air Conditioning		Unit	Material	Labor	Equip.	Total	Specification
Motor							
	Demolish	Ea.		51.50		51.50	Includes material and labor to install
	Install	Ea.	240	164		404	an evaporative cooler motor.
	Demolish and Install	Ea.	240	215.50		455.50	
	Reinstall	Ea.		130.88		130.88	
	Clean	Ea.	1.31	30		31.31	
	Minimum Charge	Job		187		187	
Water Pump							
	Demolish	Ea.		51.50		51.50	Includes material and labor to install
	Install	Ea.	44.50	93.50		138	an evaporative cooler water pump.
	Demolish and Install	Ea.	44.50	145		189.50	
	Reinstall	Ea.		74.80		74.80	
	Clean	Ea.	1.31	30		31.31	
	Minimum Charge	Job		187		187	
Fan Belt							
	Demolish	Ea.		10.30		10.30	Includes material and labor to install
	Install	Ea.	30.50	18.70		49.20	an evaporative cooler fan belt.
	Demolish and Install	Ea.	30.50	29		59.50	
	Reinstall	Ea.		14.96		14.96	
	Minimum Charge	Job		187		187	
Pads (4 ea.)							
	Demolish	Ea.		10.30		10.30	Includes material and labor to install
	Install	Ea.	36	47		83	pre-fabricated roof pads for an
	Demolish and Install	Ea.	36	57.30		93.30	evaporative cooler.
	Reinstall	Ea.		37.40		37.40	
	Clean	Ea.	1.31	30		31.31	
	Minimum Charge	Job		187		187	
Service call							
	Install	Job		187		187	Includes 4 hour minimum labor charge for sheetmetal worker.
Window Mounted							
Low Output BTU							
	Install	Ea.	224	42.50		266.50	Includes material and labor to install a
	Reinstall	Ea.		33.80		33.80	window air conditioning unit.
	Clean	Ea.	.50	7.55		8.05	
	Minimum Charge	Job		187		187	
Mid-range Output BTU							
	Demolish	Ea.		29.50		29.50	Includes material and labor to install a
	Install	Ea.	320	56.50		376.50	window air conditioning unit.
	Demolish and Install	Ea.	320	86		406	
	Reinstall	Ea.		45.07		45.07	
	Clean	Ea.	.50	7.55		8.05	
	Minimum Charge	Job		187		187	
High Output BTU							
	Demolish	Ea.		37.50		37.50	Includes material and labor to install a
	Install	Ea.	385	56.50		441.50	window air conditioning unit.
	Demolish and Install	Ea.	385	94		479	
	Reinstall	Ea.		45.07		45.07	
	Clean	Ea.	.50	7.55		8.05	
	Minimum Charge	Job		187		187	
Hotel-type Hot / Cold							
	Demolish	Ea.		62.50		62.50	Includes material and labor to install a
	Install	Ea.	1425	224		1649	packaged air conditioning unit,
	Demolish and Install	Ea.	1425	286.50		1711.50	15,000 BTU cooling, 13,900 BTU
	Reinstall	Ea.		179.33		179.33	heating.
	Clean	Ea.	.66	15.10		15.76	
	Minimum Charge	Job		187		187	

Rough Mechanical

Air Conditioning

Air Conditioning		Unit	Material	Labor	Equip.	Total	Specification
Thermostat							
	Demolish	Ea.		10.40		10.40	Includes material and labor to install
	Install	Ea.	196	47		243	an electric thermostat, 2 set back.
	Demolish and Install	Ea.	196	57.40		253.40	
	Reinstall	Ea.		37.40		37.40	
	Clean	Ea.	.07	7.55		7.62	
	Minimum Charge	Job		187		187	
Concrete Pad							
	Demolish	S.F.		1.68		1.68	Includes material and labor to install
	Install	S.F.	2.23	3.90		6.13	air conditioner pad.
	Demolish and Install	S.F.	2.23	5.58		7.81	
	Reinstall	S.F.		3.12		3.12	
	Clean	S.F.	.07	.48		.55	
	Paint	S.F.	.13	.30		.43	
	Minimum Charge	Job		187		187	
A/C Junction Box							
	Install	Ea.	186	105		291	Includes labor and materials to install
	Minimum Charge	Job		184		184	an air conditioning circuit including up to 40 feet of non-metallic cable.

Sprinkler System

Sprinkler System		Unit	Material	Labor	Equip.	Total	Specification
Exposed							
Wet Type							
	Demolish	SF Flr.		.17		.17	Includes material and labor to install
	Install	SF Flr.	1.27	1.60		2.87	an exposed wet pipe fire sprinkler
	Demolish and Install	SF Flr.	1.27	1.77		3.04	system.
	Reinstall	SF Flr.		1.60		1.60	
	Minimum Charge	Job		186		186	
Dry Type							
	Demolish	SF Flr.		.17		.17	Includes material and labor to install
	Install	SF Flr.	1.27	1.60		2.87	an exposed dry pipe fire sprinkler
	Demolish and Install	SF Flr.	1.27	1.77		3.04	system.
	Reinstall	SF Flr.		1.60		1.60	
	Minimum Charge	Job		186		186	
Concealed							
Wet Type							
	Demolish	SF Flr.		.33		.33	Includes material and labor to install a
	Install	SF Flr.	1.27	2.19		3.46	concealed wet pipe fire sprinkler
	Demolish and Install	SF Flr.	1.27	2.52		3.79	system.
	Reinstall	SF Flr.		2.19		2.19	
	Minimum Charge	Job		186		186	
Dry Type							
	Demolish	SF Flr.		.33		.33	Includes material and labor to install a
	Install	SF Flr.	.98	2.19		3.17	concealed dry pipe fire sprinkler
	Demolish and Install	SF Flr.	.98	2.52		3.50	system.
	Reinstall	SF Flr.		2.19		2.19	
	Minimum Charge	Job		186		186	

Junction Boxes	Cable	Switch Box

Electrical Per S.F.	Unit	Material	Labor	Equip.	Total	Specification
Residential						
Wiring, Lighting and Fixtures						
Install	S.F.	1.39	1.53		2.92	Includes labor and material to replace house wiring, lighting fixtures, and built-in appliances. Apply S.F. cost to floor area including garages but not basements.
Minimum Charge	Job		184		184	
Wiring Only						
Install	S.F.	.15	.60		.75	Includes labor and material to replace house wiring only. Apply S.F. cost to floor area including garages but not basements.

Main Service	Unit	Material	Labor	Equip.	Total	Specification
100 Amp.						
Demolish	Ea.		151		151	Cost includes material and labor to install 100 Amp. residential service panel including 12 breaker interior panel, 10 single pole breakers and hook-up. Wiring is not included.
Install	Ea.	540	310		850	
Demolish and Install	Ea.	540	461		1001	
Reinstall	Ea.		246.86		246.86	
Clean	Ea.	.33	60.50		60.83	
Minimum Charge	Job		184		184	
150 Amp.						
Demolish	Ea.		151		151	Cost includes material and labor to install 150 Amp. residential service, incl. 20 breaker exterior panel, main switch, 1 GFI and 14 single pole breakers, meter socket including hookup. Wiring is not included.
Install	Ea.	835	355		1190	
Demolish and Install	Ea.	835	506		1341	
Reinstall	Ea.		285.20		285.20	
Clean	Ea.	.33	60.50		60.83	
Minimum Charge	Job		184		184	
200 Amp.						
Demolish	Ea.		151		151	Cost includes material and labor to install 200 Amp. residential service, incl. 20 breaker exterior panel, main switch, 1 GFI and 18 single pole breakers, meter socket including hookup. Wiring is not included.
Install	Ea.	1100	410		1510	
Demolish and Install	Ea.	1100	561		1661	
Reinstall	Ea.		326.40		326.40	
Clean	Ea.	.33	60.50		60.83	
Minimum Charge	Job		184		184	

Rough Electrical

Main Service	Unit	Material	Labor	Equip.	Total	Specification
Masthead						
Demolish	Ea.		25.50		25.50	Includes material and labor to install
Install	Ea.	111	153		264	galvanized steel conduit including
Demolish and Install	Ea.	111	178.50		289.50	masthead.
Reinstall	Ea.		122.40		122.40	
Minimum Charge	Job		184		184	
Allowance to Repair						
Install	Job		184		184	Minimum labor charge for residential
Minimum Charge	Job		184		184	wiring.

Insulation

Wall Insulation

Floor Insulation	Unit	Material	Labor	Equip.	Total	Specification
Kraft Faced Batt						
3-1/2″ (R-11)						
Demolish	S.F.		.11		.11	Cost includes material and labor to
Install	S.F.	.40	.48		.88	install 3-1/2″ (R-11) kraft paper or foil
Demolish and Install	S.F.	.40	.59		.99	faced roll and batt fiberglass
Minimum Charge	Job		169		169	insulation, joists 16″ O.C.
6″ (R-19)						
Demolish	S.F.		.11		.11	Cost includes material and labor to
Install	S.F.	.54	.56		1.10	install 6″ (R-19) kraft paper or foil
Demolish and Install	S.F.	.54	.67		1.21	faced roll and batt fiberglass
Minimum Charge	Job		169		169	insulation, joists 16″ O.C.
10″ (R-30)						
Demolish	S.F.		.11		.11	Cost includes material and labor to
Install	S.F.	.77	.68		1.45	install 10″ (R-30) kraft paper-faced roll
Demolish and Install	S.F.	.77	.79		1.56	and batt fiberglass insulation, joists
Minimum Charge	Job		169		169	16″ O.C.
12″ (R-38)						
Demolish	S.F.		.11		.11	Cost includes material and labor to
Install	S.F.	.83	.71		1.54	install 12″ (R-38) kraft paper-faced roll
Demolish and Install	S.F.	.83	.82		1.65	and batt fiberglass insulation, joists
Minimum Charge	Job		169		169	16″ O.C.
Foil Faced Batt						
3-1/2″ (R-11)						
Demolish	S.F.		.11		.11	Cost includes material and labor to
Install	S.F.	.40	.48		.88	install 3-1/2″ (R-11) kraft paper or foil
Demolish and Install	S.F.	.40	.59		.99	faced roll and batt fiberglass
Minimum Charge	Job		169		169	insulation, joists 16″ O.C.
6″ (R-19)						
Demolish	S.F.		.11		.11	Cost includes material and labor to
Install	S.F.	.54	.56		1.10	install 6″ (R-19) kraft paper or foil
Demolish and Install	S.F.	.54	.67		1.21	faced roll and batt fiberglass
Minimum Charge	Job		169		169	insulation, joists 16″ O.C.
Unfaced Batt						
3-1/2″ (R-11)						
Demolish	S.F.		.11		.11	Cost includes material and labor to
Install	S.F.	.36	.56		.92	install 3-1/2″ (R-11) unfaced roll and
Demolish and Install	S.F.	.36	.67		1.03	batt fiberglass insulation, joists 16″
Minimum Charge	Job		169		169	O.C.

Insulation

Floor Insulation		Unit	Material	Labor	Equip.	Total	Specification
6″ (R-19)							
	Demolish	S.F.		.11		.11	Cost includes material and labor to
	Install	S.F.	.62	.68		1.30	install 6″ (R-19) unfaced roll and batt
	Demolish and Install	S.F.	.62	.79		1.41	fiberglass insulation, joists 16″ O.C.
	Minimum Charge	Job		169		169	
10″ (R-30)							
	Demolish	S.F.		.11		.11	Cost includes material and labor to
	Install	S.F.	.77	.75		1.52	install 10″ (R-30) unfaced roll and batt
	Demolish and Install	S.F.	.77	.86		1.63	fiberglass insulation, joists 16″ O.C.
	Minimum Charge	Job		169		169	
12″ (R-38)							
	Demolish	S.F.		.11		.11	Cost includes material and labor to
	Install	S.F.	.91	.80		1.71	install 12″ (R-38) unfaced roll and batt
	Demolish and Install	S.F.	.91	.91		1.82	fiberglass insulation, joists 16″ O.C.
	Minimum Charge	Job		169		169	
Rigid Board							
1/2″							
	Demolish	S.F.		.11		.11	Includes material and labor to install
	Install	S.F.	.32	.42		.74	rigid foam insulation board foil faced
	Demolish and Install	S.F.	.32	.53		.85	two sides.
	Minimum Charge	Job		169		169	
3/4″							
	Demolish	S.F.		.11		.11	Includes material and labor to install
	Install	S.F.	.42	.42		.84	rigid foam insulation board foil faced
	Demolish and Install	S.F.	.42	.53		.95	two sides.
	Minimum Charge	Job		169		169	
1-1/2″							
	Demolish	S.F.		.11		.11	Includes material and labor to install
	Install	S.F.	.62	.45		1.07	rigid foam insulation board foil faced
	Demolish and Install	S.F.	.62	.56		1.18	two sides.
	Minimum Charge	Job		169		169	
Support Wire							
	Demolish	S.F.		.04		.04	Includes material and labor to install
	Install	S.F.	.02	.09		.11	friction fit support wire.
	Demolish and Install	S.F.	.02	.13		.15	
	Minimum Charge	Job		169		169	
Vapor Barrier							
	Demolish	S.F.		.06		.06	Cost includes material and labor to
	Install	S.F.	.03	.09		.12	install 4 mil polyethylene vapor barrier
	Demolish and Install	S.F.	.03	.15		.18	10′ wide sheets with 6″ overlaps.
	Minimum Charge	Job		169		169	

Wall Insulation		Unit	Material	Labor	Equip.	Total	Specification
Kraft Faced Batt							
3-1/2″ (R-11)							
	Demolish	S.F.		.11		.11	Cost includes material and labor to
	Install	S.F.	.32	.21		.53	install 3-1/2″ (R-11) kraft paper-faced
	Demolish and Install	S.F.	.32	.32		.64	roll and batt fiberglass insulation, joist
	Minimum Charge	Job		169		169	or studs 16″ O.C.
6″ (R-19)							
	Demolish	S.F.		.11		.11	Cost includes material and labor to
	Install	S.F.	.46	.25		.71	install 6″ (R-19) kraft paper-faced roll
	Demolish and Install	S.F.	.46	.36		.82	and batt fiberglass insulation, joist or
	Minimum Charge	Job		169		169	studs 16″ O.C.

Insulation

Wall Insulation		Unit	Material	Labor	Equip.	Total	Specification
10″ (R-30)							
	Demolish	S.F.		.11		.11	Cost includes material and labor to
	Install	S.F.	.78	.29		1.07	install 10″ (R-30) kraft paper-faced roll
	Demolish and Install	S.F.	.78	.40		1.18	and batt fiberglass insulation, joist or
	Minimum Charge	Job		169		169	studs 16″ O.C.
12″ (R-38)							
	Demolish	S.F.		.11		.11	Cost includes material and labor to
	Install	S.F.	.94	.34		1.28	install 12″ (R-38) kraft paper-faced roll
	Demolish and Install	S.F.	.94	.45		1.39	and batt fiberglass insulation, joist or
	Minimum Charge	Job		169		169	studs 16″ O.C.
Foil Faced Batt							
3-1/2″ (R-11)							
	Demolish	S.F.		.11		.11	Cost includes material and labor to
	Install	S.F.	.52	.21		.73	install 3-1/2″ (R-11) foil faced roll and
	Demolish and Install	S.F.	.52	.32		.84	batt fiberglass insulation, joist or studs
	Minimum Charge	Job		169		169	16″ O.C.
6″ (R-19)							
	Demolish	S.F.		.11		.11	Cost includes material and labor to
	Install	S.F.	.56	.25		.81	install 6″ (R-19) foil faced roll and batt
	Demolish and Install	S.F.	.56	.36		.92	fiberglass insulation, joist or studs 16″
	Minimum Charge	Job		169		169	O.C.
Unfaced Batt							
3-1/2″ (R-11)							
	Demolish	S.F.		.11		.11	Cost includes material and labor to
	Install	S.F.	.36	.25		.61	install 3-1/2″ (R-11) unfaced roll and
	Demolish and Install	S.F.	.36	.36		.72	batt fiberglass insulation, joist or studs
	Minimum Charge	Job		169		169	16″ O.C.
6″ (R-19)							
	Demolish	S.F.		.11		.11	Cost includes material and labor to
	Install	S.F.	.62	.29		.91	install 6″ (R-19) unfaced roll and batt
	Demolish and Install	S.F.	.62	.40		1.02	fiberglass insulation, joist or studs 16″
	Minimum Charge	Job		169		169	O.C.
10″ (R-30)							
	Demolish	S.F.		.11		.11	Cost includes material and labor to
	Install	S.F.	.73	.34		1.07	install 10″ (R-30) unfaced roll and batt
	Demolish and Install	S.F.	.73	.45		1.18	fiberglass insulation, joist or studs 16″
	Minimum Charge	Job		169		169	O.C.
12″ (R-38)							
	Demolish	S.F.		.11		.11	Cost includes material and labor to
	Install	S.F.	.91	.34		1.25	install 12″ (R-38) unfaced roll and batt
	Demolish and Install	S.F.	.91	.45		1.36	fiberglass insulation, joist or studs 16″
	Minimum Charge	Job		169		169	O.C.
Blown							
R-11							
	Demolish	S.F.		.11		.11	Includes material and labor to install
	Install	S.F.	.25	1.09	.36	1.70	fiberglass blown-in insulation in wall
	Demolish and Install	S.F.	.25	1.20	.36	1.81	with wood siding.
	Minimum Charge	Job		375	125	500	
R-19							
	Demolish	S.F.		.11		.11	Includes material and labor to install
	Install	S.F.	.39	1.25	.42	2.06	fiberglass blown-in insulation in wall
	Demolish and Install	S.F.	.39	1.36	.42	2.17	with wood siding.
	Minimum Charge	Job		375	125	500	

Insulation

Wall Insulation

	Unit	Material	Labor	Equip.	Total	Specification
Rigid Board						
1/2"						
Demolish	S.F.		.13		.13	Includes material and labor to install
Install	S.F.	.32	.42		.74	rigid foam insulation board, foil faced,
Demolish and Install	S.F.	.32	.55		.87	two sides.
Minimum Charge	Job		169		169	
3/4"						
Demolish	S.F.		.13		.13	Includes material and labor to install
Install	S.F.	.42	.42		.84	rigid foam insulation board, foil faced,
Demolish and Install	S.F.	.42	.55		.97	two sides.
Minimum Charge	Job		169		169	
1-1/2"						
Demolish	S.F.		.13		.13	Includes material and labor to install
Install	S.F.	.62	.46		1.08	rigid foam insulation board, foil faced,
Demolish and Install	S.F.	.62	.59		1.21	two sides.
Minimum Charge	Job		169		169	
Vapor Barrier						
Demolish	S.F.		.04		.04	Cost includes material and labor to
Install	S.F.	.03	.09		.12	install 4 mil polyethylene vapor barrier
Demolish and Install	S.F.	.03	.13		.16	10' wide sheets with 6" overlaps.
Minimum Charge	Job		169		169	

Ceiling Insulation

	Unit	Material	Labor	Equip.	Total	Specification
Kraft Faced Batt						
3-1/2" (R-11)						
Demolish	S.F.		.11		.11	Cost includes material and labor to
Install	S.F.	.32	.21		.53	install 3-1/2" (R-11) kraft paper-faced
Demolish and Install	S.F.	.32	.32		.64	roll and batt fiberglass insulation, joist
Minimum Charge	Job		169		169	or studs 16" O.C.
6" (R-19)						
Demolish	S.F.		.11		.11	Cost includes material and labor to
Install	S.F.	.46	.25		.71	install 6" (R-19) kraft paper-faced roll
Demolish and Install	S.F.	.46	.36		.82	and batt fiberglass insulation, joist or
Minimum Charge	Job		169		169	studs 16" O.C.
10" (R-30)						
Demolish	S.F.		.11		.11	Cost includes material and labor to
Install	S.F.	.78	.29		1.07	install 10" (R-30) kraft paper-faced roll
Demolish and Install	S.F.	.78	.40		1.18	and batt fiberglass insulation, joist or
Minimum Charge	Job		169		169	studs 16" O.C.
12" (R-38)						
Demolish	S.F.		.11		.11	Cost includes material and labor to
Install	S.F.	.94	.34		1.28	install 12" (R-38) kraft paper-faced roll
Demolish and Install	S.F.	.94	.45		1.39	and batt fiberglass insulation, joist or
Minimum Charge	Job		169		169	studs 16" O.C.
Foil Faced Batt						
3-1/2" (R-11)						
Demolish	S.F.		.11		.11	Cost includes material and labor to
Install	S.F.	.52	.21		.73	install 3-1/2" (R-11) foil faced roll and
Demolish and Install	S.F.	.52	.32		.84	batt fiberglass insulation, joist or studs
Minimum Charge	Job		169		169	16" O.C.

Insulation

Ceiling Insulation		Unit	Material	Labor	Equip.	Total	Specification
6" (R-19)							
	Demolish	S.F.		.11		.11	Cost includes material and labor to
	Install	S.F.	.56	.25		.81	install 6" (R-19) foil faced roll and batt
	Demolish and Install	S.F.	.56	.36		.92	fiberglass insulation, joist or studs 16"
	Minimum Charge	Job		169		169	O.C.
Unfaced Batt							
3-1/2" (R-11)							
	Demolish	S.F.		.11		.11	Cost includes material and labor to
	Install	S.F.	.36	.25		.61	install 3-1/2" (R-11) unfaced roll and
	Demolish and Install	S.F.	.36	.36		.72	batt fiberglass insulation, joist or studs
	Minimum Charge	Job		169		169	16" O.C.
6" (R-19)							
	Demolish	S.F.		.11		.11	Cost includes material and labor to
	Install	S.F.	.62	.29		.91	install 6" (R-19) unfaced roll and batt
	Demolish and Install	S.F.	.62	.40		1.02	fiberglass insulation, joist or studs 16"
	Minimum Charge	Job		169		169	O.C.
10" (R-30)							
	Demolish	S.F.		.11		.11	Cost includes material and labor to
	Install	S.F.	.73	.34		1.07	install 10" (R-30) unfaced roll and batt
	Demolish and Install	S.F.	.73	.45		1.18	fiberglass insulation, joist or studs 16"
	Minimum Charge	Job		169		169	O.C.
12" (R-38)							
	Demolish	S.F.		.11		.11	Cost includes material and labor to
	Install	S.F.	.91	.34		1.25	install 12" (R-38) unfaced roll and batt
	Demolish and Install	S.F.	.91	.45		1.36	fiberglass insulation, joist or studs 16"
	Minimum Charge	Job		169		169	O.C.
Blown							
5"							
	Demolish	S.F.		.34		.34	Includes material and labor to install
	Install	S.F.	.19	.20	.07	.46	fiberglass blown-in insulation.
	Demolish and Install	S.F.	.19	.54	.07	.80	
	Minimum Charge	Job		375	125	500	
6"							
	Demolish	S.F.		.34		.34	Includes material and labor to install
	Install	S.F.	.23	.25	.08	.56	fiberglass blown-in insulation.
	Demolish and Install	S.F.	.23	.59	.08	.90	
	Minimum Charge	Job		375	125	500	
9"							
	Demolish	S.F.		.34		.34	Includes material and labor to install
	Install	S.F.	.32	.34	.11	.77	fiberglass blown-in insulation.
	Demolish and Install	S.F.	.32	.68	.11	1.11	
	Minimum Charge	Job		375	125	500	
12"							
	Demolish	S.F.		.34		.34	Includes material and labor to install
	Install	S.F.	.45	.50	.17	1.12	fiberglass blown-in insulation.
	Demolish and Install	S.F.	.45	.84	.17	1.46	
	Minimum Charge	Job		375	125	500	
Rigid Board							
1/2"							
	Demolish	S.F.		.19		.19	Includes material and labor to install
	Install	S.F.	.32	.42		.74	rigid foam insulation board, foil faced,
	Demolish and Install	S.F.	.32	.61		.93	two sides.
	Minimum Charge	Job		169		169	

Insulation

Ceiling Insulation

Ceiling Insulation		Unit	Material	Labor	Equip.	Total	Specification
3/4"							
	Demolish	S.F.		.19		.19	Includes material and labor to install
	Install	S.F.	.42	.42		.84	rigid foam insulation board, foil faced,
	Demolish and Install	S.F.	.42	.61		1.03	two sides.
	Minimum Charge	Job		169		169	
1-1/2"							
	Demolish	S.F.		.19		.19	Includes material and labor to install
	Install	S.F.	.62	.46		1.08	rigid foam insulation board, foil faced,
	Demolish and Install	S.F.	.62	.65		1.27	two sides.
	Minimum Charge	Job		169		169	

Exterior Insulation

Exterior Insulation		Unit	Material	Labor	Equip.	Total	Specification
Rigid Board							
1/2"							
	Demolish	S.F.		.10		.10	Includes material and labor to install
	Install	S.F.	.32	.42		.74	rigid foam insulation board foil faced
	Demolish and Install	S.F.	.32	.52		.84	two sides.
	Minimum Charge	Job		169		169	
3/4"							
	Demolish	S.F.		.10		.10	Includes material and labor to install
	Install	S.F.	.42	.42		.84	rigid foam insulation board foil faced
	Demolish and Install	S.F.	.42	.52		.94	two sides.
	Minimum Charge	Job		169		169	
1-1/2"							
	Demolish	S.F.		.10		.10	Includes material and labor to install
	Install	S.F.	.62	.45		1.07	rigid foam insulation board foil faced
	Demolish and Install	S.F.	.62	.55		1.17	two sides.
	Minimum Charge	Job		169		169	
Vapor Barrier							
	Demolish	S.F.		.06		.06	Cost includes material and labor to
	Install	S.F.	.03	.09		.12	install 4 mil polyethylene vapor barrier
	Demolish and Install	S.F.	.03	.15		.18	10' wide sheets with 6" overlaps.
	Minimum Charge	Job		169		169	

Roof Insulation Board

Roof Insulation Board		Unit	Material	Labor	Equip.	Total	Specification
Fiberglass							
3/4" Fiberglass							
	Demolish	Sq.		35		35	Includes material and labor to install
	Install	Sq.	55	25		80	fiberglass board insulation 3/4" thick,
	Demolish and Install	Sq.	55	60		115	R-2.80 and C-0.36 values.
	Minimum Charge	Job		169		169	
1" Fiberglass							
	Demolish	Sq.		35		35	Includes material and labor to install
	Install	Sq.	78.50	28		106.50	fiberglass board insulation 1" thick,
	Demolish and Install	Sq.	78.50	63		141.50	R-4.20 and C-0.24 values.
	Minimum Charge	Job		169		169	
1-3/8" Fiberglass							
	Demolish	Sq.		35		35	Includes material and labor to install
	Install	Sq.	127	30.50		157.50	fiberglass board insulation 1-3/8"
	Demolish and Install	Sq.	127	65.50		192.50	thick, R-5.30 and C-0.19 values.
	Minimum Charge	Job		169		169	

Roof Insulation Board		Unit	Material	Labor	Equip.	Total	Specification
1-5/8" Fiberglass							
	Demolish	Sq.		35		35	Includes material and labor to install
	Install	Sq.	129	32		161	fiberglass board insulation 1-5/8"
	Demolish and Install	Sq.	129	67		196	thick, R-6.70 and C-0.15 values.
	Minimum Charge	Job		169		169	
2-1/4" Fiberglass							
	Demolish	Sq.		35		35	Includes material and labor to install
	Install	Sq.	154	35.50		189.50	fiberglass board insulation 2-1/4"
	Demolish and Install	Sq.	154	70.50		224.50	thick, R-8.30 and C-0.12 values.
	Minimum Charge	Job		169		169	
Perlite							
1" Perlite							
	Demolish	Sq.		35		35	Includes material and labor to install
	Install	Sq.	35	28		63	perlite board insulation 1" thick,
	Demolish and Install	Sq.	35	63		98	R-2.80 and C-0.36 values.
	Minimum Charge	Job		169		169	
4" Perlite							
	Demolish	Sq.		35		35	Includes material and labor to install
	Install	Sq.	141	54		195	perlite board insulation 4" thick,
	Demolish and Install	Sq.	141	89		230	R-10.0 and C-0.10 values.
	Minimum Charge	Job		169		169	
1-1/2" Perlite							
	Demolish	Sq.		35		35	Includes material and labor to install
	Install	Sq.	44	30.50		74.50	perlite board insulation 1-1/2" thick,
	Demolish and Install	Sq.	44	65.50		109.50	R-4.20 and C-0.24 values.
	Minimum Charge	Job		169		169	
2" Perlite							
	Demolish	Sq.		35		35	Includes material and labor to install
	Install	Sq.	70.50	37.50		108	perlite board insulation 2" thick,
	Demolish and Install	Sq.	70.50	72.50		143	R-5.30 and C-0.19 values.
	Minimum Charge	Job		169		169	
2-1/2" Perlite							
	Demolish	Sq.		35		35	Includes material and labor to install
	Install	Sq.	87	42.50		129.50	perlite board insulation 2-1/2" thick,
	Demolish and Install	Sq.	87	77.50		164.50	R-6.70 and C-0.15 values.
	Minimum Charge	Job		169		169	
3" Perlite							
	Demolish	Sq.		35		35	Includes material and labor to install
	Install	Sq.	106	45		151	perlite board insulation 3" thick,
	Demolish and Install	Sq.	106	80		186	R-8.30 and C-0.12 values.
	Minimum Charge	Job		169		169	
Polystyrene							
2" Polystyrene							
	Demolish	Sq.		35		35	Includes material and labor to install
	Install	Sq.	37.50	35.50		73	extruded polystyrene board, R5.0/ 2"
	Demolish and Install	Sq.	37.50	70.50		108	in thickness @25PSI compressive
	Minimum Charge	Job		169		169	strength.
1" Polystyrene							
	Demolish	Sq.		35		35	Includes material and labor to install
	Install	Sq.	36	24		60	extruded polystyrene board, R5.0/ 1"
	Demolish and Install	Sq.	36	59		95	in thickness @25PSI compressive
	Minimum Charge	Job		169		169	strength.
1-1/2" Polystyrene							
	Demolish	Sq.		35		35	Includes material and labor to install
	Install	Sq.	36.50	28		64.50	extruded polystyrene board, R5.0/
	Demolish and Install	Sq.	36.50	63		99.50	1-1/2" in thickness @25PSI
	Minimum Charge	Job		169		169	compressive strength.

Insulation

Roof Insulation Board	Unit	Material	Labor	Equip.	Total	Specification
Urethane						
3/4" Urethane						
Demolish	Sq.		35		35	Includes material and labor to install
Install	Sq.	53	21		74	urethane board insulation 3/4" thick,
Demolish and Install	Sq.	53	56		109	R-5.30 and C-0.19 values.
Minimum Charge	Job		169		169	
1" Urethane						
Demolish	Sq.		35		35	Includes material and labor to install
Install	Sq.	58	22.50		80.50	urethane board insulation 1" thick,
Demolish and Install	Sq.	58	57.50		115.50	R-6.70 and C-0.15 values.
Minimum Charge	Job		169		169	
1-1/2" Urethane						
Demolish	Sq.		35		35	Includes material and labor to install
Install	Sq.	61	26		87	urethane board insulation 1-1/2"
Demolish and Install	Sq.	61	61		122	thick, R-10.1 and C-0.09 values.
Minimum Charge	Job		169		169	
2" Urethane						
Demolish	Sq.		35		35	Includes material and labor to install
Install	Sq.	77	30.50		107.50	urethane board insulation 2" thick,
Demolish and Install	Sq.	77	65.50		142.50	R-13.4 and C-0.06 values.
Minimum Charge	Job		169		169	

Gypsum Wallboard   **Tile Ceiling**   **Gypsum Wallboard Ceiling**

Wall Plaster	Unit	Material	Labor	Equip.	Total	Specification
Interior						
Cement						
Demolish	S.Y.		4.47		4.47	Includes material and labor to install
Install	S.Y.	13.95	19.15	1.57	34.67	cement plaster on interior wood
Demolish and Install	S.Y.	13.95	23.62	1.57	39.14	framing with metal lath.
Clean	S.Y.		2.20		2.20	
Paint	S.Y.	1.64	3.95		5.59	
Minimum Charge	Job		150		150	
Acoustic						
Demolish	S.Y.		4.47		4.47	Includes material and labor to install
Install	S.Y.	9.55	23	1.57	34.12	acoustical cement plaster on interior
Demolish and Install	S.Y.	9.55	27.47	1.57	38.59	wood framing with metal lath.
Clean	S.Y.		2.20		2.20	
Paint	S.Y.	1.64	3.95		5.59	
Minimum Charge	Job		150		150	
Gypsum						
Demolish	S.Y.		4.47		4.47	Includes material and labor to install
Install	S.Y.	8.25	19.70	1.34	29.29	gypsum wall or ceiling plaster on
Demolish and Install	S.Y.	8.25	24.17	1.34	33.76	interior wood framing with metal lath.
Clean	S.Y.		2.20		2.20	
Paint	S.Y.	1.64	3.95		5.59	
Minimum Charge	Job		150		150	
Stucco						
Demolish	S.Y.		10.75		10.75	Includes material and labor to install
Install	S.Y.	5.90	27	1.84	34.74	stucco on wire mesh over wood
Demolish and Install	S.Y.	5.90	37.75	1.84	45.49	framing.
Clean	S.Y.		2.20		2.20	
Paint	S.Y.	1.64	3.95		5.59	
Minimum Charge	Job		150		150	
Thin Coat						
Veneer						
Demolish	S.Y.		4.47		4.47	Includes materials and labor to install
Install	S.Y.	.65	3.54	.29	4.48	thin coat veneer plaster.
Demolish and Install	S.Y.	.65	8.01	.29	8.95	
Clean	S.Y.		2.20		2.20	
Paint	S.Y.	1.64	3.95		5.59	
Minimum Charge	Job		150		150	

Walls / Ceilings

Wall Plaster	Unit	Material	Labor	Equip.	Total	Specification
Lath						
Metal						
Demolish	S.Y.		1.05		1.05	Includes material and labor to install
Install	S.Y.	3.61	3.98		7.59	wire lath on steel framed walls.
Demolish and Install	S.Y.	3.61	5.03		8.64	
Clean	S.Y.		2.20		2.20	
Minimum Charge	Job		70.50		70.50	
Gypsum						
Demolish	S.Y.		1.05		1.05	Includes material and labor to install
Install	S.Y.	4.65	3.74		8.39	gypsum lath on walls.
Demolish and Install	S.Y.	4.65	4.79		9.44	
Clean	S.Y.		2.20		2.20	
Minimum Charge	Job		150		150	
Repairs						
Fill in Small Cracks						
Install	L.F.	.06	2.40		2.46	Includes labor and material to fill-in
Minimum Charge	Job		150		150	hairline cracks up to 1/8" with filler,
						sand and prep for paint.
Repair up to 1" Crack						
Install	L.F.	.22	3		3.22	Includes labor and material to fill-in
Minimum Charge	Job		150		150	hairline cracks up to 1" with filler,
						sand and prep for paint.
Patch Small Hole (2")						
Install	Ea.	.06	3.16		3.22	Includes labor and material to patch
Minimum Charge	Job		150		150	up to a 2" diameter hole in plaster.
Patch Large Hole						
Install	Ea.	6.45	50		56.45	Includes labor and material to patch
Minimum Charge	Job		150		150	up to a 20" diameter hole in plaster.
Patch Plaster Section						
Install	Ea.	2.18	3		5.18	Includes labor and material to repair
Minimum Charge	Job		150		150	section of plaster wall, area less than
						20 S.F.
Repair Inside Corner						
Install	L.F.	2.57	7.50		10.07	Includes material and labor to patch
Minimum Charge	Job		150		150	plaster corners.

Ceiling Plaster	Unit	Material	Labor	Equip.	Total	Specification
Interior						
Cement						
Demolish	S.Y.		4.47		4.47	Includes material and labor to install
Install	S.Y.	13.95	19.15	1.57	34.67	cement plaster on interior wood
Demolish and Install	S.Y.	13.95	23.62	1.57	39.14	framing with metal lath.
Clean	S.Y.		2.30		2.30	
Paint	S.Y.	1.64	3.95		5.59	
Minimum Charge	Job		150		150	
Acoustic						
Demolish	S.Y.		4.47		4.47	Includes material and labor to install
Install	S.Y.	9.55	23	1.57	34.12	acoustical cement plaster on interior
Demolish and Install	S.Y.	9.55	27.47	1.57	38.59	wood framing with metal lath.
Clean	S.Y.		2.30		2.30	
Paint	S.Y.	1.64	3.95		5.59	
Minimum Charge	Job		150		150	

Walls / Ceilings

Ceiling Plaster	Unit	Material	Labor	Equip.	Total	Specification
Gypsum						
Demolish	S.Y.		4.47		4.47	Includes material and labor to install
Install	S.Y.	8.25	19.70	1.34	29.29	gypsum wall or ceiling plaster on
Demolish and Install	S.Y.	8.25	24.17	1.34	33.76	interior wood framing with metal lath.
Clean	S.Y.		2.30		2.30	
Paint	S.Y.	1.64	3.95		5.59	
Minimum Charge	Job		150		150	
Stucco						
Demolish	S.Y.		10.75		10.75	Includes material and labor to install
Install	S.Y.	5.90	27	1.84	34.74	stucco on wire mesh over wood
Demolish and Install	S.Y.	5.90	37.75	1.84	45.49	framing.
Clean	S.Y.		2.30		2.30	
Paint	S.Y.	1.64	3.95		5.59	
Minimum Charge	Job		150		150	
Thin Coat						
Veneer						
Demolish	S.Y.		4.47		4.47	Includes materials and labor to install
Install	S.Y.	.65	3.54	.29	4.48	thin coat veneer plaster.
Demolish and Install	S.Y.	.65	8.01	.29	8.95	
Clean	S.Y.		2.30		2.30	
Paint	S.Y.	1.64	3.95		5.59	
Minimum Charge	Job		150		150	
Lath						
Metal						
Demolish	S.Y.		1.05		1.05	Includes material and labor to install
Install	S.Y.	3.61	4.98		8.59	wire lath on steel framed ceilings.
Demolish and Install	S.Y.	3.61	6.03		9.64	
Minimum Charge	Job		150		150	
Gypsum						
Demolish	S.Y.		1.05		1.05	Includes material and labor to install
Install	S.Y.	4.65	3.11		7.76	gypsum lath on ceilings.
Demolish and Install	S.Y.	4.65	4.16		8.81	
Minimum Charge	Job		150		150	
Repairs						
Fill in Small Cracks						
Install	S.F.	.06	.55		.61	Includes labor and material to fill in
Minimum Charge	Job		300		300	hairline cracks up to 1/8" with filler, sand and prep for paint.

Gypsum Wallboard	Unit	Material	Labor	Equip.	Total	Specification
Finished						
1/4"						
Demolish	S.F.		.27		.27	Cost includes material and labor to
Install	S.F.	.24	1.01		1.25	install 1/4" drywall, including joint
Demolish and Install	S.F.	.24	1.28		1.52	taping, premixed compound and
Clean	S.F.		.22		.22	screws.
Paint	S.F.	.18	.26		.44	
Minimum Charge	Job		169		169	
3/8"						
Demolish	S.F.		.27		.27	Cost includes material and labor to
Install	S.F.	.22	1.03		1.25	install 3/8" drywall including joint
Demolish and Install	S.F.	.22	1.30		1.52	tape, premixed compound and
Clean	S.F.		.22		.22	screws.
Paint	S.F.	.18	.26		.44	
Minimum Charge	Job		169		169	

175

Walls / Ceilings

Gypsum Wallboard	Unit	Material	Labor	Equip.	Total	Specification
1/2"						
Demolish	S.F.		.27		.27	Cost includes material and labor to
Install	S.F.	.25	1.04		1.29	install 1/2" drywall including joint
Demolish and Install	S.F.	.25	1.31		1.56	tape, premixed compound and
Clean	S.F.		.22		.22	screws.
Paint	S.F.	.18	.26		.44	
Minimum Charge	Job		169		169	
1/2" Water Resistant						
Demolish	S.F.		.27		.27	Cost includes material and labor to
Install	S.F.	.17	1.04		1.21	install 1/2" WR drywall including joint
Demolish and Install	S.F.	.17	1.31		1.48	tape, premixed compound and
Clean	S.F.		.22		.22	screws.
Paint	S.F.	.18	.26		.44	
Minimum Charge	Job		169		169	
1/2" Fire-rated						
Demolish	S.F.		.27		.27	Cost includes material and labor to
Install	S.F.	.27	1.04		1.31	install 1/2" FR drywall including
Demolish and Install	S.F.	.27	1.31		1.58	screws. Taping and finishing not
Clean	S.F.		.22		.22	included.
Paint	S.F.	.18	.26		.44	
Minimum Charge	Job		169		169	
5/8"						
Demolish	S.F.		.27		.27	Cost includes material and labor to
Install	S.F.	.30	1.18		1.48	install 5/8" drywall including joint
Demolish and Install	S.F.	.30	1.45		1.75	tape, premixed compound and
Clean	S.F.		.22		.22	screws.
Paint	S.F.	.18	.26		.44	
Minimum Charge	Job		169		169	
5/8" Fire-rated						
Demolish	S.F.		.27		.27	Cost includes material and labor to
Install	S.F.	.30	1.18		1.48	install 5/8" FR drywall, including joint
Demolish and Install	S.F.	.30	1.45		1.75	taping, premixed compound and
Clean	S.F.		.22		.22	screws.
Paint	S.F.	.18	.26		.44	
Minimum Charge	Job		169		169	
2-hour fire-rated						
Demolish	S.F.		.67		.67	Cost includes material and labor to
Install	S.F.	.69	2.34		3.03	install 2 layers of 5/8" type X drywall
Demolish and Install	S.F.	.69	3.01		3.70	including joint taping, premixed
Clean	S.F.		.22		.22	compound and screws.
Paint	S.F.	.18	.26		.44	
Minimum Charge	Job		169		169	
Unfinished						
1/4"						
Demolish	S.F.		.27		.27	Cost includes material and labor to
Install	S.F.	.21	.35		.56	install 1/4" drywall including screws.
Demolish and Install	S.F.	.21	.62		.83	Taping and finishing not included.
Clean	S.F.		.22		.22	
Minimum Charge	Job		169		169	
3/8"						
Demolish	S.F.		.27		.27	Cost includes material and labor to
Install	S.F.	.20	.35		.55	install 3/8" drywall including screws.
Demolish and Install	S.F.	.20	.62		.82	Tape, finish or texture is not included.
Clean	S.F.		.22		.22	
Minimum Charge	Job		169		169	

Gypsum Wallboard		Unit	Material	Labor	Equip.	Total	Specification
1/2″							
	Demolish	S.F.		.27		.27	Cost includes material and labor to
	Install	S.F.	.22	.36		.58	install 1/2″ drywall including screws.
	Demolish and Install	S.F.	.22	.63		.85	Taping and finishing not included.
	Clean	S.F.		.22		.22	
	Minimum Charge	Job		169		169	
1/2″ Water Resistant							
	Demolish	S.F.		.27		.27	Cost includes material and labor to
	Install	S.F.	.15	.36		.51	install 1/2″ WR drywall including
	Demolish and Install	S.F.	.15	.63		.78	screws. Taping and finishing is not
	Clean	S.F.		.22		.22	included.
	Minimum Charge	Job		169		169	
1/2″ Fire-rated							
	Demolish	S.F.		.27		.27	Cost includes labor and material to
	Install	S.F.	.22	.24		.46	install 1/2″ FR drywall including
	Demolish and Install	S.F.	.22	.51		.73	screws. Tape, finish or texture is not
	Clean	S.F.		.22		.22	included.
	Minimum Charge	Job		169		169	
5/8″							
	Demolish	S.F.		.27		.27	Cost includes material and labor to
	Install	S.F.	.27	.40		.67	install 5/8″ drywall including screws.
	Demolish and Install	S.F.	.27	.67		.94	Taping and finishing not included.
	Clean	S.F.		.22		.22	
	Minimum Charge	Job		169		169	
5/8″ Fire-rated							
	Demolish	S.F.		.27		.27	Cost includes material and labor to
	Install	S.F.	.27	.40		.67	install 5/8″ FR drywall including
	Demolish and Install	S.F.	.27	.67		.94	screws. Taping and finishing not
	Clean	S.F.		.22		.22	included.
	Minimum Charge	Job		169		169	
2-hour Fire-rated							
	Demolish	S.F.		.67		.67	Cost includes material and labor to
	Install	S.F.	.69	2.34		3.03	install 2 layers of 5/8″ type X drywall
	Demolish and Install	S.F.	.69	3.01		3.70	including joint taping, premixed
	Clean	S.F.		.22		.22	compound and screws.
	Minimum Charge	Job		169		169	
Accessories							
Tape & Finish							
	Install	S.F.	.03	.34		.37	Includes labor and material for taping,
	Minimum Charge	Job		169		169	sanding and finish.
Texture							
Sprayed							
	Install	S.F.	.06	.37		.43	Includes labor and material to install
	Minimum Charge	Job		169		169	by spray, texture finish.
Trowel							
	Install	S.F.	.01	.20		.21	Includes labor and material to apply
	Minimum Charge	Job		169		169	hand troweled texture.
Patch Hole in Gypsum Wallboard							
	Install	S.F.	.41	45		45.41	Includes labor and material to patch a
	Minimum Charge	Job		169		169	one square foot area of damaged drywall.
Furring Strips							
	Demolish	S.F.		.07		.07	Cost includes material and labor to
	Install	S.F.	.17	.75		.92	install 1″ x 2″ furring strips, 16″ O.C.
	Demolish and Install	S.F.	.17	.82		.99	per S.F. of surface area to be covered.
	Minimum Charge	Job		84.50		84.50	

Walls / Ceilings

Ceramic Wall Tile		Unit	Material	Labor	Equip.	Total	Specification
Thin Set							
2″ x 2″							
	Demolish	S.F.		.89		.89	Cost includes material and labor to
	Install	S.F.	6.15	5.15		11.30	install 2″ x 2″ ceramic tile in organic
	Demolish and Install	S.F.	6.15	6.04		12.19	adhesive including cutting and grout.
	Clean	S.F.		.46		.46	
	Minimum Charge	Job		154		154	
4-1/4″ x 4-1/4″							
	Demolish	S.F.		.89		.89	Cost includes material and labor to
	Install	S.F.	2.19	2.88		5.07	install 4-1/4″ x 4-1/4″ ceramic tile in
	Demolish and Install	S.F.	2.19	3.77		5.96	organic adhesive including grout.
	Clean	S.F.		.46		.46	
	Minimum Charge	Job		154		154	
6″ x 6″							
	Demolish	S.F.		.89		.89	Cost includes material and labor to
	Install	S.F.	3	2.74		5.74	install 6″ x 6″ ceramic tile in organic
	Demolish and Install	S.F.	3	3.63		6.63	adhesive including and grout.
	Clean	S.F.		.46		.46	
	Minimum Charge	Job		154		154	
8″ x 8″							
	Demolish	S.F.		.89		.89	Cost includes material and labor to
	Install	S.F.	4	2.43		6.43	install 8″ x 8″ ceramic tile in organic
	Demolish and Install	S.F.	4	3.32		7.32	adhesive including cutting and grout.
	Clean	S.F.		.46		.46	
	Minimum Charge	Job		154		154	
Thick Set							
2″ x 2″							
	Demolish	S.F.		1.07		1.07	Cost includes material and labor to
	Install	S.F.	5.05	6.20		11.25	install 2″ x 2″ ceramic tile in mortar
	Demolish and Install	S.F.	5.05	7.27		12.32	bed including cutting and grout.
	Clean	S.F.		.46		.46	
	Minimum Charge	Job		154		154	
4-1/4″ x 4-1/4″							
	Demolish	S.F.		1.07		1.07	Cost includes material and labor to
	Install	S.F.	6.80	6.20		13	install 4-1/4″ x 4-1/4″ ceramic tile in
	Demolish and Install	S.F.	6.80	7.27		14.07	mortar bed including cutting and
	Clean	S.F.		.46		.46	grout.
	Minimum Charge	Job		154		154	
6″ x 6″							
	Demolish	S.F.		1.07		1.07	Cost includes material and labor to
	Install	S.F.	7.65	5.60		13.25	install 6″ x 6″ ceramic tile in mortar
	Demolish and Install	S.F.	7.65	6.67		14.32	bed including cutting and grout.
	Clean	S.F.		.46		.46	
	Minimum Charge	Job		154		154	
8″ x 8″							
	Demolish	S.F.		1.07		1.07	Cost includes material and labor to
	Install	S.F.	10	5.15		15.15	install 8″ x 8″ ceramic tile in mortar
	Demolish and Install	S.F.	10	6.22		16.22	bed including cutting and grout.
	Clean	S.F.		.46		.46	
	Minimum Charge	Job		154		154	
Re-grout							
	Install	S.F.	.14	3.09		3.23	Includes material and labor to regrout
	Minimum Charge	Job		154		154	ceramic tile walls.

Walls / Ceilings

Wall Tile		Unit	Material	Labor	Equip.	Total	Specification
Mirror							
Custom Cut Wall							
	Demolish	S.F.		.51		.51	Includes material and labor to install
	Install	S.F.	1.75	3.29		5.04	distortion-free float glass mirror, 1/8"
	Demolish and Install	S.F.	1.75	3.80		5.55	to 3/16" thick with cut and polished
	Clean	S.F.	.02	.14		.16	edges.
	Minimum Charge	Job		161		161	
Beveled Wall							
	Demolish	S.F.		.51		.51	Includes material and labor to install
	Install	S.F.	4.50	5.85		10.35	distortion-free float glass mirror, 1/8"
	Demolish and Install	S.F.	4.50	6.36		10.86	to 3/16" thick with beveled edges.
	Clean	S.F.	.02	.14		.16	
	Minimum Charge	Job		161		161	
Smoked Wall							
	Demolish	S.F.		.51		.51	Includes material and labor to install
	Install	S.F.	7.15	5.85		13	distortion-free smoked float glass
	Demolish and Install	S.F.	7.15	6.36		13.51	mirror, 1/8" to 3/16" thick with cut
	Clean	S.F.	.02	.14		.16	and polished edges.
	Minimum Charge	Job		161		161	
Marble							
Good Grade							
	Demolish	S.F.		1.68		1.68	Cost includes material and labor to
	Install	S.F.	7.60	13.45		21.05	install 3/8" x 12" x 12" marble tile in
	Demolish and Install	S.F.	7.60	15.13		22.73	mortar bed with grout.
	Clean	S.F.		.46		.46	
	Minimum Charge	Job		154		154	
Better Grade							
	Demolish	S.F.		1.68		1.68	Cost includes material and labor to
	Install	S.F.	9.80	13.45		23.25	install 3/8" x 12" x 12" marble, in
	Demolish and Install	S.F.	9.80	15.13		24.93	mortar bed with grout.
	Clean	S.F.		.46		.46	
	Minimum Charge	Job		154		154	
Premium Grade							
	Demolish	S.F.		1.68		1.68	Cost includes material and labor to
	Install	S.F.	18.90	13.45		32.35	install 3/8" x 12" x 12" marble, in
	Demolish and Install	S.F.	18.90	15.13		34.03	mortar bed with grout.
	Clean	S.F.		.46		.46	
	Minimum Charge	Job		154		154	
Glass Block							
	Demolish	S.F.		1.34		1.34	Cost includes material and labor to
	Install	S.F.	19.75	13.40		33.15	install 3-7/8" thick, 6" x 6" smooth
	Demolish and Install	S.F.	19.75	14.74		34.49	face glass block including mortar and
	Clean	S.F.		.46		.46	cleaning.
	Minimum Charge	Job		300		300	
Plastic							
	Demolish	S.F.		.41		.41	Cost includes material and labor to
	Install	S.F.	.95	1.76		2.71	install 4" plastic tile.
	Demolish and Install	S.F.	.95	2.17		3.12	
	Clean	S.F.		.46		.46	
	Minimum Charge	Job		154		154	
Cork							
	Demolish	S.F.		.21		.21	Cost includes material and labor to
	Install	S.F.	4.11	1.24		5.35	install 3/16" thick cork tile and
	Demolish and Install	S.F.	4.11	1.45		5.56	installation.
	Clean	S.F.		.46		.46	
	Minimum Charge	Job		149		149	

Walls / Ceilings

Wall Tile		Unit	Material	Labor	Equip.	Total	Specification
Re-grout							
	Install	S.F.	.14	3.09		3.23	Includes material and labor to regrout
	Minimum Charge	Job		154		154	ceramic tile walls.

Wall Paneling		Unit	Material	Labor	Equip.	Total	Specification
Plywood							
Good Grade							
	Demolish	S.F.		.27		.27	Cost includes material and labor to
	Install	S.F.	.92	1.35		2.27	install 4' x 8' x 1/4" paneling,
	Demolish and Install	S.F.	.92	1.62		2.54	including nails and adhesive.
	Clean	S.F.		.22		.22	
	Paint	S.F.	.17	.39		.56	
	Minimum Charge	Job		169		169	
Better Grade							
	Demolish	S.F.		.27		.27	Cost includes material and labor to
	Install	S.F.	1.41	1.61		3.02	install 4' x 8' x 1/4" plywood wall
	Demolish and Install	S.F.	1.41	1.88		3.29	panel, including nails and adhesive.
	Clean	S.F.		.22		.22	
	Paint	S.F.	.17	.39		.56	
	Minimum Charge	Job		169		169	
Premium Grade							
	Demolish	S.F.		.27		.27	Cost includes material and labor to
	Install	S.F.	2.06	1.93		3.99	install 4' x 8' x 1/4" plywood wall
	Demolish and Install	S.F.	2.06	2.20		4.26	panel, including nails and adhesive.
	Clean	S.F.		.22		.22	
	Paint	S.F.	.17	.39		.56	
	Minimum Charge	Job		169		169	
Hardboard							
Embossed Paneling							
	Demolish	S.F.		.27		.27	Includes material and labor to install
	Install	S.F.	.64	1.35		1.99	hardboard embossed paneling
	Demolish and Install	S.F.	.64	1.62		2.26	including adhesives.
	Clean	S.F.		.22		.22	
	Minimum Charge	Job		169		169	
Plastic Molding							
	Demolish	L.F.		.27		.27	Includes material and labor to install
	Install	L.F.	.57	.85		1.42	plastic trim for paneling.
	Demolish and Install	L.F.	.57	1.12		1.69	
	Clean	L.F.	.02	.14		.16	
	Minimum Charge	Job		169		169	
Hardboard							
Embossed Brick							
	Demolish	S.F.		.27		.27	Cost includes material and labor to
	Install	S.F.	2.10	1.50		3.60	install 4' x 8' non-ceramic mineral
	Demolish and Install	S.F.	2.10	1.77		3.87	brick-like veneer wall panels including
	Clean	S.F.		.22		.22	adhesive.
	Paint	S.F.	.18	.45		.63	
	Minimum Charge	Job		169		169	
Embossed Ceramic Tile							
	Demolish	S.F.		.27		.27	Includes material and labor to install
	Install	S.F.	.64	1.35		1.99	hardboard embossed paneling
	Demolish and Install	S.F.	.64	1.62		2.26	including adhesives.
	Clean	S.F.		.22		.22	
	Paint	S.F.	.18	.45		.63	
	Minimum Charge	Job		169		169	

Wall Paneling	Unit	Material	Labor	Equip.	Total	Specification
Pegboard						
Demolish	S.F.		.27		.27	Cost includes material and labor to
Install	S.F.	.43	1.35		1.78	install 4' x 8' x 1/4" pegboard panel,
Demolish and Install	S.F.	.43	1.62		2.05	including adhesive.
Clean	S.F.		.22		.22	
Paint	S.F.	.18	.45		.63	
Minimum Charge	Job		169		169	
Furring Strips						
Demolish	S.F.		.07		.07	Cost includes material and labor to
Install	S.F.	.17	.75		.92	install 1" x 2" furring strips, 16" O.C.
Demolish and Install	S.F.	.17	.82		.99	per S.F. of surface area to be covered.
Minimum Charge	Job		169		169	
Molding						
Unfinished Base						
Demolish	L.F.		.45		.45	Cost includes material and labor to
Install	L.F.	.59	1.04		1.63	install 1/2" x 2-1/2" pine or birch
Demolish and Install	L.F.	.59	1.49		2.08	base molding.
Reinstall	L.F.		.83		.83	
Clean	L.F.	.02	.14		.16	
Paint	L.F.	.03	.32		.35	
Minimum Charge	Job		169		169	
Unfinished Corner						
Demolish	L.F.		.45		.45	Includes material and labor to install
Install	L.F.	1.82	1.41		3.23	unfinished pine corner molding.
Demolish and Install	L.F.	1.82	1.86		3.68	
Reinstall	L.F.		1.13		1.13	
Clean	Ea.	.83	21.50		22.33	
Paint	L.F.	.12	.47		.59	
Minimum Charge	Job		169		169	
Unfinished Cove						
Demolish	L.F.		.45		.45	Cost includes material and labor to
Install	L.F.	1.56	1.25		2.81	install 3/4" x 3-1/2" unfinished pine
Demolish and Install	L.F.	1.56	1.70		3.26	cove molding.
Reinstall	L.F.		1		1	
Clean	Ea.	.83	21.50		22.33	
Paint	L.F.	.12	.47		.59	
Minimum Charge	Job		169		169	
Unfinished Crown						
Demolish	L.F.		.54		.54	Includes material and labor to install
Install	L.F.	2.05	1.35		3.40	unfinished pine crown molding.
Demolish and Install	L.F.	2.05	1.89		3.94	
Reinstall	L.F.		1.08		1.08	
Clean	L.F.	.03	.19		.22	
Paint	L.F.	.04	.32		.36	
Minimum Charge	Job		169		169	
Prefinished Base						
Demolish	L.F.		.45		.45	Cost includes material and labor to
Install	L.F.	1.69	1.08		2.77	install 9/16" x 3-5/16" prefinished
Demolish and Install	L.F.	1.69	1.53		3.22	base molding.
Reinstall	L.F.		.87		.87	
Clean	L.F.	.02	.14		.16	
Paint	L.F.	.03	.32		.35	
Minimum Charge	Job		169		169	

Walls / Ceilings

Wall Paneling	Unit	Material	Labor	Equip.	Total	Specification
Prefinished Corner						
Demolish	L.F.		.45		.45	Includes material and labor to install
Install	L.F.	2.09	1.41		3.50	inside or outside prefinished corner
Demolish and Install	L.F.	2.09	1.86		3.95	molding.
Reinstall	L.F.		1.13		1.13	
Clean	L.F.		.09		.09	
Paint	L.F.	.12	.47		.59	
Minimum Charge	Job		169		169	
Prefinished Crown						
Demolish	L.F.		.54		.54	Cost includes material and labor to
Install	L.F.	2.24	1.35		3.59	install 3/4" x 3-13/16" prefinished
Demolish and Install	L.F.	2.24	1.89		4.13	crown molding.
Reinstall	L.F.		1.08		1.08	
Clean	L.F.	.03	.19		.22	
Paint	L.F.	.04	.32		.36	
Minimum Charge	Job		169		169	

Wallcovering	Unit	Material	Labor	Equip.	Total	Specification
Surface Prep						
Patch Walls						
Install	S.F.	.11	3		3.11	Includes labor and material to repair
Minimum Charge	Job		149		149	section of plaster wall, area less than
						20 S.F.
Sizing						
Install	S.F.	.07	.15		.22	Includes labor and materials to apply
Minimum Charge	Job		149		149	wall sizing.
Good Grade						
Walls						
Demolish	S.F.		.55		.55	Includes material and labor to install
Install	S.F.	.83	.62		1.45	pre-pasted, vinyl-coated wallcovering.
Demolish and Install	S.F.	.83	1.17		2	
Clean	S.F.	.01	.05		.06	
Minimum Charge	Job		149		149	
Ceiling						
Demolish	S.F.		.55		.55	Includes material and labor to install
Install	S.F.	.83	.62		1.45	pre-pasted, vinyl-coated wallcovering.
Demolish and Install	S.F.	.83	1.17		2	
Clean	S.F.	.02	.14		.16	
Minimum Charge	Job		149		149	
Better Grade						
Walls						
Demolish	S.F.		.55		.55	Includes material and labor to install
Install	S.F.	.92	.62		1.54	pre-pasted, vinyl-coated wallcovering.
Demolish and Install	S.F.	.92	1.17		2.09	
Clean	S.F.	.01	.05		.06	
Minimum Charge	Job		149		149	
Ceiling						
Demolish	S.F.		.55		.55	Includes material and labor to install
Install	S.F.	.92	.62		1.54	pre-pasted, vinyl-coated wallcovering.
Demolish and Install	S.F.	.92	1.17		2.09	
Clean	S.F.	.02	.14		.16	
Minimum Charge	Job		149		149	

Walls / Ceilings

Wallcovering	Unit	Material	Labor	Equip.	Total	Specification
Premium Grade						
Walls						
Demolish	S.F.		.69		.69	Includes material and labor to install
Install	S.F.	1.18	.62		1.80	pre-pasted, vinyl-coated wallcovering.
Demolish and Install	S.F.	1.18	1.31		2.49	
Clean	S.F.	.01	.05		.06	
Minimum Charge	Job		149		149	
Ceiling						
Demolish	S.F.		.69		.69	Includes material and labor to install
Install	S.F.	1.18	.62		1.80	pre-pasted, vinyl-coated wallcovering.
Demolish and Install	S.F.	1.18	1.31		2.49	
Clean	S.F.	.02	.14		.16	
Minimum Charge	Job		149		149	
Mural						
Pre-printed						
Demolish	Ea.		84		84	Includes material and labor to install
Install	Ea.	220	224		444	wallcovering with a pre-printed mural.
Demolish and Install	Ea.	220	308		528	
Clean	Ea.		6.05		6.05	
Minimum Charge	Job		149		149	
Hand-painted						
Demolish	Ea.		84		84	Includes material and labor to install
Install	Ea.	440	450		890	wallcovering with a hand-painted
Demolish and Install	Ea.	440	534		974	mural.
Clean	Ea.		6.05		6.05	
Minimum Charge	Job		149		149	
Vinyl-coated						
Good Grade						
Demolish	Ea.		14.90		14.90	Includes material and labor to install
Install	Ea.	18.30	21		39.30	pre-pasted, vinyl-coated wallcovering.
Demolish and Install	Ea.	18.30	35.90		54.20	Based upon 30 S.F. rolls.
Clean	Ea.		6.05		6.05	
Minimum Charge	Job		149		149	
Better Grade						
Demolish	Ea.		14.90		14.90	Includes material and labor to install
Install	Ea.	24	21		45	pre-pasted, vinyl-coated wallcovering.
Demolish and Install	Ea.	24	35.90		59.90	Based upon 30 S.F. rolls.
Clean	Ea.		6.05		6.05	
Minimum Charge	Job		149		149	
Premium Grade						
Demolish	Ea.		14.90		14.90	Includes material and labor to install
Install	Ea.	43	21		64	pre-pasted, vinyl-coated wallcovering.
Demolish and Install	Ea.	43	35.90		78.90	Based upon 30 S.F. rolls.
Clean	Ea.		6.05		6.05	
Minimum Charge	Job		149		149	
Blank Underliner						
Demolish	Ea.		14.90		14.90	Includes material and labor to install
Install	Ea.	10.85	21.50		32.35	blank stock liner paper or bridging
Demolish and Install	Ea.	10.85	36.40		47.25	paper. Based upon 30 S.F. rolls.
Clean	Ea.		6.05		6.05	
Minimum Charge	Job		149		149	
Solid Vinyl						
Good Grade						
Demolish	Roll		18.35		18.35	Includes material and labor to install
Install	Ea.	25	21		46	pre-pasted, vinyl wallcovering. Based
Demolish and Install	Ea.	25	39.35		64.35	upon 30 S.F. rolls.
Clean	Ea.		6.05		6.05	
Minimum Charge	Job		149		149	

Wallcovering		Unit	Material	Labor	Equip.	Total	Specification
Better Grade							
	Demolish	Roll		18.35		18.35	Includes material and labor to install
	Install	Ea.	34	21		55	pre-pasted, vinyl wallcovering. Based
	Demolish and Install	Ea.	34	39.35		73.35	upon 30 S.F. rolls.
	Clean	Ea.		6.05		6.05	
	Minimum Charge	Job		149		149	
Premium Grade							
	Demolish	Roll		18.35		18.35	Includes material and labor to install
	Install	Ea.	43.50	21		64.50	pre-pasted, vinyl wallcovering. Based
	Demolish and Install	Ea.	43.50	39.35		82.85	upon 30 S.F. rolls.
	Clean	Ea.		6.05		6.05	
	Minimum Charge	Job		149		149	
Blank Underliner							
	Demolish	Ea.		14.90		14.90	Includes material and labor to install
	Install	Ea.	10.85	21.50		32.35	blank stock liner paper or bridging
	Demolish and Install	Ea.	10.85	36.40		47.25	paper. Based upon 30 S.F. rolls.
	Clean	Ea.		6.05		6.05	
	Minimum Charge	Job		149		149	
Grasscloth / String							
Good Grade							
	Demolish	Ea.		14.90		14.90	Includes material and labor to install
	Install	Ea.	34.50	27		61.50	grasscloth or string wallcovering.
	Demolish and Install	Ea.	34.50	41.90		76.40	Based upon 30 S.F. rolls.
	Clean	Ea.		8.05		8.05	
	Minimum Charge	Job		149		149	
Better Grade							
	Demolish	Ea.		14.90		14.90	Includes material and labor to install
	Install	Ea.	44.50	27		71.50	grasscloth or string wallcovering.
	Demolish and Install	Ea.	44.50	41.90		86.40	Based upon 30 S.F. rolls.
	Clean	Ea.		8.05		8.05	
	Minimum Charge	Job		149		149	
Premium Grade							
	Demolish	Ea.		14.90		14.90	Includes material and labor to install
	Install	Ea.	57	27		84	grasscloth or string wallcovering.
	Demolish and Install	Ea.	57	41.90		98.90	Based upon 30 S.F. rolls.
	Clean	Ea.		8.05		8.05	
	Minimum Charge	Job		149		149	
Blank Underliner							
	Demolish	Ea.		14.90		14.90	Includes material and labor to install
	Install	Ea.	10.85	21.50		32.35	blank stock liner paper or bridging
	Demolish and Install	Ea.	10.85	36.40		47.25	paper. Based upon 30 S.F. rolls.
	Clean	Ea.		6.05		6.05	
	Minimum Charge	Job		149		149	
Designer							
Premium Grade							
	Demolish	S.F.		.69		.69	Includes material and labor to install
	Install	Ea.	107	31.50		138.50	paperback vinyl, untrimmed and
	Demolish and Install	Ea.	107	32.19		139.19	hand-painted. Based upon 30 S.F.
	Clean	Ea.		6.05		6.05	rolls.
	Minimum Charge	Job		149		149	
Blank Underliner							
	Demolish	Ea.		14.90		14.90	Includes material and labor to install
	Install	Ea.	10.85	21.50		32.35	blank stock liner paper or bridging
	Demolish and Install	Ea.	10.85	36.40		47.25	paper. Based upon 30 S.F. rolls.
	Clean	Ea.		6.05		6.05	
	Minimum Charge	Job		149		149	

Walls / Ceilings

Wallcovering

Wallcovering		Unit	Material	Labor	Equip.	Total	Specification
Foil							
	Demolish	Ea.		14.90		14.90	Includes material and labor to install
	Install	Ea.	37	27		64	foil coated wallcovering. Based upon
	Demolish and Install	Ea.	37	41.90		78.90	30 S.F. rolls.
	Clean	Ea.		6.05		6.05	
	Minimum Charge	Job		149		149	
Wallcovering Border							
Good Grade							
	Demolish	L.F.		.98		.98	Includes material and labor install
	Install	L.F.	1.83	1.14		2.97	vinyl-coated border 4" to 6" wide.
	Demolish and Install	L.F.	1.83	2.12		3.95	
	Clean	L.F.		.09		.09	
	Minimum Charge	Job		149		149	
Better Grade							
	Demolish	L.F.		.98		.98	Includes material and labor install
	Install	L.F.	2.43	1.14		3.57	vinyl-coated border 4" to 6" wide.
	Demolish and Install	L.F.	2.43	2.12		4.55	
	Clean	L.F.		.09		.09	
	Minimum Charge	Job		149		149	
Premium Grade							
	Demolish	L.F.		.98		.98	Includes material and labor install
	Install	L.F.	3.05	1.14		4.19	vinyl-coated border 4" to 6" wide.
	Demolish and Install	L.F.	3.05	2.12		5.17	
	Clean	L.F.		.09		.09	
	Minimum Charge	Job		149		149	

Acoustical Ceiling

Acoustical Ceiling		Unit	Material	Labor	Equip.	Total	Specification
Complete System							
2' x 2' Square Edge							
	Demolish	S.F.		.45		.45	Includes material and labor to install
	Install	S.F.	1.28	1.04		2.32	T-bar grid acoustical ceiling system,
	Demolish and Install	S.F.	1.28	1.49		2.77	random pinhole 5/8" thick square
	Reinstall	S.F.		.83		.83	edge tile.
	Clean	S.F.	.03	.18		.21	
	Minimum Charge	Job		169		169	
2' x 2' Reveal Edge							
	Demolish	S.F.		.45		.45	Includes material and labor to install
	Install	S.F.	2.08	1.35		3.43	T-bar grid acoustical ceiling system,
	Demolish and Install	S.F.	2.08	1.80		3.88	random pinhole 5/8" thick reveal
	Reinstall	S.F.		1.08		1.08	edge tile.
	Clean	S.F.	.03	.18		.21	
	Minimum Charge	Job		169		169	
2' x 2' Fire-rated							
	Demolish	S.F.		.45		.45	Includes material and labor to install
	Install	S.F.	2.46	1.04		3.50	T-bar grid FR acoustical ceiling system,
	Demolish and Install	S.F.	2.46	1.49		3.95	random pinhole 5/8" thick square
	Reinstall	S.F.		.83		.83	edge tile.
	Clean	S.F.	.03	.18		.21	
	Minimum Charge	Job		169		169	
2' x 4' Square Edge							
	Demolish	S.F.		.45		.45	Cost includes material and labor to
	Install	S.F.	1.07	.89		1.96	install 2' x 4' T-bar grid acoustical
	Demolish and Install	S.F.	1.07	1.34		2.41	ceiling system, random pinhole 5/8"
	Reinstall	S.F.		.71		.71	thick square edge tile.
	Clean	S.F.	.03	.18		.21	
	Minimum Charge	Job		169		169	

Acoustical Ceiling		Unit	Material	Labor	Equip.	Total	Specification
2' x 4' Reveal Edge							
	Demolish	S.F.		.45		.45	Includes material and labor to install
	Install	S.F.	2.08	1.23		3.31	T-bar grid acoustical ceiling system,
	Demolish and Install	S.F.	2.08	1.68		3.76	random pinhole 5/8" thick reveal
	Reinstall	S.F.		.98		.98	edge tile.
	Clean	S.F.	.03	.18		.21	
	Minimum Charge	Job		169		169	
2' x 4' Fire-rated							
	Demolish	S.F.		.45		.45	Includes material and labor to install
	Install	S.F.	2.46	1.04		3.50	T-bar grid FR acoustical ceiling system,
	Demolish and Install	S.F.	2.46	1.49		3.95	random pinhole 5/8" thick square
	Reinstall	S.F.		.83		.83	edge tile.
	Clean	S.F.	.03	.18		.21	
	Minimum Charge	Job		169		169	
T-bar Grid							
2' x 2'							
	Demolish	S.F.		.53		.53	Includes material and labor to install
	Install	S.F.	.73	.52		1.25	T-bar grid acoustical ceiling
	Demolish and Install	S.F.	.73	1.05		1.78	suspension system.
	Reinstall	S.F.		.42		.42	
	Clean	S.F.	.03	.18		.21	
	Minimum Charge	Job		150		150	
2' x 4'							
	Demolish	S.F.		.53		.53	Includes material and labor to install
	Install	S.F.	.57	.42		.99	T-bar grid acoustical ceiling
	Demolish and Install	S.F.	.57	.95		1.52	suspension system.
	Reinstall	S.F.		.34		.34	
	Clean	S.F.	.03	.18		.21	
	Minimum Charge	Job		150		150	
Tile Panels							
2' x 2' Square Edge							
	Demolish	S.F.		.15		.15	Includes material and labor to install
	Install	S.F.	.50	.54		1.04	square edge panels 5/8" thick with
	Demolish and Install	S.F.	.50	.69		1.19	pinhole pattern. T-bar grid not
	Reinstall	S.F.		.43		.43	included.
	Clean	S.F.	.03	.18		.21	
	Minimum Charge	Job		169		169	
2' x 2' Reveal Edge							
	Demolish	S.F.		.15		.15	Includes material and labor to install
	Install	S.F.	1.19	.72		1.91	reveal edge panels 5/8" thick with
	Demolish and Install	S.F.	1.19	.87		2.06	pinhole pattern. T-bar grid not
	Reinstall	S.F.		.58		.58	included.
	Clean	S.F.	.03	.18		.21	
	Minimum Charge	Job		169		169	
2' x 2' Fire-rated							
	Demolish	S.F.		.15		.15	Includes material and labor to install
	Install	S.F.	.96	.50		1.46	FR square edge panels 5/8" thick with
	Demolish and Install	S.F.	.96	.65		1.61	pinhole pattern. T-bar grid not
	Reinstall	S.F.		.40		.40	included.
	Clean	S.F.	.03	.18		.21	
	Minimum Charge	Job		169		169	
2' x 4' Square Edge							
	Demolish	S.F.		.15		.15	Includes material and labor to install
	Install	S.F.	.50	.54		1.04	square edge panels 5/8" thick with
	Demolish and Install	S.F.	.50	.69		1.19	pinhole pattern. T-bar grid not
	Reinstall	S.F.		.43		.43	included.
	Clean	S.F.	.03	.18		.21	
	Minimum Charge	Job		169		169	

Walls / Ceilings

Acoustical Ceiling	Unit	Material	Labor	Equip.	Total	Specification
2' x 4' Reveal Edge						
Demolish	S.F.		.15		.15	Includes material and labor to install
Install	S.F.	1.19	.72		1.91	reveal edge panels 5/8" thick with
Demolish and Install	S.F.	1.19	.87		2.06	pinhole pattern. T-bar grid not
Reinstall	S.F.		.58		.58	included.
Clean	S.F.	.03	.18		.21	
Minimum Charge	Job		169		169	
2' x 4' Fire-rated						
Demolish	S.F.		.15		.15	Includes material and labor to install
Install	S.F.	.96	.50		1.46	FR square edge panels 5/8" thick with
Demolish and Install	S.F.	.96	.65		1.61	pinhole pattern. T-bar grid not
Reinstall	S.F.		.40		.40	included.
Clean	S.F.	.03	.18		.21	
Minimum Charge	Job		169		169	
Bleach Out Stains						
Install	S.F.	.09	.24		.33	Includes labor and material to apply
Minimum Charge	Job		169		169	laundry type chlorine bleach to
						remove light staining of acoustic tile
						material.
Adhesive Tile						
Good Grade						
Demolish	S.F.		.60		.60	Cost includes material and labor to
Install	S.F.	.85	1.13		1.98	install 12" x 12" tile with embossed
Demolish and Install	S.F.	.85	1.73		2.58	texture pattern. Furring strips not
Clean	S.F.	.02	.14		.16	included.
Minimum Charge	Job		169		169	
Better Grade						
Demolish	S.F.		.60		.60	Cost includes material and labor to
Install	S.F.	1.13	1.13		2.26	install 12" x 12" tile with embossed
Demolish and Install	S.F.	1.13	1.73		2.86	texture pattern. Furring strips not
Clean	S.F.	.02	.14		.16	included.
Minimum Charge	Job		169		169	
Premium Grade						
Demolish	S.F.		.60		.60	Cost includes material and labor to
Install	S.F.	1.75	1.13		2.88	install 12" x 12" tile with embossed
Demolish and Install	S.F.	1.75	1.73		3.48	texture pattern. Furring strips not
Clean	S.F.	.02	.14		.16	included.
Minimum Charge	Job		169		169	
Bleach Out Stains						
Install	S.F.	.09	.24		.33	Includes labor and material to apply
Minimum Charge	Job		169		169	laundry type chlorine bleach to
						remove light staining of acoustic tile
						material.
Stapled Tile						
Good Grade						
Demolish	S.F.		.36		.36	Cost includes material and labor to
Install	S.F.	.85	1.13		1.98	install 12" x 12" tile with embossed
Demolish and Install	S.F.	.85	1.49		2.34	texture pattern. Furring strips not
Clean	S.F.	.02	.14		.16	included.
Minimum Charge	Job		169		169	
Better Grade						
Demolish	S.F.		.36		.36	Cost includes material and labor to
Install	S.F.	1.13	1.13		2.26	install 12" x 12" tile with embossed
Demolish and Install	S.F.	1.13	1.49		2.62	texture pattern. Furring strips not
Clean	S.F.	.02	.14		.16	included.
Minimum Charge	Job		169		169	

Walls / Ceilings

Acoustical Ceiling		Unit	Material	Labor	Equip.	Total	Specification
Premium Grade							
	Demolish	S.F.		.36		.36	Cost includes material and labor to
	Install	S.F.	1.75	1.13		2.88	install 12″ x 12″ tile with embossed
	Demolish and Install	S.F.	1.75	1.49		3.24	texture pattern. Furring strips not
	Clean	S.F.	.02	.14		.16	included.
	Minimum Charge	Job		169		169	
Bleach Out Stains							
	Install	S.F.	.09	.24		.33	Includes labor and material to apply
	Minimum Charge	Job		169		169	laundry type chlorine bleach to
							remove light staining of acoustic tile
							material.
Blown							
Scrape Off Existing							
	Install	S.F.		.50		.50	Includes labor and material to remove
	Minimum Charge	Job		169		169	existing blown acoustical ceiling
							material.
Seal Ceiling Gypsum Wallboard							
	Install	S.F.	.06	.26		.32	Includes labor and material to seal
	Minimum Charge	Job		148		148	existing drywall with shellac-based
							material.
Reblow Existing							
	Install	S.F.	.06	.37		.43	Includes labor and material to install
	Minimum Charge	Job		400	64.50	464.50	by spray, texture finish.
Bleach Out Stains							
	Install	S.F.	.09	.24		.33	Includes labor and material to apply
	Minimum Charge	Job		169		169	laundry type chlorine bleach to
							remove light staining of acoustic tile
							material.
New Blown							
	Install	S.F.	.06	.37		.43	Includes labor and material to install
	Minimum Charge	Job		148		148	by spray, texture finish.
Apply Glitter To							
	Install	S.F.	.09	.13		.22	Includes labor and material to apply
	Minimum Charge	Job		148		148	glitter material to existing or new
							drywall.
Furring Strips							
	Demolish	S.F.		.07		.07	Cost includes material and labor to
	Install	S.F.	.17	.75		.92	install 1″ x 2″ furring strips, 16″ O.C.
	Demolish and Install	S.F.	.17	.82		.99	per S.F. of surface area to be covered.
	Minimum Charge	Job		84.50		84.50	

Finished Stair

Bi-Fold Door

Raised Panel Door

Hollow Core Door	Unit	Material	Labor	Equip.	Total	Specification
Pre-hung Wood						
1' 6" x 6' 8"						
Demolish	Ea.		31.50		31.50	Cost includes material and labor to
Install	Ea.	146	16.10		162.10	install 1' 6" x 6' 8" pre-hung 1-3/8"
Demolish and Install	Ea.	146	47.60		193.60	lauan door with split pine jamb
Reinstall	Ea.		16.10		16.10	including casing, stops, and hinges.
Clean	Ea.	3.28	10.05		13.33	
Paint	Ea.	8.05	42.50		50.55	
Minimum Charge	Job		169		169	
2' x 6' 8"						
Demolish	Ea.		31.50		31.50	Cost includes material and labor to
Install	Ea.	151	16.90		167.90	install 2' x 6' 8" pre-hung 1-3/8"
Demolish and Install	Ea.	151	48.40		199.40	lauan door with split pine jamb
Reinstall	Ea.		16.90		16.90	including casing, stops, and hinges.
Clean	Ea.	3.28	10.05		13.33	
Paint	Ea.	8.05	42.50		50.55	
Minimum Charge	Job		169		169	
2' 4" x 6' 8"						
Demolish	Ea.		31.50		31.50	Cost includes material and labor to
Install	Ea.	157	16.90		173.90	install 2' 4" x 6' 8" pre-hung 1-3/8"
Demolish and Install	Ea.	157	48.40		205.40	lauan door with split pine jamb
Reinstall	Ea.		16.90		16.90	including casing, stops, and hinges.
Clean	Ea.	3.28	10.05		13.33	
Paint	Ea.	8.05	42.50		50.55	
Minimum Charge	Job		169		169	
2' 6" x 6' 8"						
Demolish	Ea.		31.50		31.50	Cost includes material and labor to
Install	Ea.	157	16.90		173.90	install 2' 6" x 6' 8" pre-hung 1-3/8"
Demolish and Install	Ea.	157	48.40		205.40	lauan door with split pine jamb
Reinstall	Ea.		16.90		16.90	including casing, stops, and hinges.
Clean	Ea.	3.28	10.05		13.33	
Paint	Ea.	8.05	42.50		50.55	
Minimum Charge	Job		169		169	
2' 8" x 6' 8"						
Demolish	Ea.		31.50		31.50	Cost includes material and labor to
Install	Ea.	163	16.90		179.90	install 2' 8" x 6' 8" pre-hung 1-3/8"
Demolish and Install	Ea.	163	48.40		211.40	lauan door with split pine jamb,
Reinstall	Ea.		16.90		16.90	including casing, stop and hinges.
Clean	Ea.	3.28	10.05		13.33	
Paint	Ea.	8.05	42.50		50.55	
Minimum Charge	Job		169		169	

Finish Carpentry

Hollow Core Door

	Unit	Material	Labor	Equip.	Total	Specification
3' x 6' 8"						
Demolish	Ea.		31.50		31.50	Cost includes material and labor to
Install	Ea.	168	17.80		185.80	install 3' x 6' 8" pre-hung 1-3/8"
Demolish and Install	Ea.	168	49.30		217.30	lauan door with split pine jamb
Reinstall	Ea.		17.79		17.79	including casing, stops, and hinges.
Clean	Ea.	3.28	10.05		13.33	
Paint	Ea.	8.05	42.50		50.55	
Minimum Charge	Job		169		169	
Pocket Type						
Demolish	Ea.		47		47	Cost includes material and labor to
Install	Ea.	35	37.50		72.50	install 2' 8" x 6' 8" lauan door.
Demolish and Install	Ea.	35	84.50		119.50	
Reinstall	Ea.		37.56		37.56	
Clean	Ea.	3.28	10.05		13.33	
Paint	Ea.	8.05	42.50		50.55	
Minimum Charge	Job		169		169	
Stain						
Install	Ea.	4.22	24.50		28.72	Includes labor and material to stain
Minimum Charge	Job		148		148	single door and trim on both sides.
Pre-hung Masonite						
1' 6" x 6' 8"						
Demolish	Ea.		31.50		31.50	Cost includes material and labor to
Install	Ea.	196	16.10		212.10	install 1' 6" x 6' 8" pre-hung 1-3/8"
Demolish and Install	Ea.	196	47.60		243.60	masonite door with split pine jamb
Reinstall	Ea.		16.10		16.10	including casing, stops, and hinges.
Clean	Ea.	3.28	10.05		13.33	
Paint	Ea.	8.05	42.50		50.55	
Minimum Charge	Job		169		169	
2' x 6' 8"						
Demolish	Ea.		31.50		31.50	Cost includes material and labor to
Install	Ea.	205	16.90		221.90	install 2' x 6' 8" pre-hung 1-3/8"
Demolish and Install	Ea.	205	48.40		253.40	masonite door with split pine jamb
Reinstall	Ea.		16.90		16.90	including casing, stops, and hinges.
Clean	Ea.	3.28	10.05		13.33	
Paint	Ea.	8.05	42.50		50.55	
Minimum Charge	Job		169		169	
2' 4" x 6' 8"						
Demolish	Ea.		31.50		31.50	Cost includes material and labor to
Install	Ea.	207	16.90		223.90	install 2' 4" x 6' 8" pre-hung 1-3/8"
Demolish and Install	Ea.	207	48.40		255.40	masonite door with split pine jamb
Reinstall	Ea.		16.90		16.90	including casing, stops, and hinges.
Clean	Ea.	3.28	10.05		13.33	
Paint	Ea.	8.05	42.50		50.55	
Minimum Charge	Job		169		169	
2' 8" x 6' 8"						
Demolish	Ea.		31.50		31.50	Cost includes material and labor to
Install	Ea.	209	16.90		225.90	install 2' 8" x 6' 8" pre-hung 1-3/8"
Demolish and Install	Ea.	209	48.40		257.40	masonite door with split pine jamb
Reinstall	Ea.		16.90		16.90	including casing, stops, and hinges.
Clean	Ea.	3.28	10.05		13.33	
Paint	Ea.	8.05	42.50		50.55	
Minimum Charge	Job		169		169	
3' x 6' 8"						
Demolish	Ea.		31.50		31.50	Cost includes material and labor to
Install	Ea.	213	17.80		230.80	install 3' x 6' 8" pre-hung 1-3/8"
Demolish and Install	Ea.	213	49.30		262.30	masonite door with split pine jamb
Reinstall	Ea.		17.79		17.79	including casing, stops, and hinges.
Clean	Ea.	3.28	10.05		13.33	
Paint	Ea.	8.05	42.50		50.55	
Minimum Charge	Job		169		169	

Finish Carpentry

Hollow Core Door

Hollow Core Door	Unit	Material	Labor	Equip.	Total	Specification
Pocket Type						
Demolish	Ea.		47		47	Cost includes material and labor to
Install	Ea.	35	37.50		72.50	install 2' 8" x 6' 8" lauan door.
Demolish and Install	Ea.	35	84.50		119.50	
Reinstall	Ea.		37.56		37.56	
Clean	Ea.	3.28	10.05		13.33	
Paint	Ea.	8.05	42.50		50.55	
Minimum Charge	Job		169		169	
Stain						
Install	Ea.	4.22	24.50		28.72	Includes labor and material to stain
Minimum Charge	Job		148		148	single door and trim on both sides.
Hollow Core Door Only						
1' 6" x 6' 8'						
Demolish	Ea.		6.70		6.70	Cost includes material and labor to
Install	Ea.	38	17.80		55.80	install 1' 6" x 6' 8" lauan hollow core
Demolish and Install	Ea.	38	24.50		62.50	door slab.
Reinstall	Ea.		17.79		17.79	
Clean	Ea.	1.36	7.55		8.91	
Paint	Ea.	2.41	37		39.41	
Minimum Charge	Job		169		169	
2' x 6' 8'						
Demolish	Ea.		6.70		6.70	Cost includes material and labor to
Install	Ea.	37	18.80		55.80	install 2' x 6' 8" lauan hollow core
Demolish and Install	Ea.	37	25.50		62.50	door slab.
Reinstall	Ea.		18.78		18.78	
Clean	Ea.	1.36	7.55		8.91	
Paint	Ea.	2.41	37		39.41	
Minimum Charge	Job		169		169	
2' 4" x 6' 8'						
Demolish	Ea.		6.70		6.70	Cost includes material and labor to
Install	Ea.	40	18.80		58.80	install 2' 4" x 6' 8" lauan hollow core
Demolish and Install	Ea.	40	25.50		65.50	door slab.
Reinstall	Ea.		18.78		18.78	
Clean	Ea.	1.36	7.55		8.91	
Paint	Ea.	2.41	37		39.41	
Minimum Charge	Job		169		169	
2' 8" x 6' 8'						
Demolish	Ea.		6.70		6.70	Cost includes material and labor to
Install	Ea.	42.50	18.80		61.30	install 2' 8" x 6' 8" lauan hollow core
Demolish and Install	Ea.	42.50	25.50		68	door slab.
Reinstall	Ea.		18.78		18.78	
Clean	Ea.	1.36	7.55		8.91	
Paint	Ea.	2.41	37		39.41	
Minimum Charge	Job		169		169	
3' x 6' 8'						
Demolish	Ea.		6.70		6.70	Cost includes material and labor to
Install	Ea.	44	19.90		63.90	install 3' x 6' 8" lauan hollow core
Demolish and Install	Ea.	44	26.60		70.60	door slab.
Reinstall	Ea.		19.88		19.88	
Clean	Ea.	1.36	7.55		8.91	
Paint	Ea.	2.41	37		39.41	
Minimum Charge	Job		169		169	
Pocket Type						
Demolish	Ea.		6.70		6.70	Cost includes material and labor to
Install	Ea.	35	37.50		72.50	install 2' 8" x 6' 8" lauan door.
Demolish and Install	Ea.	35	44.20		79.20	
Reinstall	Ea.		37.56		37.56	
Clean	Ea.	1.36	7.55		8.91	
Paint	Ea.	2.41	37		39.41	
Minimum Charge	Job		169		169	

Finish Carpentry

Hollow Core Door

	Unit	Material	Labor	Equip.	Total	Specification
Stain						
Install	Ea.	2.33	33		35.33	Includes labor and materials to stain a flush style door on both sides by brush.
Minimum Charge	Job		148		148	
Casing Trim						
Single Width						
Demolish	Ea.		5.95		5.95	Cost includes material and labor to install 11/16" x 2-1/2" pine ranch style casing for one side of a standard door opening.
Install	Opng.	24	57.50		81.50	
Demolish and Install	Opng.	24	63.45		87.45	
Reinstall	Opng.		45.83		45.83	
Clean	Opng.	.46	2.42		2.88	
Paint	Ea.	3.15	5.95		9.10	
Minimum Charge	Job		169		169	
Double Width						
Demolish	Ea.		7.10		7.10	Cost includes material and labor to install 11/16" x 2-1/2" pine ranch style door casing for one side of a double door opening.
Install	Opng.	28	67.50		95.50	
Demolish and Install	Opng.	28	74.60		102.60	
Reinstall	Opng.		54.08		54.08	
Clean	Opng.	.55	4.83		5.38	
Paint	Ea.	3.15	6.75		9.90	
Minimum Charge	Job		169		169	
Hardware						
Doorknob w / Lock						
Demolish	Ea.		12.50		12.50	Includes material and labor to install privacy lockset.
Install	Ea.	21	21		42	
Demolish and Install	Ea.	21	33.50		54.50	
Reinstall	Ea.		16.90		16.90	
Clean	Ea.	.33	7.55		7.88	
Minimum Charge	Job		169		169	
Doorknob						
Demolish	Ea.		12.50		12.50	Includes material and labor to install residential passage lockset, keyless bored type.
Install	Ea.	18.20	21		39.20	
Demolish and Install	Ea.	18.20	33.50		51.70	
Reinstall	Ea.		16.90		16.90	
Clean	Ea.	.33	7.55		7.88	
Minimum Charge	Job		169		169	
Deadbolt						
Demolish	Ea.		11.75		11.75	Includes material and labor to install a deadbolt.
Install	Ea.	38.50	24		62.50	
Demolish and Install	Ea.	38.50	35.75		74.25	
Reinstall	Ea.		19.31		19.31	
Clean	Ea.	.33	7.55		7.88	
Minimum Charge	Job		169		169	
Lever Handle						
Demolish	Ea.		12.50		12.50	Includes material and labor to install residential passage lockset, keyless bored type with lever handle.
Install	Ea.	28.50	34		62.50	
Demolish and Install	Ea.	28.50	46.50		75	
Reinstall	Ea.		27.04		27.04	
Clean	Ea.	.33	7.55		7.88	
Minimum Charge	Job		169		169	
Closer						
Demolish	Ea.		7.05		7.05	Includes material and labor to install pneumatic light duty closer for interior type doors.
Install	Ea.	95.50	52		147.50	
Demolish and Install	Ea.	95.50	59.05		154.55	
Reinstall	Ea.		41.60		41.60	
Clean	Ea.	.17	10.05		10.22	
Minimum Charge	Job		169		169	

Hollow Core Door		Unit	Material	Labor	Equip.	Total	Specification
Push Plate							
	Demolish	Ea.		10.45		10.45	Includes material and labor to install
	Install	Ea.	19.10	28		47.10	bronze push-pull plate.
	Demolish and Install	Ea.	19.10	38.45		57.55	
	Reinstall	Ea.		22.53		22.53	
	Clean	Ea.	.33	7.55		7.88	
	Minimum Charge	Job		169		169	
Kickplate							
	Demolish	Ea.		10.45		10.45	Includes material and labor to install
	Install	Ea.	22	22.50		44.50	aluminum kickplate, 10" x 28".
	Demolish and Install	Ea.	22	32.95		54.95	
	Reinstall	Ea.		18.03		18.03	
	Clean	Ea.	.33	7.55		7.88	
	Minimum Charge	Job		169		169	
Exit Sign							
	Demolish	Ea.		18.55		18.55	Includes material and labor to install a
	Install	Ea.	50.50	46		96.50	wall mounted interior electric exit sign.
	Demolish and Install	Ea.	50.50	64.55		115.05	
	Reinstall	Ea.		36.72		36.72	
	Clean	Ea.	.02	5.45		5.47	
	Minimum Charge	Job		169		169	
Jamb							
	Demolish	Ea.		23.50		23.50	Includes material and labor to install
	Install	Ea.	98.50	9.15		107.65	flat pine jamb with square cut heads
	Demolish and Install	Ea.	98.50	32.65		131.15	and rabbeted sides for 6' 8" high and
	Clean	Ea.	.10	6.05		6.15	3-9/16" door including trim sets for
	Paint	Ea.	5.70	19.75		25.45	both sides.
	Minimum Charge	Job		169		169	
Shave & Refit							
	Install	Ea.		28		28	Includes labor to shave and rework
	Minimum Charge	Job		169		169	door to fit opening at the job site.

Solid Core Door		Unit	Material	Labor	Equip.	Total	Specification
Pre-hung Wood							
2' 4" x 6' 8"							
	Demolish	Ea.		31.50		31.50	Cost includes material and labor to
	Install	Ea.	190	16.90		206.90	install 2' 4" x 6' 8" pre-hung 1-3/4"
	Demolish and Install	Ea.	190	48.40		238.40	door, 4-1/2" split jamb, including
	Reinstall	Ea.		16.90		16.90	casing and stop, hinges, aluminum sill,
	Clean	Ea.	3.28	10.05		13.33	weatherstripping.
	Paint	Ea.	9.30	42.50		51.80	
	Minimum Charge	Job		169		169	
2' 6" x 6' 8"							
	Demolish	Ea.		31.50		31.50	Cost includes material and labor to
	Install	Ea.	199	16.90		215.90	install 2' 6" x 6' 8" pre-hung 1-3/4"
	Demolish and Install	Ea.	199	48.40		247.40	door, 4-1/2" split jamb, including
	Reinstall	Ea.		16.90		16.90	casing and stop, hinges, aluminum sill,
	Clean	Ea.	3.28	10.05		13.33	weatherstripping.
	Paint	Ea.	9.30	42.50		51.80	
	Minimum Charge	Job		169		169	

Finish Carpentry

Solid Core Door		Unit	Material	Labor	Equip.	Total	Specification
2′ 8″ x 6′ 8″							
	Demolish	Ea.		31.50		31.50	Cost includes material and labor to
	Install	Ea.	211	16.90		227.90	install 2′ 8″ x 6′ 8″ pre-hung 1-3/4″
	Demolish and Install	Ea.	211	48.40		259.40	door, 4-1/2″ split jamb, including
	Reinstall	Ea.		16.90		16.90	casing and stop, hinges, aluminum sill,
	Clean	Ea.	3.28	10.05		13.33	weatherstripping.
	Paint	Ea.	9.30	42.50		51.80	
	Minimum Charge	Job		169		169	
3′ x 6′ 8″							
	Demolish	Ea.		31.50		31.50	Cost includes material and labor to
	Install	Ea.	219	17.80		236.80	install 3′ x 6′ 8″ pre-hung 1-3/4″
	Demolish and Install	Ea.	219	49.30		268.30	door, including casing and stop,
	Reinstall	Ea.		17.79		17.79	hinges, jamb, aluminum sill,
	Clean	Ea.	3.28	10.05		13.33	weatherstripped.
	Paint	Ea.	9.30	42.50		51.80	
	Minimum Charge	Job		169		169	
3′ 6″ x 6′ 8″							
	Demolish	Ea.		31.50		31.50	Cost includes material and labor to
	Install	Ea.	370	18.80		388.80	install 3′ 6″ x 6′ 8″ pre-hung 1-3/4″
	Demolish and Install	Ea.	370	50.30		420.30	door, 4-1/2″ split jamb, including
	Reinstall	Ea.		18.78		18.78	casing and stop, hinges, aluminum sill,
	Clean	Ea.	3.28	10.05		13.33	weatherstripping.
	Paint	Ea.	9.30	42.50		51.80	
	Minimum Charge	Job		169		169	
Raised Panel							
	Demolish	Ea.		31.50		31.50	Cost includes material and labor to
	Install	Ea.	390	18.80		408.80	install 3′ x 6′ 8″ pre-hung 1-3/4″
	Demolish and Install	Ea.	390	50.30		440.30	door, jamb, including casing and
	Reinstall	Ea.		18.78		18.78	stop, hinges, aluminum sill,
	Clean	Ea.	3.28	10.05		13.33	weatherstripping.
	Paint	Ea.	9.30	99		108.30	
	Minimum Charge	Job		169		169	
Stain							
	Install	Ea.	4.22	24.50		28.72	Includes labor and material to stain
	Minimum Charge	Job		148		148	single door and trim on both sides.

Bifold Door		Unit	Material	Labor	Equip.	Total	Specification
Wood, Single							
3′							
	Demolish	Ea.		23.50		23.50	Cost includes material and labor to
	Install	Ea.	58	52		110	install 3′ wide solid wood bi-fold door
	Demolish and Install	Ea.	58	75.50		133.50	not including jamb, trim, track, and
	Reinstall	Ea.		52		52	hardware.
	Clean	Ea.	3.28	10.05		13.33	
	Paint	Ea.	8.05	42.50		50.55	
	Minimum Charge	Job		169		169	
5′							
	Demolish	Ea.		23.50		23.50	Includes material and labor to install
	Install	Ea.	107	61.50		168.50	pair of 2′ 6″ wide solid wood bi-fold
	Demolish and Install	Ea.	107	85		192	doors not including jamb, trim, track,
	Reinstall	Ea.		61.45		61.45	and hardware.
	Clean	Ea.	1.43	17.25		18.68	
	Paint	Ea.	16.70	74		90.70	
	Minimum Charge	Job		169		169	

Finish Carpentry

Bifold Door		Unit	Material	Labor	Equip.	Total	Specification
6'							
	Demolish	Ea.		23.50		23.50	Includes material and labor to install
	Install	Ea.	113	67.50		180.50	pair of 3' wide solid wood bi-fold
	Demolish and Install	Ea.	113	91		204	doors not including jamb, trim, track,
	Reinstall	Ea.		67.60		67.60	and hardware.
	Clean	Ea.	1.64	20		21.64	
	Paint	Ea.	16.70	74		90.70	
	Minimum Charge	Job		169		169	
8'							
	Demolish	Ea.		23.50		23.50	Includes material and labor to install
	Install	Ea.	211	37.50		248.50	pair of 4' wide solid wood bi-fold
	Demolish and Install	Ea.	211	61		272	doors not including jamb, trim, track,
	Reinstall	Ea.		37.56		37.56	and hardware.
	Clean	Ea.	2.26	24		26.26	
	Paint	Ea.	8.70	74		82.70	
	Minimum Charge	Job		169		169	
Mirrored							
5'							
	Demolish	Ea.		23.50		23.50	Includes material and labor to install
	Install	Ea.	395	30.50		425.50	pair of 2' 6" wide mirrored bi-fold
	Demolish and Install	Ea.	395	54		449	doors including jamb, trim, track, and
	Reinstall	Ea.		30.73		30.73	hardware.
	Clean	Ea.	1.43	8.65		10.08	
	Minimum Charge	Job		169		169	
6'							
	Demolish	Ea.		23.50		23.50	Includes material and labor to install
	Install	Ea.	435	37.50		472.50	pair of 3' wide mirrored bi-fold doors
	Demolish and Install	Ea.	435	61		496	including jamb, trim, track, and
	Reinstall	Ea.		37.56		37.56	hardware.
	Clean	Ea.	1.64	10.05		11.69	
	Minimum Charge	Job		169		169	
8'							
	Demolish	Ea.		23.50		23.50	Includes material and labor to install
	Install	Ea.	705	56.50		761.50	pair of 4' wide mirrored bi-fold doors
	Demolish and Install	Ea.	705	80		785	including jamb, trim, track, and
	Reinstall	Ea.		56.33		56.33	hardware.
	Clean	Ea.	2.26	12.10		14.36	
	Minimum Charge	Job		169		169	
Louvered							
3'							
	Demolish	Ea.		23.50		23.50	Cost includes material and labor to
	Install	Ea.	173	52		225	install 3' wide louvered bi-fold door
	Demolish and Install	Ea.	173	75.50		248.50	not including jamb, trim, track, and
	Reinstall	Ea.		52		52	hardware.
	Clean	Ea.	3.28	10.05		13.33	
	Paint	Ea.	8.05	42.50		50.55	
	Minimum Charge	Job		169		169	
5'							
	Demolish	Ea.		23.50		23.50	Includes material and labor to install
	Install	Ea.	242	61.50		303.50	pair of 2' 6" wide louvered wood
	Demolish and Install	Ea.	242	85		327	bi-fold doors not including jamb, trim,
	Reinstall	Ea.		61.45		61.45	track, and hardware.
	Clean	Ea.	1.43	17.25		18.68	
	Paint	Ea.	16.70	74		90.70	
	Minimum Charge	Job		169		169	

Finish Carpentry

Bifold Door	Unit	Material	Labor	Equip.	Total	Specification
6'						
Demolish	Ea.		23.50		23.50	Includes material and labor to install
Install	Ea.	267	67.50		334.50	pair of 3' wide louvered wood bi-fold
Demolish and Install	Ea.	267	91		358	doors not including jamb, trim, track,
Reinstall	Ea.		67.60		67.60	and hardware.
Clean	Ea.	1.64	20		21.64	
Paint	Ea.	16.70	74		90.70	
Minimum Charge	Job		169		169	
8'						
Demolish	Ea.		23.50		23.50	Includes material and labor to install
Install	Ea.	425	34		459	pair of 4' wide solid wood bi-fold
Demolish and Install	Ea.	425	57.50		482.50	doors not including jamb, trim, track,
Reinstall	Ea.		33.80		33.80	and hardware.
Clean	Ea.	7.80	17.25		25.05	
Paint	Ea.	8.70	74		82.70	
Minimum Charge	Job		169		169	
Accordion Type						
Demolish	S.F.		.79		.79	Includes material and labor to install
Install	S.F.	1.91	1.69		3.60	vinyl folding accordion closet doors,
Demolish and Install	S.F.	1.91	2.48		4.39	including track and frame.
Reinstall	S.F.		1.69		1.69	
Clean	S.F.		.29		.29	
Minimum Charge	Job		169		169	
Casing Trim						
Single Width						
Demolish	Ea.		5.95		5.95	Cost includes material and labor to
Install	Opng.	24	57.50		81.50	install 11/16" x 2-1/2" pine ranch
Demolish and Install	Opng.	24	63.45		87.45	style casing for one side of a standard
Reinstall	Opng.		45.83		45.83	door opening.
Clean	Opng.	.46	2.42		2.88	
Paint	Ea.	3.15	5.95		9.10	
Minimum Charge	Job		169		169	
Double Width						
Demolish	Ea.		7.10		7.10	Cost includes material and labor to
Install	Opng.	28	67.50		95.50	install 11/16" x 2-1/2" pine ranch
Demolish and Install	Opng.	28	74.60		102.60	style door casing for one side of a
Reinstall	Opng.		54.08		54.08	double door opening.
Clean	Opng.	.55	4.83		5.38	
Paint	Ea.	3.15	6.75		9.90	
Minimum Charge	Job		169		169	
Jamb						
Demolish	Ea.		23.50		23.50	Includes material and labor to install
Install	Ea.	98.50	9.15		107.65	flat pine jamb with square cut heads
Demolish and Install	Ea.	98.50	32.65		131.15	and rabbeted sides for 6' 8" high and
Clean	Ea.	.10	6.05		6.15	3-9/16" door including trim sets for
Paint	Ea.	5.70	19.75		25.45	both sides.
Minimum Charge	Job		84.50		84.50	
Stain						
Install	Ea.	4.22	24.50		28.72	Includes labor and material to stain
Minimum Charge	Job		148		148	single door and trim on both sides.
Shave & Refit						
Install	Ea.		28		28	Includes labor to shave and rework
Minimum Charge	Job		84.50		84.50	door to fit opening at the job site.

Finish Carpentry

Louver Door

	Unit	Material	Labor	Equip.	Total	Specification
Full						
2' x 6' 8"						
Demolish	Ea.		6.70		6.70	Cost includes material and labor to
Install	Ea.	184	34		218	install 2' x 6' 8" pine full louver door,
Demolish and Install	Ea.	184	40.70		224.70	jamb, hinges.
Reinstall	Ea.		33.80		33.80	
Clean	Ea.	6.05	30		36.05	
Paint	Ea.	8.05	42.50		50.55	
Minimum Charge	Job		169		169	
2' 8" x 6' 8"						
Demolish	Ea.		6.70		6.70	Cost includes material and labor to
Install	Ea.	205	34		239	install 2' 8" x 6' 8" pine full louver
Demolish and Install	Ea.	205	40.70		245.70	door, jamb, hinges.
Reinstall	Ea.		33.80		33.80	
Clean	Ea.	6.05	30		36.05	
Paint	Ea.	8.05	42.50		50.55	
Minimum Charge	Job		169		169	
3' x 6' 8"						
Demolish	Ea.		6.70		6.70	Cost includes material and labor to
Install	Ea.	216	35.50		251.50	install 3' x 6' 8" pine full louver door,
Demolish and Install	Ea.	216	42.20		258.20	jamb, hinges.
Reinstall	Ea.		35.58		35.58	
Clean	Ea.	6.05	30		36.05	
Paint	Ea.	8.05	42.50		50.55	
Minimum Charge	Job		169		169	
Half						
2' x 6' 8"						
Demolish	Ea.		6.70		6.70	Cost includes material and labor to
Install	Ea.	296	34		330	install 2' x 6' 8" pine half louver door,
Demolish and Install	Ea.	296	40.70		336.70	jamb, hinges.
Reinstall	Ea.		33.80		33.80	
Clean	Ea.	6.05	30		36.05	
Paint	Ea.	8.05	42.50		50.55	
Minimum Charge	Job		169		169	
2' 8" x 6' 8"						
Demolish	Ea.		6.70		6.70	Cost includes material and labor to
Install	Ea.	320	34		354	install 2' 8" x 6' 8" pine half louver
Demolish and Install	Ea.	320	40.70		360.70	door, jamb, hinges.
Reinstall	Ea.		33.80		33.80	
Clean	Ea.	6.05	30		36.05	
Paint	Ea.	8.05	42.50		50.55	
Minimum Charge	Job		169		169	
3' x 6' 8"						
Demolish	Ea.		6.70		6.70	Cost includes material and labor to
Install	Ea.	335	35.50		370.50	install 3' x 6' 8" pine half louver door,
Demolish and Install	Ea.	335	42.20		377.20	jamb, hinges.
Reinstall	Ea.		35.58		35.58	
Clean	Ea.	6.05	30		36.05	
Paint	Ea.	8.05	42.50		50.55	
Minimum Charge	Job		169		169	
Jamb						
Demolish	Ea.		23.50		23.50	Includes material and labor to install
Install	Ea.	98.50	9.15		107.65	flat pine jamb with square cut heads
Demolish and Install	Ea.	98.50	32.65		131.15	and rabbeted sides for 6' 8" high and
Clean	Ea.	.10	6.05		6.15	3-9/16" door including trim sets for
Paint	Ea.	5.70	19.75		25.45	both sides.
Minimum Charge	Job		84.50		84.50	

Finish Carpentry

Louver Door		Unit	Material	Labor	Equip.	Total	Specification
Stain							
	Install	Ea.	4.22	24.50		28.72	Includes labor and material to stain
	Minimum Charge	Job		148		148	single door and trim on both sides.
Shave & Refit							
	Install	Ea.		28		28	Includes labor to shave and rework
	Minimum Charge	Job		148		148	door to fit opening at the job site.

Bypass Sliding Door		Unit	Material	Labor	Equip.	Total	Specification
Wood							
5'							
	Demolish	Ea.		31.50		31.50	Includes material and labor to install
	Install	Opng.	198	61.50		259.50	pair of 2' 6" wide lauan hollow core
	Demolish and Install	Opng.	198	93		291	wood bi-pass doors with jamb, track
	Reinstall	Opng.		61.45		61.45	and hardware.
	Clean	Ea.	1.43	17.25		18.68	
	Paint	Ea.	16.70	74		90.70	
	Minimum Charge	Job		169		169	
6'							
	Demolish	Ea.		31.50		31.50	Includes material and labor to install
	Install	Opng.	212	67.50		279.50	pair of 3' wide lauan hollow core
	Demolish and Install	Opng.	212	99		311	wood bi-pass doors with jamb, track
	Reinstall	Opng.		67.60		67.60	and hardware.
	Clean	Ea.	1.64	20		21.64	
	Paint	Ea.	16.70	74		90.70	
	Minimum Charge	Job		169		169	
8'							
	Demolish	Ea.		31.50		31.50	Includes material and labor to install
	Install	Opng.	380	84.50		464.50	pair of 4' wide lauan hollow core
	Demolish and Install	Opng.	380	116		496	wood bi-pass doors with jamb, track
	Reinstall	Opng.		84.50		84.50	and hardware.
	Clean	Ea.	2.26	24		26.26	
	Paint	Ea.	8.70	74		82.70	
	Minimum Charge	Job		169		169	
Mirrored							
5'							
	Demolish	Ea.		31.50		31.50	Includes material and labor to install
	Install	Opng.	223	67.50		290.50	pair of 2' 6" wide mirrored bi-pass
	Demolish and Install	Opng.	223	99		322	doors with jamb, trim, track and
	Reinstall	Opng.		67.60		67.60	hardware.
	Clean	Ea.	1.43	8.65		10.08	
	Minimum Charge	Job		169		169	
6'							
	Demolish	Ea.		31.50		31.50	Includes material and labor to install
	Install	Opng.	254	67.50		321.50	pair of 3' wide mirrored bi-pass doors
	Demolish and Install	Opng.	254	99		353	with jamb, trim, track and hardware.
	Reinstall	Opng.		67.60		67.60	
	Clean	Ea.	1.64	10.05		11.69	
	Minimum Charge	Job		169		169	
8'							
	Demolish	Ea.		31.50		31.50	Includes material and labor to install
	Install	Opng.	535	75		610	pair of 4' wide mirrored bi-pass doors
	Demolish and Install	Opng.	535	106.50		641.50	with jamb, trim, track and hardware.
	Reinstall	Opng.		75.11		75.11	
	Clean	Ea.	2.26	12.10		14.36	
	Minimum Charge	Job		169		169	

Finish Carpentry

Bypass Sliding Door

	Unit	Material	Labor	Equip.	Total	Specification
Casing Trim						
Single Width						
Demolish	Ea.		5.95		5.95	Cost includes material and labor to
Install	Opng.	24	57.50		81.50	install 11/16" x 2-1/2" pine ranch
Demolish and Install	Opng.	24	63.45		87.45	style casing for one side of a standard
Reinstall	Opng.		45.83		45.83	door opening.
Clean	Opng.	.46	2.42		2.88	
Paint	Ea.	3.15	5.95		9.10	
Minimum Charge	Job		169		169	
Double Width						
Demolish	Ea.		7.10		7.10	Cost includes material and labor to
Install	Opng.	28	67.50		95.50	install 11/16" x 2-1/2" pine ranch
Demolish and Install	Opng.	28	74.60		102.60	style door casing for one side of a
Reinstall	Opng.		54.08		54.08	double door opening.
Clean	Opng.	.55	4.83		5.38	
Paint	Ea.	3.15	6.75		9.90	
Minimum Charge	Job		169		169	
Jamb						
Demolish	Ea.		23.50		23.50	Includes material and labor to install
Install	Ea.	98.50	9.15		107.65	flat pine jamb with square cut heads
Demolish and Install	Ea.	98.50	32.65		131.15	and rabbeted sides for 6' 8" high and
Clean	Ea.	.10	6.05		6.15	3-9/16" door including trim sets for
Paint	Ea.	5.70	19.75		25.45	both sides.
Minimum Charge	Job		84.50		84.50	
Stain						
Install	Ea.	4.22	24.50		28.72	Includes labor and material to stain
Minimum Charge	Job		148		148	single door and trim on both sides.
Shave & Refit						
Install	Ea.		28		28	Includes labor to shave and rework
Minimum Charge	Job		84.50		84.50	door to fit opening at the job site.

Base Molding

	Unit	Material	Labor	Equip.	Total	Specification
One Piece						
1-5/8"						
Demolish	L.F.		.45		.45	Includes material and labor to install
Install	L.F.	.39	1		1.39	pine or birch base molding.
Demolish and Install	L.F.	.39	1.45		1.84	
Reinstall	L.F.		.80		.80	
Clean	L.F.	.02	.14		.16	
Paint	L.F.	.03	.32		.35	
Minimum Charge	Job		169		169	
2-1/2"						
Demolish	L.F.		.45		.45	Cost includes material and labor to
Install	L.F.	.59	1.04		1.63	install 1/2" x 2-1/2" pine or birch
Demolish and Install	L.F.	.59	1.49		2.08	base molding.
Reinstall	L.F.		.83		.83	
Clean	L.F.	.02	.14		.16	
Paint	L.F.	.03	.32		.35	
Minimum Charge	Job		169		169	

Finish Carpentry

Base Molding

	Unit	Material	Labor	Equip.	Total	Specification
3-1/2"						
Demolish	L.F.		.45		.45	Cost includes material and labor to
Install	L.F.	1.70	1.08		2.78	install 9/16" x 3-1/2" pine or birch
Demolish and Install	L.F.	1.70	1.53		3.23	base molding.
Reinstall	L.F.		.87		.87	
Clean	L.F.	.02	.14		.16	
Paint	L.F.	.03	.35		.38	
Minimum Charge	Job		169		169	
6"						
Demolish	L.F.		.45		.45	Cost includes material and labor to
Install	L.F.	1.16	1.18		2.34	install 1" x 6" pine or birch base
Demolish and Install	L.F.	1.16	1.63		2.79	molding.
Reinstall	L.F.		.95		.95	
Clean	L.F.	.03	.20		.23	
Paint	L.F.	.06	.45		.51	
Minimum Charge	Job		169		169	
8"						
Demolish	L.F.		.54		.54	Cost includes material and labor to
Install	L.F.	1.46	1.30		2.76	install 1" x 8" pine or birch base
Demolish and Install	L.F.	1.46	1.84		3.30	molding.
Reinstall	L.F.		1.04		1.04	
Clean	L.F.	.03	.24		.27	
Paint	L.F.	.06	.49		.55	
Minimum Charge	Job		169		169	
Premium / Custom Grade						
Demolish	L.F.		.54		.54	Includes material and labor to install
Install	L.F.	4.20	3.25		7.45	custom 2 or 3 piece oak or other
Demolish and Install	L.F.	4.20	3.79		7.99	hardwood, stain grade trim.
Reinstall	L.F.		2.60		2.60	
Clean	L.F.	.03	.24		.27	
Paint	L.F.	.06	.49		.55	
Minimum Charge	Job		169		169	

One Piece Oak

	Unit	Material	Labor	Equip.	Total	Specification
1-5/8"						
Demolish	L.F.		.45		.45	Cost includes material and labor to
Install	L.F.	.70	1		1.70	install 1/2" x 1-5/8" oak base
Demolish and Install	L.F.	.70	1.45		2.15	molding.
Reinstall	L.F.		.80		.80	
Clean	L.F.	.02	.14		.16	
Paint	L.F.	.08	.44		.52	
Minimum Charge	Job		169		169	
2-1/2"						
Demolish	L.F.		.45		.45	Cost includes material and labor to
Install	L.F.	1.07	1.02		2.09	install 1/2" x 2-1/2" oak base
Demolish and Install	L.F.	1.07	1.47		2.54	molding.
Reinstall	L.F.		.82		.82	
Clean	L.F.	.02	.14		.16	
Paint	L.F.	.08	.45		.53	
Minimum Charge	Job		169		169	
3-1/2"						
Demolish	L.F.		.45		.45	Cost includes material and labor to
Install	L.F.	2.62	1.04		3.66	install 1/2" x 3-1/2" oak base
Demolish and Install	L.F.	2.62	1.49		4.11	molding.
Reinstall	L.F.		.83		.83	
Clean	L.F.	.02	.14		.16	
Paint	L.F.	.09	.46		.55	
Minimum Charge	Job		169		169	

Finish Carpentry

Base Molding

Base Molding		Unit	Material	Labor	Equip.	Total	Specification
6"							
	Demolish	L.F.		.45		.45	Cost includes material and labor to
	Install	L.F.	5.55	1.18		6.73	install 1" x 6" oak base molding.
	Demolish and Install	L.F.	5.55	1.63		7.18	
	Reinstall	L.F.		.95		.95	
	Clean	L.F.	.03	.24		.27	
	Paint	L.F.	.09	.47		.56	
	Minimum Charge	Job		169		169	
8"							
	Demolish	L.F.		.54		.54	Cost includes material and labor to
	Install	L.F.	6.45	1.30		7.75	install 1" x 8" oak base molding.
	Demolish and Install	L.F.	6.45	1.84		8.29	
	Reinstall	L.F.		1.04		1.04	
	Clean	L.F.	.03	.24		.27	
	Paint	L.F.	.09	.49		.58	
	Minimum Charge	Job		169		169	

Vinyl Base Molding

		Unit	Material	Labor	Equip.	Total	Specification
2-1/2"							
	Demolish	L.F.		.45		.45	Cost includes material and labor to
	Install	L.F.	.57	.98		1.55	install 2-1/2" vinyl or rubber cove
	Demolish and Install	L.F.	.57	1.43		2	base including adhesive.
	Reinstall	L.F.		.98		.98	
	Clean	L.F.		.22		.22	
	Minimum Charge	Job		154		154	
4"							
	Demolish	L.F.		.45		.45	Cost includes material and labor to
	Install	L.F.	.58	.98		1.56	install 4" vinyl or rubber cove base
	Demolish and Install	L.F.	.58	1.43		2.01	including adhesive.
	Reinstall	L.F.		.98		.98	
	Clean	L.F.		.22		.22	
	Minimum Charge	Job		154		154	
6"							
	Demolish	L.F.		.45		.45	Cost includes material and labor to
	Install	L.F.	1.01	.98		1.99	install 6" vinyl or rubber cove base
	Demolish and Install	L.F.	1.01	1.43		2.44	including adhesive.
	Reinstall	L.F.		.98		.98	
	Clean	L.F.		.22		.22	
	Minimum Charge	Job		154		154	

Prefinished

		Unit	Material	Labor	Equip.	Total	Specification
	Demolish	L.F.		.45		.45	Cost includes material and labor to
	Install	L.F.	1.69	1.08		2.77	install 9/16" x 3-5/16" prefinished
	Demolish and Install	L.F.	1.69	1.53		3.22	base molding.
	Reinstall	L.F.		.87		.87	
	Clean	L.F.	.02	.14		.16	
	Paint	L.F.	.03	.35		.38	
	Minimum Charge	Job		169		169	

Shoe

		Unit	Material	Labor	Equip.	Total	Specification
	Demolish	L.F.		.45		.45	Includes material and labor to install
	Install	L.F.	.39	1		1.39	pine or birch base molding.
	Demolish and Install	L.F.	.39	1.45		1.84	
	Reinstall	L.F.		.80		.80	
	Clean	L.F.	.02	.14		.16	
	Paint	L.F.	.12	.47		.59	
	Minimum Charge	Job		169		169	

Finish Carpentry

Base Molding

Base Molding	Unit	Material	Labor	Equip.	Total	Specification
Carpeted						
Demolish	L.F.		.18		.18	Includes material and labor to install
Install	L.F.	1.25	1.87		3.12	carpet baseboard with adhesive.
Demolish and Install	L.F.	1.25	2.05		3.30	
Reinstall	L.F.		1.50		1.50	
Clean	L.F.		.09		.09	
Minimum Charge	Job		154		154	
Ceramic Tile						
Demolish	L.F.		.98		.98	Includes material and labor to install
Install	L.F.	3.53	4.28		7.81	thin set ceramic tile cove base
Demolish and Install	L.F.	3.53	5.26		8.79	including grout.
Clean	L.F.		.46		.46	
Minimum Charge	Job		154		154	
Quarry Tile						
Demolish	L.F.		.98		.98	Includes material and labor to install
Install	L.F.	5.05	4.98		10.03	quarry tile base including grout.
Demolish and Install	L.F.	5.05	5.96		11.01	
Clean	L.F.		.46		.46	
Minimum Charge	Job		154		154	
Caulk and Renail						
Install	L.F.	.01	.53		.54	Includes labor and material to renail
Minimum Charge	Job		169		169	base and caulk where necessary.

Molding

Molding	Unit	Material	Labor	Equip.	Total	Specification
Oak Base						
Demolish	L.F.		.45		.45	Cost includes material and labor to
Install	L.F.	1.07	1.02		2.09	install 1/2" x 2-1/2" oak base
Demolish and Install	L.F.	1.07	1.47		2.54	molding.
Reinstall	L.F.		.82		.82	
Clean	L.F.	.02	.14		.16	
Paint	L.F.	.08	.45		.53	
Minimum Charge	Job		169		169	
Shoe						
Demolish	L.F.		.45		.45	Includes material and labor to install
Install	L.F.	.39	1		1.39	pine or birch base molding.
Demolish and Install	L.F.	.39	1.45		1.84	
Reinstall	L.F.		.80		.80	
Clean	L.F.	.02	.14		.16	
Paint	L.F.	.12	.47		.59	
Minimum Charge	Job		169		169	
Quarter Round						
Demolish	L.F.		.45		.45	Includes material and labor to install
Install	L.F.	.70	1.33		2.03	pine base quarter round molding.
Demolish and Install	L.F.	.70	1.78		2.48	
Reinstall	L.F.		1.06		1.06	
Clean	L.F.	.02	.14		.16	
Paint	L.F.	.06	.74		.80	
Minimum Charge	Job		169		169	
Chair Rail						
Demolish	L.F.		.45		.45	Cost includes material and labor to
Install	L.F.	1.58	1.25		2.83	install 5/8" x 2-1/2" oak chair rail
Demolish and Install	L.F.	1.58	1.70		3.28	molding.
Reinstall	L.F.		1		1	
Clean	L.F.	.02	.14		.16	
Paint	L.F.	.06	.74		.80	
Minimum Charge	Job		169		169	

Finish Carpentry

Molding

	Unit	Material	Labor	Equip.	Total	Specification
Crown						
Demolish	L.F.		.45		.45	Includes material and labor to install
Install	L.F.	2.05	1.35		3.40	unfinished pine crown molding.
Demolish and Install	L.F.	2.05	1.80		3.85	
Reinstall	L.F.		1.08		1.08	
Clean	L.F.	.03	.19		.22	
Paint	L.F.	.06	.74		.80	
Minimum Charge	Job		169		169	
Cove						
Demolish	L.F.		.45		.45	Cost includes material and labor to
Install	L.F.	1.38	1.33		2.71	install 1/2" x 2-3/4" pine cove
Demolish and Install	L.F.	1.38	1.78		3.16	molding.
Reinstall	L.F.		1.06		1.06	
Clean	L.F.	.02	.14		.16	
Paint	L.F.	.06	.74		.80	
Minimum Charge	Job		169		169	
Corner						
Demolish	L.F.		.45		.45	Includes material and labor to install
Install	L.F.	.61	.53		1.14	inside or outside pine corner molding.
Demolish and Install	L.F.	.61	.98		1.59	
Reinstall	L.F.		.42		.42	
Clean	L.F.	.02	.14		.16	
Paint	L.F.	.06	.74		.80	
Minimum Charge	Job		169		169	
Picture						
Demolish	L.F.		.45		.45	Cost includes material and labor to
Install	L.F.	1.41	1.41		2.82	install 9/16" x 2-1/2" pine casing.
Demolish and Install	L.F.	1.41	1.86		3.27	
Reinstall	L.F.		1.13		1.13	
Clean	L.F.	.02	.14		.16	
Paint	L.F.	.06	.74		.80	
Minimum Charge	Job		169		169	
Single Cased Opening						
Demolish	Ea.		23.50		23.50	Includes material and labor to install
Install	Ea.	16.05	21		37.05	flat pine jamb with square cut heads
Demolish and Install	Ea.	16.05	44.50		60.55	and rabbeted sides for 6' 8" high
Clean	Ea.	.23	5.05		5.28	opening including trim sets for both
Paint	Ea.	1.02	7.40		8.42	sides.
Minimum Charge	Job		169		169	
Double Cased Opening						
Demolish	Ea.		23.50		23.50	Includes material and labor to install
Install	Ea.	32	24		56	flat pine jamb with square cut heads
Demolish and Install	Ea.	32	47.50		79.50	and rabbeted sides for 6' 8" high
Clean	Ea.	.26	5.90		6.16	opening including trim sets for both
Paint	Ea.	1.19	8.70		9.89	sides.
Minimum Charge	Job		169		169	

Window Trim Set

	Unit	Material	Labor	Equip.	Total	Specification
Single						
Demolish	Ea.		5.95		5.95	Cost includes material and labor to
Install	Opng.	38.50	34		72.50	install 11/16" x 2-1/2" pine ranch
Demolish and Install	Opng.	38.50	39.95		78.45	style window casing.
Reinstall	Opng.		27.04		27.04	
Clean	Ea.	.33	3.45		3.78	
Paint	Ea.	.96	7.05		8.01	
Minimum Charge	Job		169		169	

Finish Carpentry

Window Trim Set

	Unit	Material	Labor	Equip.	Total	Specification
Double						
Demolish	Ea.		6.70		6.70	Cost includes material and labor to
Install	Ea.	53	42.50		95.50	install 11/16" x 2-1/2" pine ranch
Demolish and Install	Ea.	53	49.20		102.20	style trim for casing one side of a
Reinstall	Ea.		33.80		33.80	double window opening.
Clean	Ea.	.58	4.03		4.61	
Paint	Ea.	1.61	11.85		13.46	
Minimum Charge	Job		169		169	
Triple						
Demolish	Ea.		7.65		7.65	Cost includes material and labor to
Install	Opng.	69	56.50		125.50	install 11/16" x 2-1/2" pine ranch
Demolish and Install	Opng.	69	64.15		133.15	style trim for casing one side of a
Reinstall	Opng.		45.07		45.07	triple window opening.
Clean	Ea.	.83	4.83		5.66	
Paint	Ea.	2.26	16.45		18.71	
Minimum Charge	Job		169		169	
Window Casing Per L.F.						
Demolish	L.F.		.45		.45	Cost includes material and labor to
Install	L.F.	1.41	1.41		2.82	install 11/16" x 2-1/2" pine ranch
Demolish and Install	L.F.	1.41	1.86		3.27	style trim for casing.
Reinstall	L.F.		1.13		1.13	
Clean	L.F.	.02	.14		.16	
Paint	L.F.	.06	.74		.80	
Minimum Charge	Job		169		169	

Window Sill

	Unit	Material	Labor	Equip.	Total	Specification
Wood						
Demolish	L.F.		.45		.45	Includes material and labor to install
Install	L.F.	2.42	1.69		4.11	flat, wood window stool.
Demolish and Install	L.F.	2.42	2.14		4.56	
Reinstall	L.F.		1.35		1.35	
Clean	L.F.	.02	.14		.16	
Paint	L.F.	.06	.74		.80	
Minimum Charge	Job		169		169	
Marble						
Demolish	L.F.		.67		.67	Includes material and labor to install
Install	L.F.	8.05	7.05		15.10	marble window sills, 6" x 3/4" thick.
Demolish and Install	L.F.	8.05	7.72		15.77	
Reinstall	L.F.		5.64		5.64	
Clean	L.F.	.02	.14		.16	
Minimum Charge	Job		154		154	

Shelving

	Unit	Material	Labor	Equip.	Total	Specification
Pine						
18"						
Demolish	L.F.		.77		.77	Cost includes material and labor to
Install	L.F.	5.10	3.56		8.66	install 1" thick custom shelving board
Demolish and Install	L.F.	5.10	4.33		9.43	18" wide.
Reinstall	L.F.		2.85		2.85	
Clean	L.F.	.04	.40		.44	
Paint	L.F.	.74	.47		1.21	
Minimum Charge	Job		169		169	

Finish Carpentry

Shelving		Unit	Material	Labor	Equip.	Total	Specification
24"							
	Demolish	L.F.		.77		.77	Cost includes material and labor to
	Install	L.F.	6.80	3.98		10.78	install 1" thick custom shelving board.
	Demolish and Install	L.F.	6.80	4.75		11.55	
	Reinstall	L.F.		3.18		3.18	
	Clean	L.F.	.07	.60		.67	
	Paint	L.F.	1	.62		1.62	
	Minimum Charge	Job		169		169	
Particle Board							
	Demolish	L.F.		.71		.71	Cost includes material and labor to
	Install	L.F.	1.08	1.78		2.86	install 3/4" thick particle shelving
	Demolish and Install	L.F.	1.08	2.49		3.57	board 18" wide.
	Reinstall	L.F.		1.42		1.42	
	Clean	L.F.	.04	.40		.44	
	Paint	S.F.	.11	.69		.80	
	Minimum Charge	Job		169		169	
Plywood							
	Demolish	L.F.		.71		.71	Cost includes material and labor to
	Install	L.F.	.89	5.20		6.09	install 3/4" thick plywood shelving
	Demolish and Install	L.F.	.89	5.91		6.80	board.
	Reinstall	L.F.		4.16		4.16	
	Clean	L.F.	.04	.40		.44	
	Paint	S.F.	.11	.69		.80	
	Minimum Charge	Job		169		169	
Hardwood (oak)							
	Demolish	L.F.		.71		.71	Includes material and labor to install
	Install	L.F.	7.05	2.33		9.38	oak shelving, 1" x 12", incl. cleats and
	Demolish and Install	L.F.	7.05	3.04		10.09	bracing.
	Reinstall	L.F.		1.86		1.86	
	Clean	L.F.	.04	.40		.44	
	Paint	S.F.	.11	.69		.80	
	Minimum Charge	Job		169		169	
Custom Bookcase							
	Demolish	L.F.		8.95		8.95	Includes material and labor to install
	Install	L.F.	23	34		57	custom modular bookcase unit with
	Demolish and Install	L.F.	23	42.95		65.95	clear pine faced frames, shelves 12"
	Reinstall	L.F.		27.04		27.04	O.C., 7' high, 8" deep.
	Clean	L.F.	1.31	8.65		9.96	
	Paint	L.F.	.98	23		23.98	
	Minimum Charge	Job		169		169	
Glass							
	Demolish	L.F.		.71		.71	Includes material and labor to install
	Install	L.F.	27	2.33		29.33	glass shelving, 1" x 12", incl. cleats
	Demolish and Install	L.F.	27	3.04		30.04	and bracing.
	Reinstall	L.F.		1.86		1.86	
	Clean	L.F.	.06	.37		.43	
	Minimum Charge	Job		161		161	

Closet Shelving		Unit	Material	Labor	Equip.	Total	Specification
Closet Shelf and Rod							
	Demolish	L.F.		.71		.71	Cost includes material and labor to
	Install	L.F.	2.40	5.65		8.05	install 1" thick custom shelving board
	Demolish and Install	L.F.	2.40	6.36		8.76	18" wide and 1" diameter clothes pole
	Reinstall	L.F.		4.51		4.51	with brackets 3' O.C.
	Clean	L.F.	.03	.30		.33	
	Paint	L.F.	.11	.74		.85	
	Minimum Charge	Job		169		169	

Finish Carpentry

Closet Shelving

	Unit	Material	Labor	Equip.	Total	Specification
Clothing Rod						
Demolish	L.F.		.47		.47	Includes material and labor to install
Install	L.F.	1.76	1.69		3.45	fir closet pole, 1-5/8" diameter.
Demolish and Install	L.F.	1.76	2.16		3.92	
Reinstall	L.F.		1.35		1.35	
Clean	L.F.	.03	.43		.46	
Paint	L.F.	.12	.81		.93	
Minimum Charge	Job		169		169	

Stair Assemblies

	Unit	Material	Labor	Equip.	Total	Specification
Straight Hardwood						
Demolish	Ea.		325		325	Includes material and labor to install
Install	Flight	955	225		1180	factory cut and assembled straight
Demolish and Install	Flight	955	550		1505	closed box stairs with oak treads and
Clean	Flight	3.28	21.50		24.78	prefinished stair rail with balusters.
Paint	Flight	10.35	39.50		49.85	
Minimum Charge	Job		169		169	
Clear Oak Tread						
Demolish	Ea.		8.55		8.55	Includes material and labor to install
Install	Ea.	32	18.80		50.80	1-1/4" thick clear oak treads per riser.
Demolish and Install	Ea.	32	27.35		59.35	
Clean	Ea.	.10	2.13		2.23	
Paint	Ea.	.34	2.20		2.54	
Minimum Charge	Job		169		169	
Landing						
Demolish	S.F.		2.80		2.80	Includes material and labor to install
Install	S.F.	11.40	18.80		30.20	clear oak landing.
Demolish and Install	S.F.	11.40	21.60		33	
Clean	S.F.		.30		.30	
Paint	S.F.	.28	.73		1.01	
Minimum Charge	Job		169		169	
Refinish / Stain						
Install	S.F.	1.18	1.86		3.04	Includes labor and material to sand
Minimum Charge	Job		148		148	and finish (3 passes) with two coats of
						urethane on a new floor.
Spiral Hardwood						
Demolish	Ea.		325		325	Cost includes material and labor to
Install	Flight	4850	450		5300	install 4' - 6' diameter spiral stairs with
Demolish and Install	Flight	4850	775		5625	oak treads, factory cut and assembled
Clean	Flight	3.28	21.50		24.78	with double handrails.
Paint	Flight	10.35	39.50		49.85	
Minimum Charge	Job		169		169	

Stairs

	Unit	Material	Labor	Equip.	Total	Specification
Disappearing						
Demolish	Ea.		16.80		16.80	Includes material and labor to install a
Install	Ea.	130	96.50		226.50	folding staircase.
Demolish and Install	Ea.	130	113.30		243.30	
Reinstall	Ea.		77.26		77.26	
Clean	Ea.	.50	15.10		15.60	
Minimum Charge	Job		169		169	

Finish Carpentry

Stair Components

	Unit	Material	Labor	Equip.	Total	Specification
Handrail w / Balusters						
Demolish	L.F.		2.24		2.24	Includes material and labor to install
Install	L.F.	23	7.05		30.05	prefinished assembled stair rail with
Demolish and Install	L.F.	23	9.29		32.29	brackets, turned balusters and newel.
Clean	L.F.	.66	1.67		2.33	
Paint	L.F.	1.40	3.49		4.89	
Minimum Charge	Job		169		169	
Bannister						
Demolish	L.F.		2.24		2.24	Includes material and labor to install
Install	L.F.	34.50	5.65		40.15	built-up oak railings.
Demolish and Install	L.F.	34.50	7.89		42.39	
Clean	L.F.	.66	1.51		2.17	
Paint	L.F.	1.40	2.69		4.09	
Minimum Charge	Job		169		169	

Wall and Base Cabinets

Countertop

Wall Cabinets	Unit	Material	Labor	Equip.	Total	Specification
Good Grade Laminated						
12″ Wide, 1 Door						
Demolish	Ea.		6.70		6.70	Includes material and labor to install
Install	Ea.	109	30.50		139.50	good modular unit with melamine
Demolish and Install	Ea.	109	37.20		146.20	laminated to particle board, in
Reinstall	Ea.		24.58		24.58	textured colors or wood grain print
Clean	Ea.	.50	4.65		5.15	finish, hinges and pulls.
Paint	Ea.	.42	10.60		11.02	
Minimum Charge	Job		169		169	
15″ Wide, 1 Door						
Demolish	Ea.		8.40		8.40	Includes material and labor to install
Install	Ea.	124	31.50		155.50	good modular unit with melamine
Demolish and Install	Ea.	124	39.90		163.90	laminated to particle board, in
Reinstall	Ea.		25.27		25.27	textured colors or wood grain print
Clean	Ea.	.62	5.75		6.37	finish, hinges and pulls.
Paint	Ea.	.53	13.45		13.98	
Minimum Charge	Job		169		169	
18″ Wide, 1 Door						
Demolish	Ea.		10.15		10.15	Includes material and labor to install
Install	Ea.	135	32.50		167.50	good modular unit with melamine
Demolish and Install	Ea.	135	42.65		177.65	laminated to particle board, in
Reinstall	Ea.		25.88		25.88	textured colors or wood grain print
Clean	Ea.	.74	6.90		7.64	finish, hinges and pulls.
Paint	Ea.	.63	15.60		16.23	
Minimum Charge	Job		169		169	
21″ Wide, 1 Door						
Demolish	Ea.		11.65		11.65	Includes material and labor to install
Install	Ea.	160	33		193	good modular unit with melamine
Demolish and Install	Ea.	160	44.65		204.65	laminated to particle board, in
Reinstall	Ea.		26.25		26.25	textured colors or wood grain print
Clean	Ea.	.86	8.05		8.91	finish, hinges and pulls.
Paint	Ea.	.74	18.55		19.29	
Minimum Charge	Job		169		169	
24″ w / Blind Corner						
Demolish	Ea.		13.40		13.40	Includes material and labor to install
Install	Ea.	153	33.50		186.50	good modular unit with melamine
Demolish and Install	Ea.	153	46.90		199.90	laminated to particle board, in
Reinstall	Ea.		26.64		26.64	textured colors or wood grain print
Clean	Ea.	.98	9.30		10.28	finish, hinges and pulls.
Paint	Ea.	.84	21		21.84	
Minimum Charge	Job		169		169	

Cabinets and Countertops

Wall Cabinets	Unit	Material	Labor	Equip.	Total	Specification
27″ Wide, 2 Door						
Demolish	Ea.		14.90		14.90	Includes material and labor to install
Install	Ea.	189	34		223	good modular unit with melamine
Demolish and Install	Ea.	189	48.90		237.90	laminated to particle board, in
Reinstall	Ea.		27.31		27.31	textured colors or wood grain print
Clean	Ea.	1.11	10.50		11.61	finish, hinges and pulls.
Paint	Ea.	.95	24.50		25.45	
Minimum Charge	Job		169		169	
30″ Wide, 2 Door						
Demolish	Ea.		16.80		16.80	Includes material and labor to install
Install	Ea.	183	35		218	good modular unit with melamine
Demolish and Install	Ea.	183	51.80		234.80	laminated to particle board, in
Reinstall	Ea.		28.02		28.02	textured colors or wood grain print
Clean	Ea.	1.23	11.50		12.73	finish, hinges and pulls.
Paint	Ea.	1.05	27		28.05	
Minimum Charge	Job		169		169	
36″ Wide, 2 Door						
Demolish	Ea.		19.90		19.90	Includes material and labor to install
Install	Ea.	209	36		245	good modular unit with melamine
Demolish and Install	Ea.	209	55.90		264.90	laminated to particle board, in
Reinstall	Ea.		28.77		28.77	textured colors or wood grain print
Clean	Ea.	1.47	14.20		15.67	finish, hinges and pulls.
Paint	Ea.	1.27	33		34.27	
Minimum Charge	Job		169		169	
48″ Wide, 2 Door						
Demolish	Ea.		27		27	Includes material and labor to install
Install	Ea.	256	36.50		292.50	good modular unit with melamine
Demolish and Install	Ea.	256	63.50		319.50	laminated to particle board, in
Reinstall	Ea.		29.39		29.39	textured colors or wood grain print
Clean	Ea.	1.97	18.60		20.57	finish, hinges and pulls.
Paint	Ea.	1.68	42.50		44.18	
Minimum Charge	Job		169		169	
Above-Appliance						
Demolish	Ea.		16.80		16.80	Includes material and labor to install
Install	Ea.	121	27.50		148.50	good modular unit with melamine
Demolish and Install	Ea.	121	44.30		165.30	laminated to particle board, in
Reinstall	Ea.		21.81		21.81	textured colors or wood grain print
Clean	Ea.	1.23	11.50		12.73	finish, hinges and pulls.
Paint	Ea.	1.05	27		28.05	
Minimum Charge	Job		169		169	
Better Grade Wood						
12″ Wide, 1 Door						
Demolish	Ea.		6.70		6.70	Includes material and labor to install
Install	Ea.	146	30.50		176.50	custom modular unit with solid
Demolish and Install	Ea.	146	37.20		183.20	hardwood faced frames, hardwood
Reinstall	Ea.		24.58		24.58	door frames and drawer fronts,
Clean	Ea.	.50	4.65		5.15	hardwood veneer on raised door
Paint	Ea.	.42	11.40		11.82	panels, hinges and pulls.
Minimum Charge	Job		169		169	
15″ Wide, 1 Door						
Demolish	Ea.		8.40		8.40	Includes material and labor to install
Install	Ea.	165	31.50		196.50	custom modular unit with solid
Demolish and Install	Ea.	165	39.90		204.90	hardwood faced frames, hardwood
Reinstall	Ea.		25.27		25.27	door frames and drawer fronts,
Clean	Ea.	.62	5.75		6.37	hardwood veneer on raised door
Paint	Ea.	.53	14.10		14.63	panels, hinges and pulls.
Minimum Charge	Job		169		169	

Cabinets and Countertops

Wall Cabinets	Unit	Material	Labor	Equip.	Total	Specification
18″ Wide, 1 Door						
Demolish	Ea.		10.15		10.15	Includes material and labor to install
Install	Ea.	180	32.50		212.50	custom modular unit with solid
Demolish and Install	Ea.	180	42.65		222.65	hardwood faced frames, hardwood
Reinstall	Ea.		25.88		25.88	door frames and drawer fronts,
Clean	Ea.	.74	6.90		7.64	hardwood veneer on raised door
Paint	Ea.	.63	17.45		18.08	panels, hinges and pulls.
Minimum Charge	Job		169		169	
21″ Wide, 1 Door						
Demolish	Ea.		11.65		11.65	Includes material and labor to install
Install	Ea.	206	33		239	custom modular unit with solid
Demolish and Install	Ea.	206	44.65		250.65	hardwood faced frames, hardwood
Reinstall	Ea.		26.25		26.25	door frames and drawer fronts,
Clean	Ea.	.86	8.05		8.91	hardwood veneer on raised door
Paint	Ea.	.74	19.75		20.49	panels, hinges and pulls.
Minimum Charge	Job		169		169	
24″ w / Blind Corner						
Demolish	Ea.		13.40		13.40	Includes material and labor to install
Install	Ea.	204	33.50		237.50	custom modular unit with solid
Demolish and Install	Ea.	204	46.90		250.90	hardwood faced frames, hardwood
Reinstall	Ea.		26.64		26.64	door frames and drawer fronts,
Clean	Ea.	.98	9.30		10.28	hardwood veneer on raised door
Paint	Ea.	.84	23		23.84	panels, hinges and pulls.
Minimum Charge	Job		169		169	
27″ Wide, 2 Door						
Demolish	Ea.		14.90		14.90	Includes material and labor to install
Install	Ea.	230	34		264	custom modular unit with solid
Demolish and Install	Ea.	230	48.90		278.90	hardwood faced frames, hardwood
Reinstall	Ea.		27.31		27.31	door frames and drawer fronts,
Clean	Ea.	1.11	10.50		11.61	hardwood veneer on raised door
Paint	Ea.	.95	24.50		25.45	panels, hinges and pulls.
Minimum Charge	Job		169		169	
30″ Wide, 2 Door						
Demolish	Ea.		16.80		16.80	Includes material and labor to install
Install	Ea.	244	35		279	custom modular unit with solid
Demolish and Install	Ea.	244	51.80		295.80	hardwood faced frames, hardwood
Reinstall	Ea.		28.02		28.02	door frames and drawer fronts,
Clean	Ea.	1.23	11.50		12.73	hardwood veneer on raised door
Paint	Ea.	1.05	29.50		30.55	panels, hinges and pulls.
Minimum Charge	Job		169		169	
36″ Wide, 2 Door						
Demolish	Ea.		19.90		19.90	Includes material and labor to install
Install	Ea.	278	36		314	custom modular unit with solid
Demolish and Install	Ea.	278	55.90		333.90	hardwood faced frames, hardwood
Reinstall	Ea.		28.77		28.77	door frames and drawer fronts,
Clean	Ea.	1.47	14.20		15.67	hardwood veneer on raised door
Paint	Ea.	1.27	33		34.27	panels, hinges and pulls.
Minimum Charge	Job		169		169	
48″ Wide, 2 Door						
Demolish	Ea.		27		27	Includes material and labor to install
Install	Ea.	340	36.50		376.50	custom modular unit with solid
Demolish and Install	Ea.	340	63.50		403.50	hardwood faced frames, hardwood
Reinstall	Ea.		29.39		29.39	door frames and drawer fronts,
Clean	Ea.	1.97	18.60		20.57	hardwood veneer on raised door
Paint	Ea.	1.68	42.50		44.18	panels, hinges and pulls.
Minimum Charge	Job		169		169	

Cabinets and Countertops

Wall Cabinets		Unit	Material	Labor	Equip.	Total	Specification
Above-appliance							
	Demolish	Ea.		16.80		16.80	Includes material and labor to install
	Install	Ea.	166	27.50		193.50	custom modular unit with solid
	Demolish and Install	Ea.	166	44.30		210.30	hardwood faced frames, hardwood
	Reinstall	Ea.		21.81		21.81	door frames and drawer fronts,
	Clean	Ea.	1.23	11.50		12.73	hardwood veneer on raised door
	Paint	Ea.	1.05	29.50		30.55	panels, hinges and pulls.
	Minimum Charge	Job		169		169	
Premium Grade Wood							
12″ Wide, 1 Door							
	Demolish	Ea.		6.70		6.70	Includes material and labor to install
	Install	Ea.	189	30.50		219.50	premium modular unit with solid
	Demolish and Install	Ea.	189	37.20		226.20	hardwood faced frames, hardwood
	Reinstall	Ea.		24.58		24.58	door frames and drawer fronts,
	Clean	Ea.	.50	4.65		5.15	hardwood veneer on raised door
	Paint	Ea.	.42	12.35		12.77	panels, hinges and pulls.
	Minimum Charge	Job		169		169	
15″ Wide, 1 Door							
	Demolish	Ea.		8.40		8.40	Includes material and labor to install
	Install	Ea.	220	31.50		251.50	premium modular unit with solid
	Demolish and Install	Ea.	220	39.90		259.90	hardwood faced frames, hardwood
	Reinstall	Ea.		25.27		25.27	door frames and drawer fronts,
	Clean	Ea.	.62	5.75		6.37	hardwood veneer on raised door
	Paint	Ea.	.53	15.60		16.13	panels, hinges and pulls.
	Minimum Charge	Job		169		169	
18″ Wide, 1 Door							
	Demolish	Ea.		10.15		10.15	Includes material and labor to install
	Install	Ea.	232	32.50		264.50	premium modular unit with solid
	Demolish and Install	Ea.	232	42.65		274.65	hardwood faced frames, hardwood
	Reinstall	Ea.		25.88		25.88	door frames and drawer fronts,
	Clean	Ea.	.74	6.90		7.64	hardwood veneer on raised door
	Paint	Ea.	.63	18.55		19.18	panels, hinges and pulls.
	Minimum Charge	Job		169		169	
21″ Wide, 1 Door							
	Demolish	Ea.		11.65		11.65	Includes material and labor to install
	Install	Ea.	246	33		279	premium modular unit with solid
	Demolish and Install	Ea.	246	44.65		290.65	hardwood faced frames, hardwood
	Reinstall	Ea.		26.25		26.25	door frames and drawer fronts,
	Clean	Ea.	.86	8.05		8.91	hardwood veneer on raised door
	Paint	Ea.	.74	21		21.74	panels, hinges and pulls.
	Minimum Charge	Job		169		169	
24″ w / Blind Corner							
	Demolish	Ea.		13.40		13.40	Includes material and labor to install
	Install	Ea.	271	33.50		304.50	premium modular unit with solid
	Demolish and Install	Ea.	271	46.90		317.90	hardwood faced frames, hardwood
	Reinstall	Ea.		26.64		26.64	door frames and drawer fronts,
	Clean	Ea.	.98	9.30		10.28	hardwood veneer on raised door
	Paint	Ea.	.84	24.50		25.34	panels, hinges and pulls.
	Minimum Charge	Job		169		169	
27″ Wide, 2 Door							
	Demolish	Ea.		14.90		14.90	Includes material and labor to install
	Install	Ea.	252	34		286	premium modular unit with solid
	Demolish and Install	Ea.	252	48.90		300.90	hardwood faced frames, hardwood
	Reinstall	Ea.		27.31		27.31	door frames and drawer fronts,
	Clean	Ea.	1.11	10.50		11.61	hardwood veneer on raised door
	Paint	Ea.	.95	27		27.95	panels, hinges and pulls.
	Minimum Charge	Job		169		169	

Cabinets and Countertops

Wall Cabinets		Unit	Material	Labor	Equip.	Total	Specification
30" Wide, 2 Door							
	Demolish	Ea.		16.80		16.80	Includes material and labor to install
	Install	Ea.	305	35		340	premium modular unit with solid
	Demolish and Install	Ea.	305	51.80		356.80	hardwood faced frames, hardwood
	Reinstall	Ea.		28.02		28.02	door frames and drawer fronts,
	Clean	Ea.	1.23	11.50		12.73	hardwood veneer on raised door
	Paint	Ea.	1.05	29.50		30.55	panels, hinges and pulls.
	Minimum Charge	Job		169		169	
36" Wide, 2 Door							
	Demolish	Ea.		19.90		19.90	Includes material and labor to install
	Install	Ea.	350	36		386	premium modular unit with solid
	Demolish and Install	Ea.	350	55.90		405.90	hardwood faced frames, hardwood
	Reinstall	Ea.		28.77		28.77	door frames and drawer fronts,
	Clean	Ea.	1.47	14.20		15.67	hardwood veneer on raised door
	Paint	Ea.	1.27	37		38.27	panels, hinges and pulls.
	Minimum Charge	Job		169		169	
48" Wide, 2 Door							
	Demolish	Ea.		27		27	Includes material and labor to install
	Install	Ea.	435	36.50		471.50	premium modular unit with solid
	Demolish and Install	Ea.	435	63.50		498.50	hardwood faced frames, hardwood
	Reinstall	Ea.		29.39		29.39	door frames and drawer fronts,
	Clean	Ea.	1.97	18.60		20.57	hardwood veneer on raised door
	Paint	Ea.	1.68	49.50		51.18	panels, hinges and pulls.
	Minimum Charge	Job		169		169	
Above-appliance							
	Demolish	Ea.		16.80		16.80	Includes material and labor to install
	Install	Ea.	206	27.50		233.50	premium modular unit with solid
	Demolish and Install	Ea.	206	44.30		250.30	hardwood faced frames, hardwood
	Reinstall	Ea.		21.81		21.81	door frames and drawer fronts,
	Clean	Ea.	1.23	11.50		12.73	hardwood veneer on raised door
	Paint	Ea.	1.05	29.50		30.55	panels, hinges, pulls.
	Minimum Charge	Job		169		169	
Premium Hardwood							
12" Wide, 1 Door							
	Demolish	Ea.		6.70		6.70	Includes material and labor to install
	Install	Ea.	220	30.50		250.50	premium hardwood modular unit with
	Demolish and Install	Ea.	220	37.20		257.20	solid hardwood faced frames, drawer
	Reinstall	Ea.		24.58		24.58	fronts, and door panels, steel drawer
	Clean	Ea.	.50	4.65		5.15	guides, hinges, pulls, laminated
	Paint	Ea.	.42	13.45		13.87	interior.
	Minimum Charge	Job		169		169	
15" Wide, 1 Door							
	Demolish	Ea.		8.40		8.40	Includes material and labor to install
	Install	Ea.	289	31.50		320.50	premium hardwood modular unit with
	Demolish and Install	Ea.	289	39.90		328.90	solid hardwood faced frames, drawer
	Reinstall	Ea.		25.27		25.27	fronts, and door panels, steel drawer
	Clean	Ea.	.62	5.75		6.37	guides, hinges, pulls, laminated
	Paint	Ea.	.53	16.45		16.98	interior.
	Minimum Charge	Job		169		169	
18" Wide, 1 Door							
	Demolish	Ea.		10.15		10.15	Includes material and labor to install
	Install	Ea.	263	32.50		295.50	premium hardwood modular unit with
	Demolish and Install	Ea.	263	42.65		305.65	solid hardwood faced frames, drawer
	Reinstall	Ea.		25.88		25.88	fronts, and door panels, steel drawer
	Clean	Ea.	.74	6.90		7.64	guides, hinges, pulls, laminated
	Paint	Ea.	.63	19.75		20.38	interior.
	Minimum Charge	Job		169		169	

Cabinets and Countertops

Wall Cabinets	Unit	Material	Labor	Equip.	Total	Specification
21" Wide, 1 Door						
Demolish	Ea.		11.65		11.65	Includes material and labor to install
Install	Ea.	270	33		303	premium hardwood modular unit with
Demolish and Install	Ea.	270	44.65		314.65	solid hardwood faced frames, drawer
Reinstall	Ea.		26.25		26.25	fronts, and door panels, steel drawer
Clean	Ea.	.86	8.05		8.91	guides, hinges, pulls, laminated
Paint	Ea.	.74	23		23.74	interior.
Minimum Charge	Job		169		169	
24" w / Blind Corner						
Demolish	Ea.		13.40		13.40	Includes material and labor to install
Install	Ea.	300	33.50		333.50	premium hardwood modular unit with
Demolish and Install	Ea.	300	46.90		346.90	solid hardwood faced frames, drawer
Reinstall	Ea.		26.64		26.64	fronts, and door panels, steel drawer
Clean	Ea.	.98	9.30		10.28	guides, hinges, pulls, laminated
Paint	Ea.	.84	27		27.84	interior.
Minimum Charge	Job		169		169	
27" Wide, 2 Door						
Demolish	Ea.		14.90		14.90	Includes material and labor to install
Install	Ea.	510	34		544	premium hardwood modular unit with
Demolish and Install	Ea.	510	48.90		558.90	solid hardwood faced frames, drawer
Reinstall	Ea.		27.31		27.31	fronts, and door panels, steel drawer
Clean	Ea.	1.11	10.50		11.61	guides, hinges, pulls, laminated
Paint	Ea.	.95	29.50		30.45	interior.
Minimum Charge	Job		169		169	
30" Wide, 2 Door						
Demolish	Ea.		16.80		16.80	Includes material and labor to install
Install	Ea.	340	35		375	premium hardwood modular unit with
Demolish and Install	Ea.	340	51.80		391.80	solid hardwood faced frames, drawer
Reinstall	Ea.		28.02		28.02	fronts, and door panels, steel drawer
Clean	Ea.	1.23	11.50		12.73	guides, hinges, pulls, laminated
Paint	Ea.	1.05	33		34.05	interior.
Minimum Charge	Job		169		169	
36" Wide, 2 Door						
Demolish	Ea.		19.90		19.90	Includes material and labor to install
Install	Ea.	385	36		421	premium hardwood modular unit with
Demolish and Install	Ea.	385	55.90		440.90	solid hardwood faced frames, drawer
Reinstall	Ea.		28.77		28.77	fronts, and door panels, steel drawer
Clean	Ea.	1.47	14.20		15.67	guides, hinges, pulls, laminated
Paint	Ea.	1.27	42.50		43.77	interior.
Minimum Charge	Job		169		169	
48" Wide, 2 Door						
Demolish	Ea.		27		27	Includes material and labor to install
Install	Ea.	470	36.50		506.50	premium hardwood modular unit with
Demolish and Install	Ea.	470	63.50		533.50	solid hardwood faced frames, drawer
Reinstall	Ea.		29.39		29.39	fronts, and door panels, steel drawer
Clean	Ea.	1.97	18.60		20.57	guides, hinges, pulls, laminated
Paint	Ea.	1.68	49.50		51.18	interior.
Minimum Charge	Job		169		169	
Above-appliance						
Demolish	Ea.		16.80		16.80	Includes material and labor to install
Install	Ea.	237	27.50		264.50	premium hardwood modular unit with
Demolish and Install	Ea.	237	44.30		281.30	solid hardwood faced frames, drawer
Reinstall	Ea.		21.81		21.81	fronts, and door panels, steel drawer
Clean	Ea.	1.23	11.50		12.73	guides, hinges, pulls, laminated
Paint	Ea.	1.05	33		34.05	interior.
Minimum Charge	Job		169		169	

Cabinets and Countertops

Wall Cabinets	Unit	Material	Labor	Equip.	Total	Specification
Medicine Cabinet						
Good Grade						
Demolish	Ea.		11.20		11.20	Cost includes material and labor to
Install	Ea.	128	24		152	install 14" x 18" recessed medicine
Demolish and Install	Ea.	128	35.20		163.20	cabinet with stainless steel frame,
Reinstall	Ea.		19.31		19.31	mirror. Light not included.
Clean	Ea.	.02	10.85		10.87	
Paint	Ea.	1	18.55		19.55	
Minimum Charge	Job		169		169	
Good Grade Lighted						
Demolish	Ea.		11.20		11.20	Cost includes material and labor to
Install	Ea.	335	28		363	install 14" x 18" recessed medicine
Demolish and Install	Ea.	335	39.20		374.20	cabinet with stainless steel frame,
Reinstall	Ea.		22.53		22.53	mirror and light.
Clean	Ea.	.02	10.85		10.87	
Paint	Ea.	1	18.55		19.55	
Minimum Charge	Job		169		169	
Better Grade						
Demolish	Ea.		11.20		11.20	Includes material and labor to install
Install	Ea.	108	24		132	recessed medicine cabinet with
Demolish and Install	Ea.	108	35.20		143.20	polished chrome frame and mirror.
Reinstall	Ea.		19.31		19.31	Light not included.
Clean	Ea.	.02	10.85		10.87	
Paint	Ea.	1	18.55		19.55	
Minimum Charge	Job		169		169	
Better Grade Lighted						
Demolish	Ea.		11.20		11.20	Includes material and labor to install
Install	Ea.	315	28		343	recessed medicine cabinet with
Demolish and Install	Ea.	315	39.20		354.20	polished chrome frame, mirror and
Reinstall	Ea.		22.53		22.53	light.
Clean	Ea.	.02	10.85		10.87	
Paint	Ea.	1	18.55		19.55	
Minimum Charge	Job		169		169	
Premium Grade						
Demolish	Ea.		11.20		11.20	Cost includes material and labor to
Install	Ea.	251	24		275	install 14" x 24" recessed medicine
Demolish and Install	Ea.	251	35.20		286.20	cabinet with beveled mirror, frameless
Reinstall	Ea.		19.31		19.31	swing door. Light not included.
Clean	Ea.	.02	10.85		10.87	
Paint	Ea.	1	18.55		19.55	
Minimum Charge	Job		169		169	
Premium Grade Light						
Demolish	Ea.		11.20		11.20	Cost includes material and labor to
Install	Ea.	460	28		488	install 36" x 30" surface mounted
Demolish and Install	Ea.	460	39.20		499.20	medicine cabinet with beveled tri-view
Reinstall	Ea.		22.53		22.53	mirror, frameless swing door and light.
Clean	Ea.	.02	10.85		10.87	
Paint	Ea.	1	18.55		19.55	
Minimum Charge	Job		169		169	
Bath Mirror						
Demolish	Ea.		8.40		8.40	Cost includes material and labor to
Install	Ea.	72.50	16.90		89.40	install 18" x 24" surface mount mirror
Demolish and Install	Ea.	72.50	25.30		97.80	with stainless steel frame.
Reinstall	Ea.		13.52		13.52	
Clean	Ea.	.02	.72		.74	
Minimum Charge	Job		169		169	

Cabinets and Countertops

Wall Cabinets	Unit	Material	Labor	Equip.	Total	Specification
Garage						
Demolish	L.F.		6.70		6.70	Includes material and labor to install
Install	L.F.	53.50	16.90		70.40	good modular unit with melamine
Demolish and Install	L.F.	53.50	23.60		77.10	laminated to particle board, in
Reinstall	L.F.		13.52		13.52	textured colors or wood grain print
Clean	L.F.	.50	4.65		5.15	finish, hinges and pulls.
Paint	L.F.	.42	10.60		11.02	
Minimum Charge	Job		169		169	
Strip and Refinish						
Install	L.F.	1.11	5.40		6.51	Includes labor and material to strip,
Minimum Charge	Job		148		148	prep and refinish exterior of cabinets.

Tall Cabinets	Unit	Material	Labor	Equip.	Total	Specification
Good Grade Laminated						
Demolish	L.F.		11.95		11.95	Includes material and labor to install
Install	L.F.	216	190		406	custom modular unit with solid
Demolish and Install	L.F.	216	201.95		417.95	hardwood faced frames, hardwood
Reinstall	L.F.		152.34		152.34	door frames and drawer fronts,
Clean	L.F.	1.31	8.65		9.96	hardwood veneer on raised door
Paint	L.F.	.98	21		21.98	panels, hinges and pulls.
Minimum Charge	Job		169		169	
Better Grade Wood						
Demolish	L.F.		11.95		11.95	Includes material and labor to install
Install	L.F.	288	190		478	custom modular unit with solid
Demolish and Install	L.F.	288	201.95		489.95	hardwood faced frames, hardwood
Reinstall	L.F.		152.34		152.34	door frames and drawer fronts,
Clean	L.F.	1.31	8.65		9.96	hardwood veneer on raised door
Paint	L.F.	.98	23		23.98	panels, hinges and pulls.
Minimum Charge	Job		169		169	
Premium Grade Wood						
Demolish	L.F.		11.95		11.95	Includes material and labor to install
Install	L.F.	385	190		575	premium modular unit with solid
Demolish and Install	L.F.	385	201.95		586.95	hardwood faced frames, hardwood
Reinstall	L.F.		152.34		152.34	door frames and drawer fronts,
Clean	L.F.	1.31	8.65		9.96	hardwood veneer on raised door
Paint	L.F.	.98	24.50		25.48	panels, hinges and pulls.
Minimum Charge	Job		169		169	
Premium Hardwood						
Demolish	L.F.		11.95		11.95	Includes material and labor to install
Install	L.F.	505	190		695	premium hardwood modular unit with
Demolish and Install	L.F.	505	201.95		706.95	solid hardwood faced frames, drawer
Reinstall	L.F.		152.34		152.34	fronts, and door panels, steel drawer
Clean	L.F.	1.31	8.65		9.96	guides, hinges and pulls, laminated
Paint	L.F.	.98	27		27.98	interior.
Minimum Charge	Job		169		169	

Cabinets and Countertops

Tall Cabinets	Unit	Material	Labor	Equip.	Total	Specification
Single Oven, 27″ Wide						
Good Grade Laminated						
Demolish	Ea.		27		27	Includes material and labor to install
Install	Ea.	485	84.50		569.50	good modular unit with melamine
Demolish and Install	Ea.	485	111.50		596.50	laminated to particle board, in
Reinstall	Ea.		67.60		67.60	textured colors or wood grain print
Clean	Ea.	4.10	15.10		19.20	finish, hinges and pulls.
Paint	Ea.	1.64	47.50		49.14	
Minimum Charge	Job		169		169	
Better Grade Veneer						
Demolish	Ea.		27		27	Includes material and labor to install
Install	Ea.	865	84.50		949.50	custom modular unit with solid
Demolish and Install	Ea.	865	111.50		976.50	hardwood faced frames, hardwood
Reinstall	Ea.		67.60		67.60	door frames and drawer fronts,
Clean	Ea.	4.10	15.10		19.20	hardwood veneer on raised door
Paint	Ea.	1.64	51		52.64	panels, hinges and pulls.
Minimum Charge	Job		169		169	
Premium Grade Veneer						
Demolish	Ea.		27		27	Includes material and labor to install
Install	Ea.	765	84.50		849.50	premium modular unit with solid
Demolish and Install	Ea.	765	111.50		876.50	hardwood faced frames, hardwood
Reinstall	Ea.		67.60		67.60	door frames and drawer fronts,
Clean	Ea.	4.10	15.10		19.20	hardwood veneer on raised door
Paint	Ea.	1.64	56		57.64	panels, hinges and pulls.
Minimum Charge	Job		169		169	
Premium Hardwood						
Demolish	Ea.		27		27	Includes material and labor to install
Install	Ea.	985	84.50		1069.50	premium hardwood modular unit with
Demolish and Install	Ea.	985	111.50		1096.50	solid hardwood faced frames, drawer
Reinstall	Ea.		67.60		67.60	fronts, and door panels, steel drawer
Clean	Ea.	4.10	15.10		19.20	guides, hinges and pulls, laminated
Paint	Ea.	1.64	60.50		62.14	interior.
Minimum Charge	Job		169		169	
Single Oven, 30″ Wide						
Good Grade Laminated						
Demolish	Ea.		27		27	Includes material and labor to install
Install	Ea.	440	84.50		524.50	good modular unit with melamine
Demolish and Install	Ea.	440	111.50		551.50	laminated to particle board, in
Reinstall	Ea.		67.60		67.60	textured colors or wood grain print
Clean	Ea.	4.10	15.10		19.20	finish, hinges and pulls.
Paint	Ea.	1.64	47.50		49.14	
Minimum Charge	Job		169		169	
Better Grade Veneer						
Demolish	Ea.		27		27	Includes material and labor to install
Install	Ea.	645	84.50		729.50	custom modular unit with solid
Demolish and Install	Ea.	645	111.50		756.50	hardwood faced frames, hardwood
Reinstall	Ea.		67.60		67.60	door frames and drawer fronts,
Clean	Ea.	4.10	15.10		19.20	hardwood veneer on raised door
Paint	Ea.	1.64	47.50		49.14	panels, hinges and pulls.
Minimum Charge	Job		169		169	

Cabinets and Countertops

Tall Cabinets	Unit	Material	Labor	Equip.	Total	Specification
Premium Grade Veneer						Includes material and labor to install
Demolish	Ea.		27		27	premium modular unit with solid
Install	Ea.	755	84.50		839.50	hardwood faced frames, hardwood
Demolish and Install	Ea.	755	111.50		866.50	door frames and drawer fronts,
Reinstall	Ea.		67.60		67.60	hardwood veneer on raised door
Clean	Ea.	4.10	15.10		19.20	panels, hinges and pulls.
Paint	Ea.	1.64	47.50		49.14	
Minimum Charge	Job		169		169	
Premium Hardwood						Includes material and labor to install
Demolish	Ea.		27		27	premium hardwood modular unit with
Install	Ea.	895	84.50		979.50	solid hardwood faced frames, drawer
Demolish and Install	Ea.	895	111.50		1006.50	fronts, and door panels, steel drawer
Reinstall	Ea.		67.60		67.60	guides, hinges and pulls, laminated
Clean	Ea.	4.10	15.10		19.20	interior.
Paint	Ea.	1.64	47.50		49.14	
Minimum Charge	Job		169		169	

Utility 18″ W x 12″ D

	Unit	Material	Labor	Equip.	Total	Specification
Good Grade Laminated						Includes material and labor to install
Demolish	Ea.		27		27	good modular unit with melamine
Install	Ea.	395	67.50		462.50	laminated to particle board, in
Demolish and Install	Ea.	395	94.50		489.50	textured colors or wood grain print
Reinstall	Ea.		54.08		54.08	finish, hinges and pulls.
Clean	Ea.	.97	7.30		8.27	
Paint	Ea.	.63	18.55		19.18	
Minimum Charge	Job		169		169	
Better Grade Veneer						Includes material and labor to install
Demolish	Ea.		27		27	custom modular unit with solid
Install	Ea.	430	67.50		497.50	hardwood faced frames, hardwood
Demolish and Install	Ea.	430	94.50		524.50	door frames and drawer fronts,
Reinstall	Ea.		54.08		54.08	hardwood veneer on raised door
Clean	Ea.	.97	7.30		8.27	panels, hinges and pulls.
Paint	Ea.	.63	19.75		20.38	
Minimum Charge	Job		169		169	
Premium Grade Veneer						Includes material and labor to install
Demolish	Ea.		27		27	premium modular unit with solid
Install	Ea.	550	67.50		617.50	hardwood faced frames, hardwood
Demolish and Install	Ea.	550	94.50		644.50	door frames and drawer fronts,
Reinstall	Ea.		54.08		54.08	hardwood veneer on raised door
Clean	Ea.	.97	7.30		8.27	panels, hinges and pulls.
Paint	Ea.	.63	23		23.63	
Minimum Charge	Job		169		169	
Premium Hardwood						Includes material and labor to install
Demolish	Ea.		27		27	premium hardwood modular unit with
Install	Ea.	700	67.50		767.50	solid hardwood faced frames, drawer
Demolish and Install	Ea.	700	94.50		794.50	fronts, and door panels, steel drawer
Reinstall	Ea.		54.08		54.08	guides, hinges and pulls, laminated
Clean	Ea.	.97	7.30		8.27	interior.
Paint	Ea.	.63	24.50		25.13	
Minimum Charge	Job		169		169	

Cabinets and Countertops

Tall Cabinets	Unit	Material	Labor	Equip.	Total	Specification
Utility 24″ W x 24″ D						
Good Grade Laminated						
Demolish	Ea.		27		27	Includes material and labor to install
Install	Ea.	480	84.50		564.50	good modular unit with melamine
Demolish and Install	Ea.	480	111.50		591.50	laminated to particle board, in
Reinstall	Ea.		67.60		67.60	textured colors or wood grain print
Clean	Ea.	1.27	9.65		10.92	finish, hinges and pulls.
Paint	Ea.	.84	24.50		25.34	
Minimum Charge	Job		169		169	
Better Grade Veneer						
Demolish	Ea.		27		27	Includes material and labor to install
Install	Ea.	585	84.50		669.50	custom modular unit with solid
Demolish and Install	Ea.	585	111.50		696.50	hardwood faced frames, hardwood
Reinstall	Ea.		67.60		67.60	door frames and drawer fronts,
Clean	Ea.	1.27	9.65		10.92	hardwood veneer on raised door
Paint	Ea.	.84	27		27.84	panels, hinges and pulls.
Minimum Charge	Job		169		169	
Premium Grade Veneer						
Demolish	Ea.		27		27	Includes material and labor to install
Install	Ea.	730	84.50		814.50	premium modular unit with solid
Demolish and Install	Ea.	730	111.50		841.50	hardwood faced frames, hardwood
Reinstall	Ea.		67.60		67.60	door frames and drawer fronts,
Clean	Ea.	1.27	9.65		10.92	hardwood veneer on raised door
Paint	Ea.	.84	29.50		30.34	panels, hinges and pulls.
Minimum Charge	Job		169		169	
Premium Hardwood						
Demolish	Ea.		27		27	Includes material and labor to install
Install	Ea.	895	84.50		979.50	premium hardwood modular unit with
Demolish and Install	Ea.	895	111.50		1006.50	solid hardwood faced frames, drawer
Reinstall	Ea.		67.60		67.60	fronts, and door panels, steel drawer
Clean	Ea.	1.27	9.65		10.92	guides, hinges and pulls, laminated
Paint	Ea.	.84	33		33.84	interior.
Minimum Charge	Job		169		169	
Plain Shelves						
Install	Ea.	29	21		50	Includes labor and material to install
Rotating Shelves						tall cabinet shelving, per set.
Install	Ea.	78	16.90		94.90	Includes labor and material to install lazy susan shelving for wall cabinet.
Strip and Refinish						
Install	L.F.	1.72	13.45		15.17	Includes labor and material to strip,
Minimum Charge	Job		148		148	prep and refinish exterior of cabinets.
Stain						
Install	L.F.	.17	2.69		2.86	Includes labor and material to replace
Minimum Charge	Job		148		148	normal prep and stain on doors and exterior cabinets.

Cabinets and Countertops

Base Cabinets	Unit	Material	Labor	Equip.	Total	Specification
Good Grade Laminated						
12" w, 1 Door, 1 Drawer						
Demolish	Ea.		6.70		6.70	Includes material and labor to install
Install	Ea.	113	27.50		140.50	good modular unit with melamine
Demolish and Install	Ea.	113	34.20		147.20	laminate, textured colors or wood
Reinstall	Ea.		21.81		21.81	grain print finish including hinges and
Clean	Ea.	.66	4.83		5.49	pulls.
Paint	Ea.	.42	12.35		12.77	
Minimum Charge	Job		169		169	
15" w, 1 Door, 1 Drawer						
Demolish	Ea.		8.40		8.40	Includes material and labor to install
Install	Ea.	155	28		183	good modular unit with melamine
Demolish and Install	Ea.	155	36.40		191.40	laminate, textured colors or wood
Reinstall	Ea.		22.53		22.53	grain print finish including hinges and
Clean	Ea.	.83	6.05		6.88	pulls.
Paint	Ea.	.53	15.60		16.13	
Minimum Charge	Job		169		169	
18" w, 1 Door, 1 Drawer						
Demolish	Ea.		9.95		9.95	Includes material and labor to install
Install	Ea.	168	29		197	good modular unit with melamine
Demolish and Install	Ea.	168	38.95		206.95	laminate, textured colors or wood
Reinstall	Ea.		23.21		23.21	grain print finish including hinges and
Clean	Ea.	.97	7.30		8.27	pulls.
Paint	Ea.	.63	18.55		19.18	
Minimum Charge	Job		169		169	
21" w, 1 Door, 1 Drawer						
Demolish	Ea.		11.65		11.65	Includes material and labor to install
Install	Ea.	176	30		206	good modular unit with melamine
Demolish and Install	Ea.	176	41.65		217.65	laminate, textured colors or wood
Reinstall	Ea.		23.82		23.82	grain print finish including hinges and
Clean	Ea.	1.17	8.65		9.82	pulls.
Paint	Ea.	.74	21		21.74	
Minimum Charge	Job		169		169	
24" w, 1 Door, 1 Drawer						
Demolish	Ea.		13.40		13.40	Includes material and labor to install
Install	Ea.	202	30.50		232.50	good modular unit with melamine
Demolish and Install	Ea.	202	43.90		245.90	laminate, textured colors or wood
Reinstall	Ea.		24.25		24.25	grain print finish including hinges and
Clean	Ea.	1.27	9.65		10.92	pulls.
Paint	Ea.	.84	24.50		25.34	
Minimum Charge	Job		169		169	
30" Wide Sink Base						
Demolish	Ea.		16.80		16.80	Includes material and labor to install
Install	Ea.	198	31.50		229.50	good modular unit with melamine
Demolish and Install	Ea.	198	48.30		246.30	laminate, textured colors or wood
Reinstall	Ea.		25.27		25.27	grain print finish including hinges and
Clean	Ea.	1.64	12.10		13.74	pulls.
Paint	Ea.	1.05	29.50		30.55	
Minimum Charge	Job		169		169	
36" Wide Sink Base						
Demolish	Ea.		20.50		20.50	Includes material and labor to install
Install	Ea.	222	33.50		255.50	good modular unit with melamine
Demolish and Install	Ea.	222	54		276	laminate, textured colors or wood
Reinstall	Ea.		26.64		26.64	grain print finish including hinges and
Clean	Ea.	1.83	14.20		16.03	pulls.
Paint	Ea.	1.27	37		38.27	
Minimum Charge	Job		169		169	

Cabinets and Countertops

Base Cabinets	Unit	Material	Labor	Equip.	Total	Specification
36″ Blind Corner						
Demolish	Ea.		20.50		20.50	Includes material and labor to install
Install	Ea.	325	37.50		362.50	good modular unit with melamine
Demolish and Install	Ea.	325	58		383	laminate, textured colors or wood
Reinstall	Ea.		30.04		30.04	grain print finish including hinges and
Clean	Ea.	1.83	14.20		16.03	pulls.
Paint	Ea.	1.27	37		38.27	
Minimum Charge	Job		169		169	
42″ w, 2 Door, 2 Drawer						
Demolish	Ea.		24.50		24.50	Includes material and labor to install
Install	Ea.	276	34		310	good modular unit with melamine
Demolish and Install	Ea.	276	58.50		334.50	laminate, textured colors or wood
Reinstall	Ea.		27.31		27.31	grain print finish including hinges and
Clean	Ea.	2.34	17.25		19.59	pulls.
Paint	Ea.	1.45	42.50		43.95	
Minimum Charge	Job		169		169	
Clean Interior						
Clean	L.F.	.66	3.66		4.32	Includes labor and materials to clean
Minimum Charge	Job		121		121	the interior of a 27″ cabinet.
Better Grade Wood						
12″ w, 1 Door, 1 Drawer						
Demolish	Ea.		6.70		6.70	Includes material and labor to install
Install	Ea.	152	27.50		179.50	custom modular unit including solid
Demolish and Install	Ea.	152	34.20		186.20	hardwood faced frames, hardwood
Reinstall	Ea.		21.81		21.81	door frames and drawer fronts,
Clean	Ea.	.66	4.83		5.49	hardwood veneer on raised door
Paint	Ea.	.42	13.45		13.87	panels, hinges and pulls.
Minimum Charge	Job		169		169	
15″ w, 1 Door, 1 Drawer						
Demolish	Ea.		8.40		8.40	Includes material and labor to install
Install	Ea.	207	28		235	custom modular unit including solid
Demolish and Install	Ea.	207	36.40		243.40	hardwood faced frames, hardwood
Reinstall	Ea.		22.53		22.53	door frames and drawer fronts,
Clean	Ea.	.83	6.05		6.88	hardwood veneer on raised door
Paint	Ea.	.53	17.45		17.98	panels, hinges and pulls.
Minimum Charge	Job		169		169	
18″ w, 1 Door, 1 Drawer						
Demolish	Ea.		9.95		9.95	Includes material and labor to install
Install	Ea.	226	29		255	custom modular unit including solid
Demolish and Install	Ea.	226	38.95		264.95	hardwood faced frames, hardwood
Reinstall	Ea.		23.21		23.21	door frames and drawer fronts,
Clean	Ea.	.97	7.30		8.27	hardwood veneer on raised door
Paint	Ea.	.63	19.75		20.38	panels, hinges and pulls.
Minimum Charge	Job		169		169	
21″ w, 1 Door, 1 Drawer						
Demolish	Ea.		11.65		11.65	Includes material and labor to install
Install	Ea.	234	30		264	custom modular unit including solid
Demolish and Install	Ea.	234	41.65		275.65	hardwood faced frames, hardwood
Reinstall	Ea.		23.82		23.82	door frames and drawer fronts,
Clean	Ea.	1.17	8.65		9.82	hardwood veneer on raised door
Paint	Ea.	.74	23		23.74	panels, hinges and pulls.
Minimum Charge	Job		169		169	

Cabinets and Countertops

Base Cabinets	Unit	Material	Labor	Equip.	Total	Specification
24″ w, 1 Door, 1 Drawer						
Demolish	Ea.		13.40		13.40	Includes material and labor to install
Install	Ea.	270	30.50		300.50	custom modular unit including solid
Demolish and Install	Ea.	270	43.90		313.90	hardwood faced frames, hardwood
Reinstall	Ea.		24.25		24.25	door frames and drawer fronts,
Clean	Ea.	1.27	9.65		10.92	hardwood veneer on raised door
Paint	Ea.	.84	27		27.84	panels, hinges and pulls.
Minimum Charge	Job		169		169	
30″ Wide Sink Base						
Demolish	Ea.		16.80		16.80	Includes material and labor to install
Install	Ea.	264	31.50		295.50	custom modular unit including solid
Demolish and Install	Ea.	264	48.30		312.30	hardwood faced frames, hardwood
Reinstall	Ea.		25.27		25.27	door frames and drawer fronts,
Clean	Ea.	1.64	12.10		13.74	hardwood veneer on raised door
Paint	Ea.	1.05	33		34.05	panels, hinges and pulls.
Minimum Charge	Job		169		169	
36″ Wide Sink Base						
Demolish	Ea.		20.50		20.50	Includes material and labor to install
Install	Ea.	297	33.50		330.50	custom modular unit including solid
Demolish and Install	Ea.	297	54		351	hardwood faced frames, hardwood
Reinstall	Ea.		26.64		26.64	door frames and drawer fronts,
Clean	Ea.	1.83	14.20		16.03	hardwood veneer on raised door
Paint	Ea.	1.27	42.50		43.77	panels, hinges and pulls.
Minimum Charge	Job		169		169	
36″ Blind Corner						
Demolish	Ea.		20.50		20.50	Includes material and labor to install
Install	Ea.	435	37.50		472.50	custom modular unit including solid
Demolish and Install	Ea.	435	58		493	hardwood faced frames, hardwood
Reinstall	Ea.		30.04		30.04	door frames and drawer fronts,
Clean	Ea.	1.83	14.20		16.03	hardwood veneer on raised door
Paint	Ea.	1.27	42.50		43.77	panels, hinges and pulls.
Minimum Charge	Job		169		169	
42″ w, 2 Door, 2 Drawer						
Demolish	Ea.		24.50		24.50	Includes material and labor to install
Install	Ea.	370	34		404	modular unit including solid hardwood
Demolish and Install	Ea.	370	58.50		428.50	faced frames, hardwood door frames
Reinstall	Ea.		27.31		27.31	and drawer fronts, hardwood veneer
Clean	Ea.	2.34	17.25		19.59	on raised door panels, hinges and
Paint	Ea.	1.45	49.50		50.95	pulls.
Minimum Charge	Job		169		169	
Clean Interior						
Clean	L.F.	.66	3.66		4.32	Includes labor and materials to clean
Minimum Charge	Job		121		121	the interior of a 27″ cabinet.
Premium Grade Wood						
12″ w, 1 Door, 1 Drawer						
Demolish	Ea.		6.70		6.70	Includes material and labor to install
Install	Ea.	261	13.65		274.65	premium hardwood modular unit with
Demolish and Install	Ea.	261	20.35		281.35	solid hardwood faced frames, drawer
Reinstall	Ea.		10.90		10.90	fronts, and door panels, steel drawer
Clean	Ea.	.66	4.83		5.49	guides, hinges, pulls and laminated
Paint	Ea.	.42	14.80		15.22	interior.
Minimum Charge	Job		169		169	

Cabinets and Countertops

Base Cabinets		Unit	Material	Labor	Equip.	Total	Specification
15" w, 1 Door, 1 Drawer							
	Demolish	Ea.		8.40		8.40	Includes material and labor to install
	Install	Ea.	271	28		299	premium modular unit including solid
	Demolish and Install	Ea.	271	36.40		307.40	hardwood faced frames, hardwood
	Reinstall	Ea.		22.53		22.53	door frames and drawer fronts,
	Clean	Ea.	.83	6.05		6.88	hardwood veneer on raised door
	Paint	Ea.	.53	18.55		19.08	panels, hinges and pulls.
	Minimum Charge	Job		169		169	
18" w, 1 Door, 1 Drawer							
	Demolish	Ea.		9.95		9.95	Includes material and labor to install
	Install	Ea.	292	29		321	premium modular unit including solid
	Demolish and Install	Ea.	292	38.95		330.95	hardwood faced frames, hardwood
	Reinstall	Ea.		23.21		23.21	door frames and drawer fronts,
	Clean	Ea.	.97	7.30		8.27	hardwood veneer on raised door
	Paint	Ea.	.63	23		23.63	panels, hinges and pulls.
	Minimum Charge	Job		169		169	
21" w, 1 Door, 1 Drawer							
	Demolish	Ea.		11.65		11.65	Includes material and labor to install
	Install	Ea.	305	30		335	premium modular unit including solid
	Demolish and Install	Ea.	305	41.65		346.65	hardwood faced frames, hardwood
	Reinstall	Ea.		23.82		23.82	door frames and drawer fronts,
	Clean	Ea.	1.17	8.65		9.82	hardwood veneer on raised door
	Paint	Ea.	.74	24.50		25.24	panels, hinges and pulls.
	Minimum Charge	Job		169		169	
24" w, 1 Door, 1 Drawer							
	Demolish	Ea.		13.40		13.40	Includes material and labor to install
	Install	Ea.	345	30.50		375.50	premium modular unit including solid
	Demolish and Install	Ea.	345	43.90		388.90	hardwood faced frames, hardwood
	Reinstall	Ea.		24.25		24.25	door frames and drawer fronts,
	Clean	Ea.	1.27	9.65		10.92	hardwood veneer on raised door
	Paint	Ea.	.84	29.50		30.34	panels, hinges and pulls.
	Minimum Charge	Job		169		169	
30" Wide Sink Base							
	Demolish	Ea.		16.80		16.80	Includes material and labor to install
	Install	Ea.	305	31.50		336.50	premium modular unit including solid
	Demolish and Install	Ea.	305	48.30		353.30	hardwood faced frames, hardwood
	Reinstall	Ea.		25.27		25.27	door frames and drawer fronts,
	Clean	Ea.	1.64	12.10		13.74	hardwood veneer on raised door
	Paint	Ea.	1.05	37		38.05	panels, hinges and pulls.
	Minimum Charge	Job		169		169	
36" Wide Sink Base							
	Demolish	Ea.		20.50		20.50	Includes material and labor to install
	Install	Ea.	345	33.50		378.50	premium modular unit including solid
	Demolish and Install	Ea.	345	54		399	hardwood faced frames, hardwood
	Reinstall	Ea.		26.64		26.64	door frames and drawer fronts,
	Clean	Ea.	1.83	14.20		16.03	hardwood veneer on raised door
	Paint	Ea.	1.27	42.50		43.77	panels, hinges and pulls.
	Minimum Charge	Job		169		169	
36" Blind Corner							
	Demolish	Ea.		20.50		20.50	Includes material and labor to install
	Install	Ea.	725	37.50		762.50	premium modular unit including solid
	Demolish and Install	Ea.	725	58		783	hardwood faced frames, hardwood
	Reinstall	Ea.		30.04		30.04	door frames and drawer fronts,
	Clean	Ea.	1.83	14.20		16.03	hardwood veneer on raised door
	Paint	Ea.	1.27	42.50		43.77	panels, hinges and pulls.
	Minimum Charge	Job		169		169	

Base Cabinets		Unit	Material	Labor	Equip.	Total	Specification
42" w, 2 Door, 2 Drawer							
	Demolish	Ea.		24.50		24.50	Includes material and labor to install
	Install	Ea.	490	34		524	premium modular unit including solid
	Demolish and Install	Ea.	490	58.50		548.50	hardwood faced frames, hardwood
	Reinstall	Ea.		27.31		27.31	door frames and drawer fronts,
	Clean	Ea.	2.34	17.25		19.59	hardwood veneer on raised door
	Paint	Ea.	1.45	49.50		50.95	panels, hinges and pulls.
	Minimum Charge	Job		169		169	
Clean Interior							
	Clean	L.F.	.66	3.66		4.32	Includes labor and materials to clean
	Minimum Charge	Job		121		121	the interior of a 27" cabinet.
Premium Hardwood							
12" w, 1 Door, 1 Drawer							
	Demolish	Ea.		6.70		6.70	Includes material and labor to install
	Install	Ea.	320	27.50		347.50	premium hardwood modular unit with
	Demolish and Install	Ea.	320	34.20		354.20	solid hardwood faced frames, drawer
	Reinstall	Ea.		21.81		21.81	fronts, and door panels, steel drawer
	Clean	Ea.	.66	4.83		5.49	guides, hinges, pulls and laminated
	Paint	Ea.	.42	16.45		16.87	interior.
	Minimum Charge	Job		169		169	
15" w, 1 Door, 1 Drawer							
	Demolish	Ea.		8.40		8.40	Includes material and labor to install
	Install	Ea.	325	28		353	premium hardwood modular unit with
	Demolish and Install	Ea.	325	36.40		361.40	solid hardwood faced frames, drawer
	Reinstall	Ea.		22.53		22.53	fronts, and door panels, steel drawer
	Clean	Ea.	.83	6.05		6.88	guides, hinges, pulls and laminated
	Paint	Ea.	.53	21		21.53	interior.
	Minimum Charge	Job		169		169	
18" w, 1 Door, 1 Drawer							
	Demolish	Ea.		9.95		9.95	Includes material and labor to install
	Install	Ea.	350	29		379	premium hardwood modular unit with
	Demolish and Install	Ea.	350	38.95		388.95	solid hardwood faced frames, drawer
	Reinstall	Ea.		23.21		23.21	fronts, and door panels, steel drawer
	Clean	Ea.	.97	7.30		8.27	guides, hinges, pulls and laminated
	Paint	Ea.	.63	24.50		25.13	interior.
	Minimum Charge	Job		169		169	
21" w, 1 Door, 1 Drawer							
	Demolish	Ea.		11.65		11.65	Includes material and labor to install
	Install	Ea.	365	30		395	premium hardwood modular unit with
	Demolish and Install	Ea.	365	41.65		406.65	solid hardwood faced frames, drawer
	Reinstall	Ea.		23.82		23.82	fronts, and door panels, steel drawer
	Clean	Ea.	1.17	8.65		9.82	guides, hinges, pulls and laminated
	Paint	Ea.	.74	27		27.74	interior.
	Minimum Charge	Job		169		169	
24" w, 1 Door, 1 Drawer							
	Demolish	Ea.		13.40		13.40	Includes material and labor to install
	Install	Ea.	405	30.50		435.50	premium hardwood modular unit with
	Demolish and Install	Ea.	405	43.90		448.90	solid hardwood faced frames, drawer
	Reinstall	Ea.		24.25		24.25	fronts, and door panels, steel drawer
	Clean	Ea.	1.27	9.65		10.92	guides, hinges and pulls, laminated
	Paint	Ea.	.84	33		33.84	interior.
	Minimum Charge	Job		169		169	

Cabinets and Countertops

Base Cabinets	Unit	Material	Labor	Equip.	Total	Specification
30" Wide Sink Base						
Demolish	Ea.		16.80		16.80	Includes material and labor to install
Install	Ea.	370	31.50		401.50	premium hardwood modular unit with
Demolish and Install	Ea.	370	48.30		418.30	solid hardwood faced frames, drawer
Reinstall	Ea.		25.27		25.27	fronts, and door panels, steel drawer
Clean	Ea.	1.64	12.10		13.74	guides, hinges, pulls and laminated
Paint	Ea.	1.05	37		38.05	interior.
Minimum Charge	Job		169		169	
36" Wide Sink Base						
Demolish	Ea.		20.50		20.50	Includes material and labor to install
Install	Ea.	405	33.50		438.50	premium hardwood modular unit with
Demolish and Install	Ea.	405	54		459	solid hardwood faced frames, drawer
Reinstall	Ea.		26.64		26.64	fronts, and door panels, steel drawer
Clean	Ea.	1.83	14.20		16.03	guides, hinges, pulls and laminated
Paint	Ea.	1.27	49.50		50.77	interior.
Minimum Charge	Job		169		169	
36" Blind Corner						
Demolish	Ea.		20.50		20.50	Includes material and labor to install
Install	Ea.	780	37.50		817.50	premium hardwood modular unit with
Demolish and Install	Ea.	780	58		838	solid hardwood faced frames, drawer
Reinstall	Ea.		30.04		30.04	fronts, and door panels, steel drawer
Clean	Ea.	1.83	14.20		16.03	guides, hinges, pulls and laminated
Paint	Ea.	1.27	49.50		50.77	interior.
Minimum Charge	Job		169		169	
42" w, 2 Door, 2 Drawer						
Demolish	Ea.		24.50		24.50	Includes material and labor to install
Install	Ea.	645	34		679	premium hardwood modular unit with
Demolish and Install	Ea.	645	58.50		703.50	solid hardwood faced frames, drawer
Reinstall	Ea.		27.31		27.31	fronts, and door panels, steel drawer
Clean	Ea.	2.34	17.25		19.59	guides, hinges, pulls and laminated
Paint	Ea.	1.45	59.50		60.95	interior.
Minimum Charge	Job		169		169	
Clean Interior						
Clean	L.F.	.66	3.66		4.32	Includes labor and materials to clean
Minimum Charge	Job		121		121	the interior of a 27" cabinet.
Island						
24" Wide						
Demolish	Ea.		13.40		13.40	Includes material and labor to install
Install	Ea.	385	28		413	custom modular unit with solid
Demolish and Install	Ea.	385	41.40		426.40	hardwood faced frames, hardwood
Reinstall	Ea.		22.53		22.53	door frames and drawer fronts,
Clean	Ea.	1.83	14.20		16.03	hardwood veneer on raised door
Paint	Ea.	1.25	42.50		43.75	panels, hinges and pulls.
Minimum Charge	Job		169		169	
30" Wide						
Demolish	Ea.		16.80		16.80	Includes material and labor to install
Install	Ea.	450	35		485	custom modular unit with solid
Demolish and Install	Ea.	450	51.80		501.80	hardwood faced frames, hardwood
Reinstall	Ea.		28.17		28.17	door frames and drawer fronts,
Clean	Ea.	2.34	17.25		19.59	hardwood veneer on raised door
Paint	Ea.	1.57	49.50		51.07	panels, hinges and pulls.
Minimum Charge	Job		169		169	

Cabinets and Countertops

Base Cabinets		Unit	Material	Labor	Equip.	Total	Specification
36" Wide							
	Demolish	Ea.		20.50		20.50	Includes material and labor to install
	Install	Ea.	515	42.50		557.50	custom modular unit with solid
	Demolish and Install	Ea.	515	63		578	hardwood faced frames, hardwood
	Reinstall	Ea.		33.80		33.80	door frames and drawer fronts,
	Clean	Ea.	2.73	22		24.73	hardwood veneer on raised door
	Paint	Ea.	1.90	59.50		61.40	panels, hinges and pulls.
	Minimum Charge	Job		169		169	
48" Wide							
	Demolish	Ea.		27		27	Includes material and labor to install
	Install	Ea.	765	56.50		821.50	custom modular unit with solid
	Demolish and Install	Ea.	765	83.50		848.50	hardwood faced frames, hardwood
	Reinstall	Ea.		45.07		45.07	door frames and drawer fronts,
	Clean	Ea.	3.28	27		30.28	hardwood veneer on raised door
	Paint	Ea.	2.18	74		76.18	panels, hinges and pulls.
	Minimum Charge	Job		169		169	
60" Wide							
	Demolish	Ea.		30		30	Includes material and labor to install
	Install	Ea.	900	70.50		970.50	custom modular unit with solid
	Demolish and Install	Ea.	900	100.50		1000.50	hardwood faced frames, hardwood
	Reinstall	Ea.		56.33		56.33	door frames and drawer fronts,
	Clean	Ea.	4.10	34.50		38.60	hardwood veneer on raised door
	Paint	Ea.	2.18	99		101.18	panels, hinges and pulls.
	Minimum Charge	Job		169		169	
Clean Interior							
	Clean	L.F.	.66	3.66		4.32	Includes labor and materials to clean
	Minimum Charge	Job		121		121	the interior of a 27" cabinet.
Liquor Bar							
	Demolish	L.F.		6.70		6.70	Includes material and labor to install
	Install	L.F.	139	34		173	custom modular unit with solid
	Demolish and Install	L.F.	139	40.70		179.70	hardwood faced frames, hardwood
	Reinstall	L.F.		27.04		27.04	door frames and drawer fronts,
	Clean	L.F.	.66	5.05		5.71	hardwood veneer on raised door
	Paint	L.F.	.42	13.45		13.87	panels, hinges and pulls.
	Minimum Charge	Job		169		169	
Clean Interior							
	Clean	L.F.	.66	3.66		4.32	Includes labor and materials to clean
	Minimum Charge	Job		121		121	the interior of a 27" cabinet.
Seal Interior							
	Install	L.F.	.32	4.86		5.18	Includes labor and materials to seal a
	Minimum Charge	Job		148		148	cabinet interior.
Back Bar							
	Demolish	L.F.		6.70		6.70	Includes material and labor to install
	Install	L.F.	139	34		173	custom modular unit with solid
	Demolish and Install	L.F.	139	40.70		179.70	hardwood faced frames, hardwood
	Reinstall	L.F.		27.04		27.04	door frames and drawer fronts,
	Clean	L.F.	.66	5.05		5.71	hardwood veneer on raised door
	Paint	L.F.	.42	13.45		13.87	panels, hinges and pulls.
	Minimum Charge	Job		169		169	
Clean Interior							
	Clean	L.F.	.66	3.66		4.32	Includes labor and materials to clean
	Minimum Charge	Job		121		121	the interior of a 27" cabinet.

Cabinets and Countertops

Base Cabinets		Unit	Material	Labor	Equip.	Total	Specification
Seal Interior							
	Install	L.F.	.32	4.86		5.18	Includes labor and materials to seal a
	Minimum Charge	Job		148		148	cabinet interior.
Storage (garage)							
	Demolish	L.F.		6.70		6.70	Includes material and labor to install
	Install	L.F.	104	34		138	good modular unit with melamine
	Demolish and Install	L.F.	104	40.70		144.70	laminate, textured colors or wood
	Reinstall	L.F.		27.04		27.04	grain print finish including hinges and
	Clean	L.F.	.66	5.05		5.71	pulls.
	Paint	L.F.	.42	13.45		13.87	
	Minimum Charge	Job		169		169	
Clean Interior							
	Clean	L.F.	.66	3.66		4.32	Includes labor and materials to clean
	Minimum Charge	Job		121		121	the interior of a 27" cabinet.
Seal Interior							
	Install	L.F.	.32	4.86		5.18	Includes labor and materials to seal a
	Minimum Charge	Job		148		148	cabinet interior.
Built-in Desk							
	Demolish	L.F.		6.70		6.70	Includes material and labor to install
	Install	L.F.	139	34		173	custom modular unit with solid
	Demolish and Install	L.F.	139	40.70		179.70	hardwood faced frames, hardwood
	Reinstall	L.F.		27.04		27.04	door frames and drawer fronts,
	Clean	L.F.	.66	5.05		5.71	hardwood veneer on raised door
	Paint	L.F.	.42	13.45		13.87	panels, hinges and pulls.
	Minimum Charge	Job		169		169	
Built-in Bookcase							
	Demolish	L.F.		8.95		8.95	Includes material and labor to install
	Install	L.F.	23	34		57	custom modular bookcase unit with
	Demolish and Install	L.F.	23	42.95		65.95	clear pine faced frames, shelves 12"
	Reinstall	L.F.		27.04		27.04	O.C., 7' high, 8" deep.
	Clean	L.F.	1.31	8.65		9.96	
	Paint	L.F.	.98	23		23.98	
	Minimum Charge	Job		169		169	
Strip and Refinish							
	Install	L.F.	1.11	5.40		6.51	Includes labor and material to strip,
	Minimum Charge	Job		148		148	prep and refinish exterior of cabinets.
Stain							
	Install	L.F.	.37	1.38		1.75	Includes labor and material to stain
	Minimum Charge	Job		169		169	door and exterior of cabinets.

Countertop		Unit	Material	Labor	Equip.	Total	Specification
Laminated							
With Splash							
	Demolish	L.F.		4.47		4.47	Includes material and labor to install
	Install	L.F.	28	12.05		40.05	one piece laminated top with 4"
	Demolish and Install	L.F.	28	16.52		44.52	backsplash.
	Reinstall	L.F.		9.66		9.66	
	Clean	L.F.	.08	.60		.68	
	Minimum Charge	Job		169		169	

Cabinets and Countertops

Countertop		Unit	Material	Labor	Equip.	Total	Specification
Roll Top							
	Demolish	L.F.		4.47		4.47	Includes material and labor to install
	Install	L.F.	10	11.25		21.25	one piece laminated top with rolled
	Demolish and Install	L.F.	10	15.72		25.72	drip edge (post formed) and
	Reinstall	L.F.		9.01		9.01	backsplash.
	Clean	L.F.	.12	.92		1.04	
	Minimum Charge	Job		169		169	
Ceramic Tile							
	Demolish	S.F.		2.24		2.24	Cost includes material and labor to
	Install	S.F.	14.45	6.75		21.20	install 4-1/4" x 4-1/4" to 6" x 6"
	Demolish and Install	S.F.	14.45	8.99		23.44	glazed tile set in mortar bed and grout
	Clean	S.F.	.06	.46		.52	on particle board substrate.
	Minimum Charge	Job		154		154	
Cultured Marble							
	Demolish	L.F.		4.47		4.47	Includes material and labor to install
	Install	L.F.	28	34		62	marble countertop, 24" wide, no
	Demolish and Install	L.F.	28	38.47		66.47	backsplash.
	Reinstall	L.F.		27.23		27.23	
	Clean	L.F.	.12	.92		1.04	
	Minimum Charge	Job		154		154	
With Splash							
	Demolish	L.F.		4.47		4.47	Includes material and labor to install
	Install	L.F.	31.50	34		65.50	marble countertop, 24" wide, with
	Demolish and Install	L.F.	31.50	38.47		69.97	backsplash.
	Reinstall	L.F.		27.23		27.23	
	Clean	L.F.	.12	.92		1.04	
	Minimum Charge	Job		154		154	
Quarry Tile							
	Demolish	S.F.		2.24		2.24	Includes material and labor to install
	Install	S.F.	7.85	6.75		14.60	quarry tile countertop with no
	Demolish and Install	S.F.	7.85	8.99		16.84	backsplash.
	Clean	S.F.	.06	.46		.52	
	Minimum Charge	Job		154		154	
With Splash							
	Demolish	S.F.		2.24		2.24	Includes material and labor to install
	Install	S.F.	10.40	7.50		17.90	quarry tile countertop with backsplash.
	Demolish and Install	S.F.	10.40	9.74		20.14	
	Clean	S.F.	.06	.46		.52	
	Minimum Charge	Job		154		154	
Butcher Block							
	Demolish	S.F.		2.24		2.24	Includes material and labor to install
	Install	S.F.	29	6.05		35.05	solid laminated maple countertop with
	Demolish and Install	S.F.	29	8.29		37.29	no backsplash.
	Reinstall	S.F.		4.83		4.83	
	Clean	S.F.	.02	.19		.21	
	Minimum Charge	Job		169		169	
Solid Surface							
	Demolish	L.F.		4.47		4.47	Includes material and labor to install
	Install	L.F.	48.50	24		72.50	solid surface counter top 22" deep.
	Demolish and Install	L.F.	48.50	28.47		76.97	
	Reinstall	L.F.		19.31		19.31	
	Clean	L.F.	.12	.92		1.04	
	Minimum Charge	Job		169		169	

Cabinets and Countertops

Vanity Cabinets	Unit	Material	Labor	Equip.	Total	Specification
Good Grade Laminated						
24" Wide, 2 Door						
Demolish	Ea.		13.40		13.40	Includes material and labor to install
Install	Ea.	152	34		186	modular unit with melamine laminated
Demolish and Install	Ea.	152	47.40		199.40	to particle board, textured colors or
Reinstall	Ea.		27.04		27.04	wood grain print finish, hinges and
Clean	Ea.	1.27	9.65		10.92	pulls.
Paint	Ea.	.84	24.50		25.34	
Minimum Charge	Job		169		169	
30" Wide, 2 Door						
Demolish	Ea.		16.80		16.80	Includes material and labor to install
Install	Ea.	174	42.50		216.50	modular unit with melamine laminated
Demolish and Install	Ea.	174	59.30		233.30	to particle board, textured colors or
Reinstall	Ea.		33.80		33.80	wood grain print finish, hinges and
Clean	Ea.	1.64	12.10		13.74	pulls.
Paint	Ea.	1.05	29.50		30.55	
Minimum Charge	Job		169		169	
36" Wide, 2 Door						
Demolish	Ea.		20.50		20.50	Includes material and labor to install
Install	Ea.	232	50.50		282.50	modular unit with melamine laminated
Demolish and Install	Ea.	232	71		303	to particle board, textured colors or
Reinstall	Ea.		40.57		40.57	wood grain print finish, hinges and
Clean	Ea.	1.83	14.20		16.03	pulls.
Paint	Ea.	1.27	37		38.27	
Minimum Charge	Job		169		169	
42" w, 3 Doors / Drawer						
Demolish	Ea.		24.50		24.50	Includes material and labor to install
Install	Ea.	335	54.50		389.50	modular unit with melamine laminated
Demolish and Install	Ea.	335	79		414	to particle board, textured colors or
Reinstall	Ea.		43.68		43.68	wood grain print finish, hinges and
Clean	Ea.	2.34	17.25		19.59	pulls.
Paint	Ea.	1.45	42.50		43.95	
Minimum Charge	Job		169		169	
48" w, 3 Doors / Drawer						
Demolish	Ea.		27		27	Includes material and labor to install
Install	Ea.	278	59		337	modular unit with melamine laminated
Demolish and Install	Ea.	278	86		364	to particle board, textured colors or
Reinstall	Ea.		47.31		47.31	wood grain print finish, hinges and
Clean	Ea.	3.28	27		30.28	pulls.
Paint	Ea.	2.18	74		76.18	
Minimum Charge	Job		169		169	
Clean Interior						
Install	L.F.	.66	5.05		5.71	Includes labor and materials to clean
Minimum Charge	Job		121		121	base cabinetry.
Better Grade Wood						
24" Wide, 2 Door						
Demolish	Ea.		13.40		13.40	Includes material and labor to install
Install	Ea.	202	34		236	custom modular unit with solid
Demolish and Install	Ea.	202	47.40		249.40	hardwood faced frames, hardwood
Reinstall	Ea.		27.04		27.04	door frames and drawer fronts,
Clean	Ea.	1.27	9.65		10.92	hardwood veneer on raised door
Paint	Ea.	.84	27		27.84	panels, hinges and pulls.
Minimum Charge	Job		169		169	

Cabinets and Countertops

Vanity Cabinets	Unit	Material	Labor	Equip.	Total	Specification
30" Wide, 2 Door						
Demolish	Ea.		16.80		16.80	Includes material and labor to install
Install	Ea.	232	42.50		274.50	custom modular unit with solid
Demolish and Install	Ea.	232	59.30		291.30	hardwood faced frames, hardwood
Reinstall	Ea.		33.80		33.80	door frames and drawer fronts,
Clean	Ea.	1.64	12.10		13.74	hardwood veneer on raised door
Paint	Ea.	1.05	33		34.05	panels, hinges and pulls.
Minimum Charge	Job		169		169	
36" Wide, 2 Door						
Demolish	Ea.		20.50		20.50	Includes material and labor to install
Install	Ea.	310	50.50		360.50	custom modular unit with solid
Demolish and Install	Ea.	310	71		381	hardwood faced frames, hardwood
Reinstall	Ea.		40.57		40.57	door frames and drawer fronts,
Clean	Ea.	1.83	14.20		16.03	hardwood veneer on raised door
Paint	Ea.	1.27	42.50		43.77	panels, hinges and pulls.
Minimum Charge	Job		169		169	
42" w, 3 Doors / Drawer						
Demolish	Ea.		24.50		24.50	Includes material and labor to install
Install	Ea.	335	54.50		389.50	custom modular unit with solid
Demolish and Install	Ea.	335	79		414	hardwood faced frames, hardwood
Reinstall	Ea.		43.68		43.68	door frames and drawer fronts,
Clean	Ea.	2.34	17.25		19.59	hardwood veneer on raised door
Paint	Ea.	1.45	49.50		50.95	panels, hinges and pulls.
Minimum Charge	Job		169		169	
48" w, 3 Doors / Drawer						
Demolish	Ea.		27		27	Includes material and labor to install
Install	Ea.	370	59		429	custom modular unit with solid
Demolish and Install	Ea.	370	86		456	hardwood faced frames, hardwood
Reinstall	Ea.		47.31		47.31	door frames and drawer fronts,
Clean	Ea.	3.28	27		30.28	hardwood veneer on raised door
Paint	Ea.	2.18	74		76.18	panels, hinges and pulls.
Minimum Charge	Job		169		169	
Clean Interior						
Install	L.F.	.66	5.05		5.71	Includes labor and materials to clean
Minimum Charge	Job		121		121	base cabinetry.
Premium Grade Wood						
24" Wide, 2 Door						
Demolish	Ea.		13.40		13.40	Includes material and labor to install
Install	Ea.	240	34		274	premium modular unit with solid
Demolish and Install	Ea.	240	47.40		287.40	hardwood faced frames, hardwood
Reinstall	Ea.		27.04		27.04	door frames and drawer fronts,
Clean	Ea.	1.27	9.65		10.92	hardwood veneer on raised door
Paint	Ea.	.84	29.50		30.34	panels, hinges and pulls.
Minimum Charge	Job		169		169	
30" Wide, 2 Door						
Demolish	Ea.		16.80		16.80	Includes material and labor to install
Install	Ea.	276	42.50		318.50	premium modular unit with solid
Demolish and Install	Ea.	276	59.30		335.30	hardwood faced frames, hardwood
Reinstall	Ea.		33.80		33.80	door frames and drawer fronts,
Clean	Ea.	1.64	12.10		13.74	hardwood veneer on raised door
Paint	Ea.	1.05	37		38.05	panels, hinges and pulls.
Minimum Charge	Job		169		169	

Cabinets and Countertops

Vanity Cabinets	Unit	Material	Labor	Equip.	Total	Specification
36" Wide, 2 Door						
Demolish	Ea.		20.50		20.50	Includes material and labor to install
Install	Ea.	430	50.50		480.50	premium modular unit with solid
Demolish and Install	Ea.	430	71		501	hardwood faced frames, hardwood
Reinstall	Ea.		40.57		40.57	door frames and drawer fronts,
Clean	Ea.	1.83	14.20		16.03	hardwood veneer on raised door
Paint	Ea.	1.27	42.50		43.77	panels, hinges and pulls.
Minimum Charge	Job		169		169	
42" w, 3 Doors / Drawer						
Demolish	Ea.		24.50		24.50	Includes material and labor to install
Install	Ea.	415	54.50		469.50	premium modular unit with solid
Demolish and Install	Ea.	415	79		494	hardwood faced frames, hardwood
Reinstall	Ea.		43.68		43.68	door frames and drawer fronts,
Clean	Ea.	2.34	17.25		19.59	hardwood veneer on raised door
Paint	Ea.	1.45	49.50		50.95	panels, hinges and pulls.
Minimum Charge	Job		169		169	
48" w, 3 Doors / Drawer						
Demolish	Ea.		27		27	Includes material and labor to install
Install	Ea.	500	59		559	premium modular unit with solid
Demolish and Install	Ea.	500	86		586	hardwood faced frames, hardwood
Reinstall	Ea.		47.31		47.31	door frames and drawer fronts,
Clean	Ea.	3.28	27		30.28	hardwood veneer on raised door
Paint	Ea.	2.18	74		76.18	panels, hinges and pulls.
Minimum Charge	Job		169		169	
Clean Interior						
Install	L.F.	.66	5.05		5.71	Includes labor and materials to clean
Minimum Charge	Job		121		121	base cabinetry.
Premium Hardwood						
24" Wide, 2 Door						
Demolish	Ea.		13.40		13.40	Includes material and labor to install
Install	Ea.	300	34		334	premium modular unit with solid
Demolish and Install	Ea.	300	47.40		347.40	hardwood faced frames, hardwood
Reinstall	Ea.		27.04		27.04	door frames and drawer fronts,
Clean	Ea.	1.27	9.65		10.92	hardwood veneer on raised door
Paint	Ea.	.84	33		33.84	panels, hinges and pulls.
Minimum Charge	Job		169		169	
30" Wide, 2 Door						
Demolish	Ea.		16.80		16.80	Includes material and labor to install
Install	Ea.	335	42.50		377.50	premium modular unit with solid
Demolish and Install	Ea.	335	59.30		394.30	hardwood faced frames, hardwood
Reinstall	Ea.		33.80		33.80	door frames and drawer fronts,
Clean	Ea.	1.64	12.10		13.74	hardwood veneer on raised door
Paint	Ea.	1.05	37		38.05	panels, hinges and pulls.
Minimum Charge	Job		169		169	
36" Wide, 2 Door						
Demolish	Ea.		20.50		20.50	Includes material and labor to install
Install	Ea.	495	50.50		545.50	premium modular unit with solid
Demolish and Install	Ea.	495	71		566	hardwood faced frames, hardwood
Reinstall	Ea.		40.57		40.57	door frames and drawer fronts,
Clean	Ea.	1.83	14.20		16.03	hardwood veneer on raised door
Paint	Ea.	1.27	49.50		50.77	panels, hinges and pulls.
Minimum Charge	Job		169		169	

Cabinets and Countertops

Vanity Cabinets	Unit	Material	Labor	Equip.	Total	Specification
42" w, 3 Doors / Drawer						
Demolish	Ea.		24.50		24.50	Includes material and labor to install
Install	Ea.	455	54.50		509.50	premium modular unit with solid
Demolish and Install	Ea.	455	79		534	hardwood faced frames, hardwood
Reinstall	Ea.		43.68		43.68	door frames and drawer fronts,
Clean	Ea.	2.34	17.25		19.59	hardwood veneer on raised door
Paint	Ea.	1.45	59.50		60.95	panels, hinges and pulls.
Minimum Charge	Job		169		169	
48" w, 3 Doors / Drawer						
Demolish	Ea.		27		27	Includes material and labor to install
Install	Ea.	565	59		624	premium modular unit with solid
Demolish and Install	Ea.	565	86		651	hardwood faced frames, hardwood
Reinstall	Ea.		47.31		47.31	door frames and drawer fronts,
Clean	Ea.	3.28	27		30.28	hardwood veneer on raised door
Paint	Ea.	2.18	74		76.18	panels, hinges and pulls.
Minimum Charge	Job		169		169	
Clean Interior						
Install	L.F.	.66	5.05		5.71	Includes labor and materials to clean
Minimum Charge	Job		121		121	base cabinetry.
Strip and Refinish						
Install	L.F.	1.11	5.40		6.51	Includes labor and material to strip,
Minimum Charge	Job		148		148	prep and refinish exterior of cabinets.

Vanity Cabinet Tops	Unit	Material	Labor	Equip.	Total	Specification
Cultured Marble Top						
Demolish	L.F.		4.47		4.47	Includes material and labor to install
Install	L.F.	36	20		56	cultured marble vanity top with
Demolish and Install	L.F.	36	24.47		60.47	integral sink, 22" deep.
Reinstall	L.F.		16.02		16.02	
Clean	L.F.	.12	.92		1.04	
Minimum Charge	Job		154		154	
Laminated Top						
Demolish	L.F.		2.68		2.68	Includes material and labor to install
Install	L.F.	28	12.05		40.05	one piece laminated top with 4"
Demolish and Install	L.F.	28	14.73		42.73	backsplash.
Reinstall	L.F.		9.66		9.66	
Clean	L.F.	.08	.60		.68	
Minimum Charge	Job		169		169	
Solid Surface						
Demolish	L.F.		4.47		4.47	Includes material and labor to install
Install	L.F.	67.50	25		92.50	solid surface vanity top with integral
Demolish and Install	L.F.	67.50	29.47		96.97	sink installed, 22" deep.
Reinstall	L.F.		20.03		20.03	
Clean	L.F.	.12	.92		1.04	
Minimum Charge	Job		169		169	

Painting

Roller Handle   **Paint Brush**

Paint Preparation, Walls		Unit	Material	Labor	Equip.	Total	Specification
Cover / Protect Floors							
	Install	S.F.	.15	.11		.26	Includes material and labor to install
	Minimum Charge	Job		149		149	plastic masking sheet on large
							uninterrupted areas.
Cover / Protect Walls							
	Install	S.F.	.15	.11		.26	Includes material and labor to install
	Minimum Charge	Job		149		149	plastic masking sheet on walls.
Clean Walls							
Light							
	Clean	S.F.		.09		.09	Includes labor to wash gypsum
	Minimum Charge	Job		149		149	drywall or plaster wall surfaces.
Heavy							
	Clean	S.F.		.12		.12	Includes labor and material for heavy
	Minimum Charge	Job		149		149	cleaning (multiple applications) with
							detergent and solvent.
Prep Walls							
	Paint	S.F.	.03	.15		.18	Includes labor and material to prepare
	Minimum Charge	Job		149		149	for painting including scraping,
							patching and puttying.
Sand Walls							
	Paint	S.F.		.10		.10	Includes labor to sand gypsum drywall
	Minimum Charge	Job		149		149	wall surfaces.
Seal Walls							
Large Area							
	Paint	S.F.	.06	.22		.28	Includes labor and material to paint
	Minimum Charge	Job		149		149	with primer / sealer by roller.
Spot							
	Paint	S.F.	.06	.26		.32	Includes labor and material to seal
	Minimum Charge	Job		149		149	existing drywall with shellac-based
							material.

Paint Preparation, Ceilings		Unit	Material	Labor	Equip.	Total	Specification
Cover / Protect Floors							
	Install	S.F.	.15	.11		.26	Includes material and labor to install
	Minimum Charge	Job		149		149	plastic masking sheet on large
							uninterrupted areas.

Painting

Paint Preparation, Ceilings		Unit	Material	Labor	Equip.	Total	Specification
Cover / Protect Walls							
	Install	S.F.	.15	.11		.26	Includes material and labor to install
	Minimum Charge	Job		149		149	plastic masking sheet on walls.
Prep Ceiling							
	Paint	S.F.	.06	.55		.61	Includes labor and material to fill in
	Minimum Charge	Job		149		149	hairline cracks up to 1/8" with filler, sand and prep for paint.
Sand Ceiling							
	Paint	S.F.		.14		.14	Includes labor to sand gypsum drywall
	Minimum Charge	Job		149		149	ceiling surfaces.
Seal Ceiling							
Large Area							
	Paint	S.F.	.06	.23		.29	Includes labor and material to paint
	Minimum Charge	Job		148		148	ceiling, one coat flat latex.
Spot							
	Paint	S.F.	.06	.26		.32	Includes labor and material to seal
	Minimum Charge	Job		148		148	existing drywall with shellac-based material.

Paint / Texture, Walls		Unit	Material	Labor	Equip.	Total	Specification
Texture Walls							
Spray							
	Install	S.F.	.06	.37		.43	Includes labor and material to install
	Minimum Charge	Job		148		148	by spray, texture finish.
Trowel							
	Install	S.F.	.01	.20		.21	Includes labor and material to apply
	Minimum Charge	Job		148		148	hand troweled texture.
Paint Walls 1 Coat							
	Paint	S.F.	.06	.23		.29	Includes labor and material to paint,
	Minimum Charge	Job		148		148	one coat flat latex.
Paint Walls 2 Coats							
	Paint	S.F.	.12	.37		.49	Includes labor and material to paint,
	Minimum Charge	Job		148		148	two coats flat latex.
Paint Walls 3 Coats							
	Paint	S.F.	.18	.46		.64	Includes labor and material to paint,
	Minimum Charge	Job		148		148	three coats flat latex.
Clean and Seal							
	Paint	S.F.	.24	.79		1.03	Includes labor and material to do light
	Minimum Charge	Job		148		148	cleaning, sealing with 2 coats of latex paint.

Paint / Texture, Ceilings		Unit	Material	Labor	Equip.	Total	Specification
Texture Ceiling							
Spray							
	Install	S.F.	.06	.37		.43	Includes labor and material to install
	Minimum Charge	Job		148		148	by spray, texture finish.
Trowel							
	Install	S.F.	.01	.20		.21	Includes labor and material to apply
	Minimum Charge	Job		148		148	hand troweled texture.

Paint / Texture, Ceilings		Unit	Material	Labor	Equip.	Total	Specification
Paint Ceiling, 1 Coat							
	Paint	S.F.	.06	.23		.29	Includes labor and material to paint
	Minimum Charge	Job		148		148	ceiling, one coat flat latex.
Paint Ceiling, 2 Coats							
	Paint	S.F.	.12	.37		.49	Includes labor and material to paint
	Minimum Charge	Job		148		148	two coats flat latex paint.
Paint Ceiling, 3 Coats							
	Paint	S.F.	.18	.44		.62	Includes labor and material to paint
	Minimum Charge	Job		148		148	ceiling, three coats flat latex.
Clean and Seal							
	Paint	S.F.	.18	.62		.80	Includes labor and material for light
	Minimum Charge	Job		148		148	cleaning, sealing and paint 3 coats of flat latex paint.

Flooring

Plank Flooring		Wood Strip Flooring		Tile Flooring		

Wood Plank Flooring	Unit	Material	Labor	Equip.	Total	Specification
Maple						
2-1/4"						
Demolish	S.F.		1.07		1.07	Includes material and labor to install
Install	S.F.	4.31	1.68		5.99	unfinished T&G maple flooring,
Demolish and Install	S.F.	4.31	2.75		7.06	25/32" thick, 2-1/4" wide, random
Reinstall	S.F.		1.35		1.35	3' to 16' lengths including felt
Clean	S.F.		.30		.30	underlayment, nailed in place over
Paint	S.F.	1.18	1.86		3.04	prepared subfloor.
Minimum Charge	Job		169		169	
3-1/4"						
Demolish	S.F.		1.07		1.07	Includes material and labor to install
Install	S.F.	3.36	1.68		5.04	unfinished T&G maple flooring,
Demolish and Install	S.F.	3.36	2.75		6.11	33/32" thick, 3-1/4" wide, random
Reinstall	S.F.		1.35		1.35	3' to 16' lengths including felt
Clean	S.F.		.30		.30	underlayment, nailed in place over
Paint	S.F.	1.18	1.86		3.04	prepared subfloor.
Minimum Charge	Job		169		169	
Sand and Finish						
Install	S.F.	.78	.60		1.38	Includes labor and material to sand (3
Minimum Charge	Job		169		169	passes) and finish and two coats of
						urethane existing floor.
Oak						
2-1/4"						
Demolish	S.F.		1.07		1.07	Includes material and labor to install
Install	S.F.	4.06	1.68		5.74	unfinished T&G oak flooring, 25/32"
Demolish and Install	S.F.	4.06	2.75		6.81	thick, 2-1/4" wide, random 3' to 16'
Reinstall	S.F.		1.35		1.35	lengths including felt underlayment,
Clean	S.F.		.30		.30	nailed in place over prepared
Paint	S.F.	1.18	1.86		3.04	subfloor.
Minimum Charge	Job		169		169	
3-1/4"						
Demolish	S.F.		1.07		1.07	Includes material and labor to install
Install	S.F.	5.50	1.19		6.69	unfinished T&G oak flooring, 25/32"
Demolish and Install	S.F.	5.50	2.26		7.76	thick, 3-1/4" wide, random 3' to 16'
Reinstall	S.F.		.95		.95	lengths including felt underlayment,
Clean	S.F.		.30		.30	nailed in place over prepared
Paint	S.F.	1.18	1.86		3.04	subfloor.
Minimum Charge	Job		169		169	

Wood Plank Flooring		Unit	Material	Labor	Equip.	Total	Specification
Sand and Finish							
	Install	S.F.	1.18	1.86		3.04	Includes labor and material to sand
	Minimum Charge	Job		169		169	and finish (3 passes) with two coats of urethane on a new floor.
Walnut							
3″ to 7″							
	Demolish	S.F.		1.07		1.07	Includes material and labor to install
	Install	S.F.	9.05	1.19		10.24	unfinished T&G walnut flooring,
	Demolish and Install	S.F.	9.05	2.26		11.31	25/32″ thick, random 3′ to 16′
	Reinstall	S.F.		.95		.95	lengths including felt underlayment,
	Clean	S.F.		.30		.30	nailed in place over prepared
	Paint	S.F.	1.18	1.86		3.04	subfloor.
	Minimum Charge	Job		169		169	
Sand and Finish							
	Install	S.F.	1.18	1.86		3.04	Includes labor and material to sand
	Minimum Charge	Job		169		169	and finish (3 passes) with two coats of urethane on a new floor.
Pine							
2-1/4″							
	Demolish	S.F.		1.07		1.07	Includes material and labor to install
	Install	S.F.	7.35	1.27		8.62	unfinished T&G pine flooring, 25/32″
	Demolish and Install	S.F.	7.35	2.34		9.69	thick, 2-1/4″ wide, random 3′ to 16′
	Reinstall	S.F.		1.02		1.02	lengths including felt underlayment,
	Clean	S.F.		.30		.30	nailed in place over prepared
	Paint	S.F.	.78	.60		1.38	subfloor.
	Minimum Charge	Job		169		169	
Sand and Finish							
	Install	S.F.	1.18	1.86		3.04	Includes labor and material to sand
	Minimum Charge	Job		169		169	and finish (3 passes) with two coats of urethane on a new floor.
Prefinished Oak							
2-1/4″							
	Demolish	S.F.		1.07		1.07	Includes material and labor to install
	Install	S.F.	5.60	1.68		7.28	prefinished T&G oak flooring, 2-1/4″
	Demolish and Install	S.F.	5.60	2.75		8.35	wide, random 3′ to 16′ lengths
	Reinstall	S.F.		1.35		1.35	including felt underlayment, nailed in
	Clean	S.F.		.30		.30	place over prepared subfloor.
	Minimum Charge	Job		169		169	
3-1/4″							
	Demolish	S.F.		1.07		1.07	Includes material and labor to install
	Install	S.F.	7.15	1.55		8.70	prefinished T&G oak flooring, 25/32″
	Demolish and Install	S.F.	7.15	2.62		9.77	thick, 3″ to 7″ wide, random 3′ to 16′
	Reinstall	S.F.		1.24		1.24	lengths including felt underlayment,
	Clean	S.F.		.30		.30	nailed in place over prepared
	Minimum Charge	Job		169		169	subfloor.
Sand and Finish New							
	Install	S.F.	.78	.82		1.60	Includes labor and material to sand (3
	Minimum Charge	Job		169		169	passes) and finish, two coats of urethane, new floor.
Refinish Existing							
	Install	S.F.	.78	.60		1.38	Includes labor and material to sand (3
	Minimum Charge	Job		169		169	passes) and finish and two coats of urethane existing floor.

Flooring

Wood Plank Flooring

Wood Plank Flooring		Unit	Material	Labor	Equip.	Total	Specification
Urethane Coat							
	Install	S.F.	.14	.16		.30	Includes labor and material to install
	Minimum Charge	Job		169		169	two coats of urethane, on existing floor.
Clean and Wax							
	Clean	S.F.	.02	.19		.21	Includes labor and material to wash
	Minimum Charge	Job		169		169	and wax a hardwood floor.
Underlayment Per S.F.							
	Demolish	S.F.		.49		.49	Cost includes material and labor to
	Install	S.F.	.77	.45		1.22	install 1/4" lauan subfloor, standard
	Demolish and Install	S.F.	.77	.94		1.71	interior grade, nailed every 6".
	Clean	S.F.		.23		.23	
	Minimum Charge	Job		169		169	

Wood Parquet Tile Floor

Wood Parquet Tile Floor		Unit	Material	Labor	Equip.	Total	Specification
Oak							
9" x 9"							
	Demolish	S.F.		.83		.83	Includes material and labor to install
	Install	S.F.	4.76	2.70		7.46	prefinished oak 9" x 9" parquet block
	Demolish and Install	S.F.	4.76	3.53		8.29	flooring, 5/16" thick, installed in
	Reinstall	S.F.		2.16		2.16	mastic.
	Clean	S.F.		.30		.30	
	Paint	S.F.	1.18	1.86		3.04	
	Minimum Charge	Job		169		169	
13" x 13"							
	Demolish	S.F.		.83		.83	Includes material and labor to install
	Install	S.F.	9	2.11		11.11	prefinished oak 13" x 13" parquet
	Demolish and Install	S.F.	9	2.94		11.94	block flooring, 5/16" thick, installed in
	Reinstall	S.F.		1.69		1.69	mastic.
	Clean	S.F.		.30		.30	
	Paint	S.F.	1.18	1.86		3.04	
	Minimum Charge	Job		169		169	
Cherry							
9" x 9"							
	Demolish	S.F.		.83		.83	Includes material and labor to install
	Install	S.F.	8.15	2.70		10.85	prefinished cherry 9" x 9" parquet
	Demolish and Install	S.F.	8.15	3.53		11.68	block flooring, 5/16" thick, installed in
	Reinstall	S.F.		2.70		2.70	mastic.
	Clean	S.F.		.30		.30	
	Paint	S.F.	1.18	1.86		3.04	
	Minimum Charge	Job		169		169	
13" x 13"							
	Demolish	S.F.		.83		.83	Includes material and labor to install
	Install	S.F.	9.30	2.11		11.41	prefinished cherry 13" x 13" parquet
	Demolish and Install	S.F.	9.30	2.94		12.24	block flooring, 5/16" thick, installed in
	Reinstall	S.F.		2.11		2.11	mastic.
	Clean	S.F.		.30		.30	
	Paint	S.F.	1.18	1.86		3.04	
	Minimum Charge	Job		169		169	

Flooring

Wood Parquet Tile Floor	Unit	Material	Labor	Equip.	Total	Specification
Walnut						
9″ x 9″						
Demolish	S.F.		.83		.83	Includes material and labor to install
Install	S.F.	5.15	2.70		7.85	prefinished walnut 9″ x 9″ parquet
Demolish and Install	S.F.	5.15	3.53		8.68	flooring in mastic.
Reinstall	S.F.		2.16		2.16	
Clean	S.F.		.30		.30	
Paint	S.F.	1.18	1.86		3.04	
Minimum Charge	Job		169		169	
13″ x 13″						
Demolish	S.F.		.83		.83	Includes material and labor to install
Install	S.F.	9.65	2.11		11.76	prefinished walnut 13″ x 13″ parquet
Demolish and Install	S.F.	9.65	2.94		12.59	block flooring, 5/16″ thick, installed in
Reinstall	S.F.		2.11		2.11	mastic.
Clean	S.F.		.30		.30	
Paint	S.F.	1.18	1.86		3.04	
Minimum Charge	Job		169		169	
Acrylic Impregnated						
Oak 12″ x 12″						
Demolish	S.F.		.83		.83	Includes material and labor to install
Install	S.F.	7.75	2.41		10.16	red oak 12″ x 12″ acrylic
Demolish and Install	S.F.	7.75	3.24		10.99	impregnated parquet block flooring,
Reinstall	S.F.		2.41		2.41	5/16″ thick, installed in mastic.
Clean	S.F.		.30		.30	
Minimum Charge	Job		169		169	
Cherry 12″ x 12″						
Demolish	S.F.		.83		.83	Includes material and labor to install
Install	S.F.	11.40	2.41		13.81	prefinished cherry 12″ x 12″ acrylic
Demolish and Install	S.F.	11.40	3.24		14.64	impregnated parquet block flooring,
Reinstall	S.F.		2.41		2.41	5/16″ thick, installed in mastic.
Clean	S.F.		.30		.30	
Minimum Charge	Job		169		169	
Ash 12″ x 12″						
Demolish	S.F.		.83		.83	Includes material and labor to install
Install	S.F.	10.25	2.41		12.66	ash 12″ x 12″ acrylic impregnated
Demolish and Install	S.F.	10.25	3.24		13.49	parquet block flooring, 5/16″ thick in
Reinstall	S.F.		2.41		2.41	mastic.
Clean	S.F.		.30		.30	
Minimum Charge	Job		169		169	
Teak						
Demolish	S.F.		.83		.83	Includes material and labor to install
Install	S.F.	7.20	2.11		9.31	prefinished teak parquet block
Demolish and Install	S.F.	7.20	2.94		10.14	flooring, 5/16″ thick, and installed in
Reinstall	S.F.		2.11		2.11	mastic.
Clean	S.F.		.30		.30	
Paint	S.F.	1.18	1.86		3.04	
Minimum Charge	Job		169		169	
Clean and Wax						
Clean	S.F.	.02	.19		.21	Includes labor and material to wash
Minimum Charge	Job		169		169	and wax a hardwood floor.

Flooring

Ceramic Tile Flooring	Unit	Material	Labor	Equip.	Total	Specification
Economy Grade						
Demolish	S.F.		.80		.80	Cost includes material and labor to
Install	S.F.	4.29	6.05		10.34	install 4-1/4" x 4-1/4" ceramic tile in
Demolish and Install	S.F.	4.29	6.85		11.14	mortar bed including grout.
Clean	S.F.		.46		.46	
Minimum Charge	Job		154		154	
Average Grade						
Demolish	S.F.		.80		.80	Cost includes material and labor to
Install	S.F.	5.70	6.20		11.90	install 4-1/4" x 4-1/4" ceramic tile in
Demolish and Install	S.F.	5.70	7		12.70	mortar bed including grout.
Clean	S.F.		.46		.46	
Minimum Charge	Job		154		154	
Premium Grade						
Demolish	S.F.		.80		.80	Cost includes material and labor to
Install	S.F.	8.95	6.20		15.15	install 4-1/4" x 4-1/4" ceramic tile in
Demolish and Install	S.F.	8.95	7		15.95	mortar bed including grout.
Clean	S.F.		.46		.46	
Minimum Charge	Job		154		154	
Re-grout						
Install	S.F.	.14	2.47		2.61	Includes material and labor to regrout
Minimum Charge	Job		154		154	tile floors.

Hard Tile Flooring	Unit	Material	Labor	Equip.	Total	Specification
Thick Set Paver						
Brick						
Demolish	S.F.		1.68		1.68	Cost includes material and labor to
Install	S.F.	.86	14.25		15.11	install 6" x 12" adobe brick paver with
Demolish and Install	S.F.	.86	15.93		16.79	1/2" mortar joints.
Clean	S.F.		.46		.46	
Minimum Charge	Job		154		154	
Mexican Red						
Demolish	S.F.		1.68		1.68	Cost includes material and labor to
Install	S.F.	1.53	6.45		7.98	install 12" x 12" red Mexican paver
Demolish and Install	S.F.	1.53	8.13		9.66	tile in mortar bed with grout.
Clean	S.F.		.46		.46	
Minimum Charge	Job		154		154	
Saltillo						
Demolish	S.F.		.80		.80	Cost includes material and labor to
Install	S.F.	1.08	6.45		7.53	install 12" x 12"saltillo tile in mortar
Demolish and Install	S.F.	1.08	7.25		8.33	bed with grout.
Clean	S.F.		.46		.46	
Minimum Charge	Job		154		154	
Marble						
Premium Grade						
Demolish	S.F.		.80		.80	Cost includes material and labor to
Install	S.F.	18.90	13.45		32.35	install 3/8" x 12" x 12" marble, in
Demolish and Install	S.F.	18.90	14.25		33.15	mortar bed with grout.
Clean	S.F.		.46		.46	
Minimum Charge	Job		154		154	
Terrazzo						
Gray Cement						
Demolish	S.F.		1.07		1.07	Cost includes material and labor to
Install	S.F.	2.81	4.26	1.99	9.06	install 1-3/4" terrazzo , #1 and #2
Demolish and Install	S.F.	2.81	5.33	1.99	10.13	chips in gray Portland cement.
Clean	S.F.		.46		.46	
Minimum Charge	Job		277	130	407	

Flooring

Hard Tile Flooring		Unit	Material	Labor	Equip.	Total	Specification
White Cement							
	Demolish	S.F.		1.07		1.07	Cost includes material and labor to
	Install	S.F.	3.19	4.26	1.99	9.44	install 1-3/4" terrazzo, #1 and #2
	Demolish and Install	S.F.	3.19	5.33	1.99	10.51	chips in white Portland cement.
	Clean	S.F.		.46		.46	
	Minimum Charge	Job		277	130	407	
Non-skid Gray							
	Demolish	S.F.		1.07		1.07	Cost includes material and labor to
	Install	S.F.	5.10	9.25	4.32	18.67	install 1-3/4" terrazzo, #1 and #2
	Demolish and Install	S.F.	5.10	10.32	4.32	19.74	chips in gray Portland cement with
	Clean	S.F.		.46		.46	light non-skid abrasive.
	Minimum Charge	Job		277	130	407	
Non-skid White							
	Demolish	S.F.		1.07		1.07	Cost includes material and labor to
	Install	S.F.	5.60	9.25	4.32	19.17	install 1-3/4" terrazzo, #1 and #2
	Demolish and Install	S.F.	5.60	10.32	4.32	20.24	chips in white Portland cement with
	Clean	S.F.		.46		.46	light non-skid abrasive.
	Minimum Charge	Job		277	130	407	
Gray w / Brass Divider							
	Demolish	S.F.		1.07		1.07	Cost includes material and labor to
	Install	S.F.	7.05	10.10	4.71	21.86	install 1-3/4" terrazzo, #1 and #2
	Demolish and Install	S.F.	7.05	11.17	4.71	22.93	chips in gray Portland cement with
	Clean	S.F.		.46		.46	brass strips 2' O.C. each way.
	Minimum Charge	Job		277	130	407	
Clean							
	Clean	S.F.	.14	.31		.45	Includes labor and material to clean
	Minimum Charge	Job		277	130	407	terrazzo flooring.
Slate							
	Demolish	S.F.		.96		.96	Includes material and labor to install
	Install	S.F.	4.69	3.04		7.73	slate tile flooring in thin set with grout.
	Demolish and Install	S.F.	4.69	4		8.69	
	Clean	S.F.		.46		.46	
	Minimum Charge	Job		154		154	
Granite							
	Demolish	S.F.		.79		.79	Includes material and labor to install
	Install	S.F.	18.55	3.09		21.64	granite tile flooring in thin set with
	Demolish and Install	S.F.	18.55	3.88		22.43	grout.
	Reinstall	S.F.		3.09		3.09	
	Clean	S.F.		.46		.46	
	Minimum Charge	Job		154		154	
Re-grout							
	Install	S.F.	.14	2.47		2.61	Includes material and labor to regrout
	Minimum Charge	Job		154		154	tile floors.

Sheet Vinyl Per S.F.		Unit	Material	Labor	Equip.	Total	Specification
Economy Grade							
	Demolish	S.F.		.38		.38	Includes material and labor to install
	Install	S.F.	1.62	.96		2.58	resilient sheet vinyl flooring.
	Demolish and Install	S.F.	1.62	1.34		2.96	
	Clean	S.F.		.30		.30	
	Minimum Charge	Job		154		154	

Flooring

Sheet Vinyl Per S.F.

	Unit	Material	Labor	Equip.	Total	Specification
Average Grade						
Demolish	S.F.		.38		.38	Includes material and labor to install
Install	S.F.	2.61	.96		3.57	resilient sheet vinyl flooring.
Demolish and Install	S.F.	2.61	1.34		3.95	
Clean	S.F.		.30		.30	
Minimum Charge	Job		154		154	
Premium Grade						
Demolish	S.F.		.38		.38	Includes material and labor to install
Install	S.F.	4.10	.96		5.06	resilient sheet vinyl flooring.
Demolish and Install	S.F.	4.10	1.34		5.44	
Clean	S.F.		.30		.30	
Minimum Charge	Job		154		154	
Clean and Wax						
Clean	S.F.	.02	.15		.17	Includes labor and material to wash
Minimum Charge	Job		154		154	and wax a vinyl tile floor.

Sheet Vinyl

	Unit	Material	Labor	Equip.	Total	Specification
Economy Grade						
Demolish	S.Y.		3.46		3.46	Includes material and labor to install
Install	S.Y.	14.55	8.70		23.25	no wax sheet vinyl flooring.
Demolish and Install	S.Y.	14.55	12.16		26.71	
Clean	S.Y.		2.10		2.10	
Minimum Charge	Job		154		154	
Average Grade						
Demolish	S.Y.		3.46		3.46	Includes material and labor to install
Install	S.Y.	23.50	8.70		32.20	no wax sheet vinyl flooring.
Demolish and Install	S.Y.	23.50	12.16		35.66	
Clean	S.Y.		2.10		2.10	
Minimum Charge	Job		154		154	
Premium Grade						
Demolish	S.Y.		3.46		3.46	Includes material and labor to install
Install	S.Y.	37	8.70		45.70	no wax sheet vinyl flooring.
Demolish and Install	S.Y.	37	12.16		49.16	
Clean	S.Y.		2.10		2.10	
Minimum Charge	Job		154		154	
Clean and Wax						
Clean	S.F.	.02	.15		.17	Includes labor and material to wash
Minimum Charge	Job		154		154	and wax a vinyl tile floor.

Vinyl Tile

	Unit	Material	Labor	Equip.	Total	Specification
Economy Grade						
Demolish	S.F.		.54		.54	Includes material and labor to install
Install	S.F.	2.31	.62		2.93	no wax 12" x 12" vinyl tile.
Demolish and Install	S.F.	2.31	1.16		3.47	
Clean	S.F.		.30		.30	
Minimum Charge	Job		154		154	
Average Grade						
Demolish	S.F.		.54		.54	Includes material and labor to install
Install	S.F.	5.55	.62		6.17	no wax 12" x 12" vinyl tile.
Demolish and Install	S.F.	5.55	1.16		6.71	
Clean	S.F.		.30		.30	
Minimum Charge	Job		154		154	

Flooring

Vinyl Tile		Unit	Material	Labor	Equip.	Total	Specification
Premium Grade							
	Demolish	S.F.		.54		.54	Includes material and labor to install
	Install	S.F.	10	.62		10.62	no wax 12″ x 12″ vinyl tile.
	Demolish and Install	S.F.	10	1.16		11.16	
	Clean	S.F.		.30		.30	
	Minimum Charge	Job		154		154	
Clean and Wax							
	Clean	S.F.	.02	.15		.17	Includes labor and material to wash
	Minimum Charge	Job		154		154	and wax a vinyl tile floor.

Carpeting Per S.F.		Unit	Material	Labor	Equip.	Total	Specification
Economy Grade							
	Demolish	S.F.		.06		.06	Includes material and labor to install
	Install	S.F.	1.14	.60		1.74	carpet including tack strips and hot
	Demolish and Install	S.F.	1.14	.66		1.80	melt tape on seams.
	Clean	S.F.		.42		.42	
	Minimum Charge	Job		154		154	
Average Grade							
	Demolish	S.F.		.06		.06	Includes material and labor to install
	Install	S.F.	1.34	.60		1.94	carpet including tack strips and hot
	Demolish and Install	S.F.	1.34	.66		2	melt tape on seams.
	Clean	S.F.		.42		.42	
	Minimum Charge	Job		154		154	
Premium Grade							
	Demolish	S.F.		.06		.06	Includes material and labor to install
	Install	S.F.	2.31	.60		2.91	carpet including tack strips and hot
	Demolish and Install	S.F.	2.31	.66		2.97	melt tape on seams.
	Clean	S.F.		.42		.42	
	Minimum Charge	Job		154		154	
Indoor/Outdoor							
	Demolish	S.F.		.06		.06	Includes material and labor to install
	Install	S.F.	1.14	.73		1.87	indoor-outdoor carpet.
	Demolish and Install	S.F.	1.14	.79		1.93	
	Clean	S.F.		.42		.42	
	Minimum Charge	Job		154		154	
100% Wool							
Average Grade							
	Demolish	S.F.		.06		.06	Includes material and labor to install
	Install	S.F.	7.10	.61		7.71	carpet including tack strips and hot
	Demolish and Install	S.F.	7.10	.67		7.77	melt tape on seams.
	Clean	S.F.		.42		.42	
	Minimum Charge	Job		154		154	
Premium Grade							
	Demolish	S.F.		.06		.06	Includes material and labor to install
	Install	S.F.	7.90	.61		8.51	carpet including tack strips and hot
	Demolish and Install	S.F.	7.90	.67		8.57	melt tape on seams.
	Clean	S.F.		.42		.42	
	Minimum Charge	Job		154		154	
Luxury Grade							
	Demolish	S.F.		.06		.06	Includes material and labor to install
	Install	S.F.	11.20	.61		11.81	carpet including tack strips and hot
	Demolish and Install	S.F.	11.20	.67		11.87	melt tape on seams.
	Clean	S.F.		.42		.42	
	Minimum Charge	Job		154		154	

Flooring

Carpeting Per S.F.

Carpeting Per S.F.	Unit	Material	Labor	Equip.	Total	Specification
Berber						
Average Grade						
Demolish	S.F.		.06		.06	Includes material and labor to install
Install	S.F.	3.05	.73		3.78	carpet including tack strips and hot
Demolish and Install	S.F.	3.05	.79		3.84	melt tape on seams.
Clean	S.F.		.42		.42	
Minimum Charge	Job		154		154	
Premium Grade						
Demolish	S.F.		.06		.06	Includes material and labor to install
Install	S.F.	5.90	.73		6.63	carpet including tack strips and hot
Demolish and Install	S.F.	5.90	.79		6.69	melt tape on seams.
Clean	S.F.		.42		.42	
Minimum Charge	Job		154		154	

Carpeting

Carpeting	Unit	Material	Labor	Equip.	Total	Specification
Indoor/Outdoor						
Demolish	S.Y.		.89		.89	Includes material and labor to install
Install	S.Y.	10.35	6.55		16.90	indoor-outdoor carpet.
Demolish and Install	S.Y.	10.35	7.44		17.79	
Reinstall	S.Y.		5.26		5.26	
Clean	S.Y.		3.78		3.78	
Minimum Charge	Job		154		154	
Economy Grade						
Demolish	S.Y.		.89		.89	Includes material and labor to install
Install	S.Y.	10.30	5.40		15.70	carpet including tack strips and hot
Demolish and Install	S.Y.	10.30	6.29		16.59	melt tape on seams.
Reinstall	S.Y.		4.33		4.33	
Clean	S.Y.		3.78		3.78	
Minimum Charge	Job		154		154	
Average Grade						
Demolish	S.Y.		.89		.89	Includes material and labor to install
Install	S.Y.	12.05	5.40		17.45	carpet including tack strips and hot
Demolish and Install	S.Y.	12.05	6.29		18.34	melt tape on seams.
Reinstall	S.Y.		4.33		4.33	
Clean	S.Y.		3.78		3.78	
Minimum Charge	Job		154		154	
Premium Grade						
Demolish	S.Y.		.89		.89	Includes material and labor to install
Install	S.Y.	21	5.40		26.40	carpet including tack strips and hot
Demolish and Install	S.Y.	21	6.29		27.29	melt tape on seams.
Reinstall	S.Y.		4.33		4.33	
Clean	S.Y.		3.78		3.78	
Minimum Charge	Job		154		154	
100% Wool						
Average Grade						
Demolish	S.Y.		.89		.89	Includes material and labor to install
Install	S.Y.	64	5.50		69.50	wool carpet including tack strips and
Demolish and Install	S.Y.	64	6.39		70.39	hot melt tape on seams.
Reinstall	S.Y.		4.41		4.41	
Clean	S.Y.		3.78		3.78	
Minimum Charge	Job		154		154	

Flooring

Carpeting		Unit	Material	Labor	Equip.	Total	Specification
Premium Grade							
	Demolish	S.Y.		.89		.89	Includes material and labor to install
	Install	S.Y.	71.50	5.50		77	carpet including tack strips and hot
	Demolish and Install	S.Y.	71.50	6.39		77.89	melt tape on seams.
	Reinstall	S.Y.		4.41		4.41	
	Clean	S.Y.		3.78		3.78	
	Minimum Charge	Job		154		154	
Luxury Grade							
	Demolish	S.Y.		.89		.89	Includes material and labor to install
	Install	S.Y.	101	5.50		106.50	carpet including tack strips and hot
	Demolish and Install	S.Y.	101	6.39		107.39	melt tape on seams.
	Reinstall	S.Y.		4.41		4.41	
	Clean	S.Y.		3.78		3.78	
	Minimum Charge	Job		154		154	
Berber							
Average Grade							
	Demolish	S.Y.		.89		.89	Includes material and labor to install
	Install	S.Y.	27.50	6.55		34.05	berber carpet including tack strips and
	Demolish and Install	S.Y.	27.50	7.44		34.94	hot melt tape on seams.
	Reinstall	S.Y.		5.26		5.26	
	Clean	S.Y.		3.78		3.78	
	Minimum Charge	Job		154		154	
Premium Grade							
	Demolish	S.Y.		.89		.89	Includes material and labor to install
	Install	S.Y.	53	6.55		59.55	carpet including tack strips and hot
	Demolish and Install	S.Y.	53	7.44		60.44	melt tape on seams.
	Reinstall	S.Y.		5.26		5.26	
	Clean	S.Y.		3.78		3.78	
	Minimum Charge	Job		154		154	

Carpet Pad Per S.F.		Unit	Material	Labor	Equip.	Total	Specification
Urethane							
	Demolish	S.F.		.02		.02	Includes material and labor to install
	Install	S.F.	.39	.23		.62	urethane carpet pad.
	Demolish and Install	S.F.	.39	.25		.64	
	Minimum Charge	Job		154		154	
Foam Rubber Slab							
	Demolish	S.F.		.02		.02	Includes material and labor to install
	Install	S.F.	.52	.23		.75	carpet pad, 3/8" thick foam rubber.
	Demolish and Install	S.F.	.52	.25		.77	
	Minimum Charge	Job		154		154	
Waffle							
	Demolish	S.F.		.02		.02	Includes material and labor to install
	Install	S.F.	.35	.03		.38	rubber waffle carpet pad.
	Demolish and Install	S.F.	.35	.05		.40	
	Minimum Charge	Job		154		154	
Jute							
	Demolish	S.F.		.02		.02	Includes material and labor to install
	Install	S.F.	.64	.20		.84	jute hair carpet pad.
	Demolish and Install	S.F.	.64	.22		.86	
	Minimum Charge	Job		154		154	
Rebound							
	Demolish	S.F.		.02		.02	Includes material and labor to install
	Install	S.F.	.30	.17		.47	rebound carpet pad.
	Demolish and Install	S.F.	.30	.19		.49	
	Minimum Charge	Job		154		154	

Flooring

Carpet Pad	Unit	Material	Labor	Equip.	Total	Specification
Urethane						
Demolish	S.Y.		.15		.15	Includes material and labor to install
Install	S.Y.	3.51	2.06		5.57	urethane carpet pad.
Demolish and Install	S.Y.	3.51	2.21		5.72	
Reinstall	S.Y.		1.65		1.65	
Minimum Charge	Job		154		154	
Foam Rubber Slab						
Demolish	S.Y.		.15		.15	Includes material and labor to install
Install	S.Y.	4.65	2.06		6.71	rubber slab carpet pad.
Demolish and Install	S.Y.	4.65	2.21		6.86	
Reinstall	S.Y.		1.65		1.65	
Minimum Charge	Job		154		154	
Waffle						
Demolish	S.Y.		.15		.15	Includes material and labor to install
Install	S.Y.	3.17	.23		3.40	rubber waffle carpet pad.
Demolish and Install	S.Y.	3.17	.38		3.55	
Reinstall	S.Y.		.18		.18	
Minimum Charge	Job		154		154	
Jute						
Demolish	S.Y.		.15		.15	Includes material and labor to install
Install	S.Y.	5.70	1.82		7.52	jute hair carpet pad.
Demolish and Install	S.Y.	5.70	1.97		7.67	
Reinstall	S.Y.		1.45		1.45	
Minimum Charge	Job		154		154	
Rebound						
Demolish	S.Y.		.15		.15	Includes material and labor to install
Install	S.Y.	2.69	1.54		4.23	rebound carpet pad.
Demolish and Install	S.Y.	2.69	1.69		4.38	
Reinstall	S.Y.		1.23		1.23	
Minimum Charge	Job		154		154	

Stair Components	Unit	Material	Labor	Equip.	Total	Specification
Carpeting						
Demolish	Ea.		1.63		1.63	Includes material and labor to install
Install	Riser	36	8.15		44.15	carpet and pad on stairs.
Demolish and Install	Riser	36	9.78		45.78	
Reinstall	Riser		6.50		6.50	
Clean	Ea.		3.78		3.78	
Minimum Charge	Job		154		154	

Finish Mechanical

Tub/Shower

Stainless Steel Double Sink

Water Closet

Commode	Unit	Material	Labor	Equip.	Total	Specification
Demolish	Ea.		37		37	Includes material and labor to install a
Install	Ea.	186	126		312	toilet with valve, seat and cover.
Demolish and Install	Ea.	186	163		349	
Reinstall	Ea.		100.83		100.83	
Clean	Ea.	.33	12.10		12.43	
Minimum Charge	Job		186		186	

Wall w / Flush Valve

	Unit	Material	Labor	Equip.	Total	Specification
Demolish	Ea.		37		37	Includes material and labor to install a
Install	Ea.	385	115		500	wall hung toilet with flush valve, seat
Demolish and Install	Ea.	385	152		537	and cover.
Reinstall	Ea.		92.14		92.14	
Clean	Ea.	.33	12.10		12.43	
Minimum Charge	Job		186		186	

Designer Floor Mount

	Unit	Material	Labor	Equip.	Total	Specification
Demolish	Ea.		37		37	Includes material and labor to install a
Install	Ea.	545	126		671	toilet with valve, seat and cover.
Demolish and Install	Ea.	545	163		708	
Reinstall	Ea.		100.83		100.83	
Clean	Ea.	.33	12.10		12.43	
Minimum Charge	Job		186		186	

Urinal

	Unit	Material	Labor	Equip.	Total	Specification
Demolish	Ea.		37		37	Includes material and labor to install a
Install	Ea.	345	223		568	wall hung urinal with flush valve.
Demolish and Install	Ea.	345	260		605	
Reinstall	Ea.		178.13		178.13	
Clean	Ea.	.33	12.10		12.43	
Minimum Charge	Job		186		186	

Bidet

Vitreous China

	Unit	Material	Labor	Equip.	Total	Specification
Demolish	Ea.		37		37	Includes material and labor to install a
Install	Ea.	650	134		784	vitreous china bidet complete with
Demolish and Install	Ea.	650	171		821	trim.
Reinstall	Ea.		106.88		106.88	
Clean	Ea.	.33	12.10		12.43	
Minimum Charge	Job		186		186	

Finish Mechanical

Commode		Unit	Material	Labor	Equip.	Total	Specification
Chrome Fittings							
	Install	Ea.	152	46.50		198.50	Includes labor and materials to install chrome fittings for a bidet.
Brass Fittings							
	Install	Ea.	167	46.50		213.50	Includes labor and materials to install brass fittings for a bidet.
Seat							
	Demolish	Ea.		15.45		15.45	Includes material and labor to install a
	Install	Ea.	23.50	15.45		38.95	toilet seat and cover.
	Demolish and Install	Ea.	23.50	30.90		54.40	
	Reinstall	Ea.		12.37		12.37	
	Clean	Ea.	.17	5.05		5.22	
	Minimum Charge	Job		186		186	
Rough-in							
	Install	Ea.	230	235		465	Includes labor and materials to install water supply pipe with valves and drain, waste and vent pipe with all couplings, hangers and fasteners necessary.

Sink (assembly)		Unit	Material	Labor	Equip.	Total	Specification
Single							
Porcelain							
	Demolish	Ea.		23		23	Includes material and labor to install a
	Install	Ea.	239	119		358	single bowl enamel finished cast iron
	Demolish and Install	Ea.	239	142		381	kitchen sink with faucet and drain.
	Reinstall	Ea.		95.43		95.43	
	Clean	Ea.	.21	8.65		8.86	
	Minimum Charge	Job		186		186	
Stainless Steel							
	Demolish	Ea.		23		23	Includes material and labor to install a
	Install	Ea.	460	119		579	single bowl stainless steel kitchen sink
	Demolish and Install	Ea.	460	142		602	with faucet and drain.
	Reinstall	Ea.		95.43		95.43	
	Clean	Ea.	.21	8.65		8.86	
	Minimum Charge	Job		186		186	
Double							
Porcelain							
	Demolish	Ea.		26.50		26.50	Includes material and labor to install a
	Install	Ea.	530	139		669	double bowl enamel finished cast iron
	Demolish and Install	Ea.	530	165.50		695.50	kitchen sink with faucet and drain.
	Reinstall	Ea.		111.33		111.33	
	Clean	Ea.	.33	12.10		12.43	
	Minimum Charge	Job		186		186	
Stainless Steel							
	Demolish	Ea.		26.50		26.50	Includes material and labor to install a
	Install	Ea.	295	139		434	double bowl stainless steel kitchen sink
	Demolish and Install	Ea.	295	165.50		460.50	with faucet and drain.
	Reinstall	Ea.		111.33		111.33	
	Clean	Ea.	.33	12.10		12.43	
	Minimum Charge	Job		186		186	
Bar							
	Demolish	Ea.		23		23	Includes material and labor to install a
	Install	Ea.	88	93		181	small single bowl stainless steel bar /
	Demolish and Install	Ea.	88	116		204	vegetable / kitchen sink with faucet
	Reinstall	Ea.		74.24		74.24	and drain.
	Clean	Ea.	.21	8.65		8.86	
	Minimum Charge	Job		186		186	

Finish Mechanical

Sink (assembly)

	Unit	Material	Labor	Equip.	Total	Specification
Floor						
Demolish	Ea.		74		74	Includes material and labor to install
Install	Ea.	645	152		797	an enameled cast iron floor mounted
Demolish and Install	Ea.	645	226		871	corner service sink with faucet and
Reinstall	Ea.		121.45		121.45	drain.
Clean	Ea.	.33	6.05		6.38	
Minimum Charge	Job		186		186	
Pedestal						
Demolish	Ea.		23		23	Includes material and labor to install a
Install	Ea.	455	104		559	vitreous china pedestal lavatory with
Demolish and Install	Ea.	455	127		582	faucet set and pop-up drain.
Reinstall	Ea.		83.50		83.50	
Clean	Ea.	.21	8.65		8.86	
Minimum Charge	Job		186		186	
Vanity Lavatory						
Demolish	Ea.		23		23	Includes material and labor to install a
Install	Ea.	257	124		381	vitreous china lavatory with faucet set
Demolish and Install	Ea.	257	147		404	and pop-up drain.
Reinstall	Ea.		98.96		98.96	
Clean	Ea.	.21	8.65		8.86	
Minimum Charge	Job		186		186	
Rough-in						
Install	Ea.	143	310		453	Includes labor and materials to install copper supply pipe with valves and drain, waste and vent pipe with all couplings, hangers and fasteners necessary.
Laundry						
Demolish	Ea.		24		24	Includes material and labor to install a
Install	Ea.	103	103		206	laundry sink (wall mounted or with
Demolish and Install	Ea.	103	127		230	legs) including faucet and pop up
Clean	Ea.	.21	8.65		8.86	drain.
Minimum Charge	Job		186		186	

Sink Only

	Unit	Material	Labor	Equip.	Total	Specification
Single						
Porcelain						
Demolish	Ea.		23		23	Includes material and labor to install a
Install	Ea.	168	133		301	single bowl enamel finished cast iron
Demolish and Install	Ea.	168	156		324	kitchen sink.
Reinstall	Ea.		106.06		106.06	
Clean	Ea.	.21	8.65		8.86	
Minimum Charge	Job		186		186	
Stainless Steel						
Demolish	Ea.		23		23	Includes material and labor to install a
Install	Ea.	390	93		483	single bowl stainless steel kitchen sink.
Demolish and Install	Ea.	390	116		506	
Reinstall	Ea.		74.24		74.24	
Clean	Ea.	.21	8.65		8.86	
Minimum Charge	Job		186		186	

Finish Mechanical

Sink Only		Unit	Material	Labor	Equip.	Total	Specification
Double							
Porcelain							
	Demolish	Ea.		26.50		26.50	Includes material and labor to install a
	Install	Ea.	263	155		418	double bowl enamel finished cast iron
	Demolish and Install	Ea.	263	181.50		444.50	kitchen sink.
	Reinstall	Ea.		123.73		123.73	
	Clean	Ea.	.33	12.10		12.43	
	Minimum Charge	Job		186		186	
Stainless Steel							
	Demolish	Ea.		26.50		26.50	Includes material and labor to install a
	Install	Ea.	590	120		710	double bowl stainless steel kitchen
	Demolish and Install	Ea.	590	146.50		736.50	sink.
	Reinstall	Ea.		95.79		95.79	
	Clean	Ea.	.33	12.10		12.43	
	Minimum Charge	Job		186		186	
Bar							
	Demolish	Ea.		23		23	Includes material and labor to install a
	Install	Ea.	46	93		139	small single bowl stainless steel bar /
	Demolish and Install	Ea.	46	116		162	vegetable / kitchen sink.
	Reinstall	Ea.		74.24		74.24	
	Clean	Ea.	.21	8.65		8.86	
	Minimum Charge	Job		186		186	
Floor							
	Demolish	Ea.		74		74	Includes material and labor to install
	Install	Ea.	550	169		719	an enameled cast iron floor mounted
	Demolish and Install	Ea.	550	243		793	corner service sink.
	Reinstall	Ea.		134.98		134.98	
	Clean	Ea.	.33	6.05		6.38	
	Minimum Charge	Job		186		186	
Pedestal							
	Demolish	Ea.		23		23	Includes material and labor to install a
	Install	Ea.	360	116		476	vitreous china pedestal lavatory.
	Demolish and Install	Ea.	360	139		499	
	Reinstall	Ea.		92.80		92.80	
	Clean	Ea.	.21	8.65		8.86	
	Minimum Charge	Job		186		186	
Vanity Lavatory							
	Demolish	Ea.		23		23	Includes material and labor to install a
	Install	Ea.	146	116		262	vitreous china lavatory.
	Demolish and Install	Ea.	146	139		285	
	Reinstall	Ea.		92.80		92.80	
	Clean	Ea.	.21	8.65		8.86	
	Minimum Charge	Job		186		186	
Rough-in							
	Install	Ea.	124	290		414	Includes labor and materials to install copper supply pipe with valves and drain, waste and vent pipe with all couplings, hangers and fasteners necessary.
Faucet Sink							
Good Grade							
	Demolish	Ea.		18.55		18.55	Includes material and labor to install a
	Install	Ea.	56.50	25.50		82	sink faucet and fittings.
	Demolish and Install	Ea.	56.50	44.05		100.55	
	Reinstall	Ea.		20.49		20.49	
	Clean	Ea.	.13	7.55		7.68	
	Minimum Charge	Job		186		186	

Finish Mechanical

Sink Only

	Unit	Material	Labor	Equip.	Total	Specification
Better Grade						
Demolish	Ea.		18.55		18.55	Includes material and labor to install a
Install	Ea.	95	25.50		120.50	sink faucet and fittings.
Demolish and Install	Ea.	95	44.05		139.05	
Reinstall	Ea.		20.49		20.49	
Clean	Ea.	.13	7.55		7.68	
Minimum Charge	Job		186		186	
Premium Grade						
Demolish	Ea.		18.55		18.55	Includes material and labor to install a
Install	Ea.	186	32		218	sink faucet and fittings.
Demolish and Install	Ea.	186	50.55		236.55	
Reinstall	Ea.		25.56		25.56	
Clean	Ea.	.13	7.55		7.68	
Minimum Charge	Job		186		186	

Drain & Basket

	Unit	Material	Labor	Equip.	Total	Specification
Demolish	Ea.		11.60		11.60	Includes material and labor to install a
Install	Ea.	13.35	23		36.35	sink drain and basket.
Demolish and Install	Ea.	13.35	34.60		47.95	
Reinstall	Ea.		18.56		18.56	
Clean	Ea.	.13	7.55		7.68	
Minimum Charge	Job		186		186	

Bathtub

	Unit	Material	Labor	Equip.	Total	Specification
Enameled						
Steel						
Demolish	Ea.		67		67	Includes material and labor to install a
Install	Ea.	330	121		451	formed steel bathtub with spout,
Demolish and Install	Ea.	330	188		518	mixing valve, shower head and
Reinstall	Ea.		97.16		97.16	pop-up drain.
Clean	Ea.	.66	20		20.66	
Minimum Charge	Job		186		186	
Cast Iron						
Demolish	Ea.		67		67	Includes material and labor to install a
Install	Ea.	805	152		957	cast iron bath tub with spout, mixing
Demolish and Install	Ea.	805	219		1024	valve, shower head and pop-up drain.
Reinstall	Ea.		121.45		121.45	
Clean	Ea.	.66	20		20.66	
Minimum Charge	Job		186		186	
Fiberglass						
Demolish	Ea.		67		67	Includes material and labor to install
Install	Ea.	915	121		1036	an acrylic soaking tub with spout,
Demolish and Install	Ea.	915	188		1103	mixing valve, shower head and
Reinstall	Ea.		97.16		97.16	pop-up drain.
Clean	Ea.	.66	20		20.66	
Minimum Charge	Job		186		186	
Institutional						
Demolish	Ea.		67		67	Includes material and labor to install a
Install	Ea.	985	223		1208	hospital / institutional type bathtub
Demolish and Install	Ea.	985	290		1275	with spout, mixing valve, shower head
Reinstall	Ea.		178.13		178.13	and pop-up drain.
Clean	Ea.	.66	20		20.66	
Minimum Charge	Job		186		186	

Bathtub

Bathtub	Unit	Material	Labor	Equip.	Total	Specification
Whirlpool (Acrylic)						
Demolish	Ea.		206		206	Includes material and labor to install a
Install	Ea.	2525	670		3195	molded fiberglass whirlpool tub.
Demolish and Install	Ea.	2525	876		3401	
Reinstall	Ea.		534.40		534.40	
Clean	Ea.	.66	20		20.66	
Minimum Charge	Job		186		186	
Pump / Motor						
Demolish	Ea.		51.50		51.50	Includes material and labor to install a
Install	Ea.	193	155		348	whirlpool tub pump / motor.
Demolish and Install	Ea.	193	206.50		399.50	
Reinstall	Ea.		123.73		123.73	
Clean	Ea.	.17	15.10		15.27	
Minimum Charge	Job		186		186	
Heater / Motor						
Demolish	Ea.		51.50		51.50	Includes material and labor to install a
Install	Ea.	265	165		430	whirlpool tub heater.
Demolish and Install	Ea.	265	216.50		481.50	
Reinstall	Ea.		131.98		131.98	
Clean	Ea.	.17	15.10		15.27	
Minimum Charge	Job		186		186	
Thermal Cover						
Demolish	Ea.		8.25		8.25	Includes material and labor to install a
Install	Ea.	178	23		201	whirlpool tub thermal cover.
Demolish and Install	Ea.	178	31.25		209.25	
Reinstall	Ea.		18.56		18.56	
Clean	Ea.	1.64	15.10		16.74	
Minimum Charge	Job		186		186	
Tub / Shower Combination						
Demolish	Ea.		41		41	Includes material and labor to install a
Install	Ea.	545	167		712	fiberglass module tub with shower
Demolish and Install	Ea.	545	208		753	surround with spout, mixing valve,
Reinstall	Ea.		133.60		133.60	shower head and pop-up drain.
Clean	Ea.	.83	30		30.83	
Minimum Charge	Job		186		186	
Accessories						
Sliding Door						
Demolish	Ea.		16.80		16.80	Includes material and labor to install a
Install	Ea.	400	78.50		478.50	48" aluminum framed tempered glass
Demolish and Install	Ea.	400	95.30		495.30	shower door.
Reinstall	Ea.		62.99		62.99	
Clean	Ea.	1.36	10.05		11.41	
Minimum Charge	Job		186		186	
Shower Head						
Demolish	Ea.		4.64		4.64	Includes material and labor to install a
Install	Ea.	71.50	15.45		86.95	water saving shower head.
Demolish and Install	Ea.	71.50	20.09		91.59	
Reinstall	Ea.		12.37		12.37	
Clean	Ea.	.01	3.02		3.03	
Minimum Charge	Job		186		186	
Faucet Set						
Demolish	Ea.		23		23	Includes material and labor to install a
Install	Ea.	76.50	46.50		123	combination spout / diverter for a
Demolish and Install	Ea.	76.50	69.50		146	bathtub.
Reinstall	Ea.		37.12		37.12	
Clean	Ea.	.13	7.55		7.68	
Minimum Charge	Job		186		186	

Finish Mechanical

Bathtub

	Unit	Material	Labor	Equip.	Total	Specification
Shower Rod						
Demolish	Ea.		2.80		2.80	Includes material and labor to install a
Install	Ea.	14.95	26		40.95	chrome plated shower rod.
Demolish and Install	Ea.	14.95	28.80		43.75	
Reinstall	Ea.		20.80		20.80	
Clean	Ea.		3.02		3.02	
Minimum Charge	Job		186		186	
Rough-in						
Install	Ea.	183	325		508	Includes labor and materials to install copper supply pipe with valves and drain, waste and vent pipe with all couplings, hangers and fasteners necessary.

Shower

	Unit	Material	Labor	Equip.	Total	Specification
Fiberglass						
32″ x 32″						
Demolish	Ea.		139		139	Includes material and labor to install a
Install	Ea.	460	121		581	fiberglass shower stall with door,
Demolish and Install	Ea.	460	260		720	mixing valve and shower head and
Reinstall	Ea.		97.16		97.16	drain fitting.
Clean	Ea.	.66	20		20.66	
Minimum Charge	Job		186		186	
36″ x 36″						
Demolish	Ea.		139		139	Includes material and labor to install a
Install	Ea.	510	121		631	fiberglass shower stall with door,
Demolish and Install	Ea.	510	260		770	mixing valve and shower head and
Reinstall	Ea.		97.16		97.16	drain fitting.
Clean	Ea.	.66	20		20.66	
Minimum Charge	Job		186		186	
35″ x 60″						
Demolish	Ea.		139		139	Includes material and labor to install a
Install	Ea.	625	167		792	handicap fiberglass shower stall with
Demolish and Install	Ea.	625	306		931	door & seat, mixing valve and shower
Reinstall	Ea.		133.60		133.60	head and drain fitting.
Clean	Ea.	.66	20		20.66	
Minimum Charge	Job		186		186	
Accessories						
Single Door						
Demolish	Ea.		16.80		16.80	Includes material and labor to install a
Install	Ea.	120	41.50		161.50	shower door.
Demolish and Install	Ea.	120	58.30		178.30	
Reinstall	Ea.		33.24		33.24	
Clean	Ea.	1.36	10.05		11.41	
Minimum Charge	Job		186		186	
Shower Head						
Demolish	Ea.		4.64		4.64	Includes material and labor to install a
Install	Ea.	71.50	15.45		86.95	water saving shower head.
Demolish and Install	Ea.	71.50	20.09		91.59	
Reinstall	Ea.		12.37		12.37	
Clean	Ea.	.01	3.02		3.03	
Minimum Charge	Job		186		186	

Finish Mechanical

Shower		Unit	Material	Labor	Equip.	Total	Specification
Faucet Set							
	Demolish	Ea.		23		23	Includes material and labor to install a
	Install	Ea.	76.50	46.50		123	combination spout / diverter for a
	Demolish and Install	Ea.	76.50	69.50		146	bathtub.
	Reinstall	Ea.		37.12		37.12	
	Clean	Ea.	.13	7.55		7.68	
	Minimum Charge	Job		186		186	
Shower Pan							
	Demolish	Ea.		17.20		17.20	Includes material and labor to install
	Install	Ea.	206	46.50		252.50	the base portion of a fiberglass
	Demolish and Install	Ea.	206	63.70		269.70	shower unit.
	Reinstall	Ea.		37.12		37.12	
	Clean	Ea.	.66	20		20.66	
	Minimum Charge	Job		186		186	
Rough-in							
	Install	Ea.	268	325		593	Includes labor and materials to install copper supply pipe with valves and drain, waste and vent pipe with all couplings, hangers and fasteners necessary.

Ductwork		Unit	Material	Labor	Equip.	Total	Specification
Return Grill							
Under 10″ Wide							
	Demolish	Ea.		8.30		8.30	Includes material and labor to install a
	Install	Ea.	37.50	21		58.50	10″ x 10″ air return register.
	Demolish and Install	Ea.	37.50	29.30		66.80	
	Reinstall	Ea.		16.62		16.62	
	Clean	Ea.	.17	6.70		6.87	
	Minimum Charge	Job		187		187	
12″ to 20″ Wide							
	Demolish	Ea.		8.30		8.30	Includes material and labor to install a
	Install	Ea.	68	22		90	16″ x 16″ air return register.
	Demolish and Install	Ea.	68	30.30		98.30	
	Reinstall	Ea.		17.60		17.60	
	Clean	Ea.	.17	6.70		6.87	
	Minimum Charge	Job		187		187	
Over 20″ Wide							
	Demolish	Ea.		12.60		12.60	Includes material and labor to install a
	Install	Ea.	132	34		166	24″ x 24″ air return register.
	Demolish and Install	Ea.	132	46.60		178.60	
	Reinstall	Ea.		27.20		27.20	
	Clean	Ea.	.17	7.55		7.72	
	Minimum Charge	Job		187		187	
Supply Grill							
Baseboard Type							
	Demolish	Ea.		5.95		5.95	Includes material and labor to install a
	Install	Ea.	18.50	18.70		37.20	14″ x 6″ baseboard air supply
	Demolish and Install	Ea.	18.50	24.65		43.15	register.
	Reinstall	Ea.		14.96		14.96	
	Clean	Ea.	.17	6.70		6.87	
	Minimum Charge	Job		187		187	

Finish Mechanical

Ductwork

	Unit	Material	Labor	Equip.	Total	Specification
10″ Wide						
Demolish	Ea.		6.95		6.95	Includes material and labor to install a
Install	Ea.	27.50	18.70		46.20	10″ x 6″ air supply register.
Demolish and Install	Ea.	27.50	25.65		53.15	
Reinstall	Ea.		14.96		14.96	
Clean	Ea.	.17	6.70		6.87	
Minimum Charge	Job		187		187	
12″ to 15″ Wide						
Demolish	Ea.		6.95		6.95	Includes material and labor to install a
Install	Ea.	38	22		60	14″ x 8″ air supply register.
Demolish and Install	Ea.	38	28.95		66.95	
Reinstall	Ea.		17.60		17.60	
Clean	Ea.	.17	6.70		6.87	
Minimum Charge	Job		187		187	
18″ to 24″ Wide						
Demolish	Ea.		8.30		8.30	Includes material and labor to install a
Install	Ea.	53	29		82	24″ x 8″ air supply register.
Demolish and Install	Ea.	53	37.30		90.30	
Reinstall	Ea.		23.02		23.02	
Clean	Ea.	.17	7.55		7.72	
Minimum Charge	Job		187		187	
30″ to 36″ Wide						
Demolish	Ea.		10.40		10.40	Includes material and labor to install a
Install	Ea.	69.50	26.50		96	30″ x 8″ air supply register.
Demolish and Install	Ea.	69.50	36.90		106.40	
Reinstall	Ea.		21.37		21.37	
Clean	Ea.	.17	15.10		15.27	
Minimum Charge	Job		187		187	

Diffuser

	Unit	Material	Labor	Equip.	Total	Specification
12″ Louver						
Demolish	Ea.		8.30		8.30	Includes material and labor to install a
Install	Ea.	36.50	26.50		63	12″ aluminum louvered diffuser.
Demolish and Install	Ea.	36.50	34.80		71.30	
Reinstall	Ea.		21.37		21.37	
Clean	Ea.	.17	6.70		6.87	
Minimum Charge	Job		187		187	
14″ to 20″ Louver						
Demolish	Ea.		8.30		8.30	Includes material and labor to install a
Install	Ea.	41.50	37.50		79	14″ to 20″ aluminum louvered diffuser.
Demolish and Install	Ea.	41.50	45.80		87.30	
Reinstall	Ea.		29.92		29.92	
Clean	Ea.	.17	7.55		7.72	
Minimum Charge	Job		187		187	
25″ to 32″ Louver						
Demolish	Ea.		12.60		12.60	Includes material and labor to install a
Install	Ea.	53.50	53.50		107	25″ to 32″ aluminum louvered diffuser.
Demolish and Install	Ea.	53.50	66.10		119.60	
Reinstall	Ea.		42.74		42.74	
Clean	Ea.	.17	10.05		10.22	
Minimum Charge	Job		187		187	
12″ Round						
Demolish	Ea.		11.55		11.55	Includes material and labor to install a
Install	Ea.	34.50	31		65.50	12″ diameter aluminum diffuser with a
Demolish and Install	Ea.	34.50	42.55		77.05	butterfly damper.
Reinstall	Ea.		24.93		24.93	
Clean	Ea.	.17	6.70		6.87	
Minimum Charge	Job		187		187	

Finish Mechanical

Ductwork	Unit	Material	Labor	Equip.	Total	Specification
14″ to 20″ Round						
Demolish	Ea.		15.40		15.40	Includes material and labor to install a
Install	Ea.	125	41.50		166.50	20″ diameter aluminum diffuser with a
Demolish and Install	Ea.	125	56.90		181.90	butterfly damper.
Reinstall	Ea.		33.24		33.24	
Clean	Ea.	.17	7.55		7.72	
Minimum Charge	Job		187		187	
Flush-out / Sanitize						
Install	L.F.	.66	3.02		3.68	Includes labor and material costs to
Minimum Charge	Job		187		187	clean ductwork.

Lighting Fixture   Duplex Receptacle

Electrical Per S.F.	Unit	Material	Labor	Equip.	Total	Specification
Residential						
Light Fixtures						
Install	S.F.	.23	.18		.41	Includes labor and material to replace house light fixtures. Apply S.F. cost to floor area including garages but not basements.

Circuits	Unit	Material	Labor	Equip.	Total	Specification
Outlet with Wiring						
Demolish	Ea.		20.50		20.50	Includes material and labor to install general purpose outlet with cover, up to 20 feet of #12/2 wire.
Install	Ea.	7.50	25		32.50	
Demolish and Install	Ea.	7.50	45.50		53	
Reinstall	Ea.		20.19		20.19	
Clean	Ea.		1.51		1.51	
Minimum Charge	Job		184		184	
GFCI Outlet w / Wiring						
Demolish	Ea.		20.50		20.50	Cost includes material and labor to install 120V GFCI receptacle and wall plate, with up to 20 feet of #12/2 wire.
Install	Ea.	39	30		69	
Demolish and Install	Ea.	39	50.50		89.50	
Reinstall	Ea.		23.86		23.86	
Clean	Ea.		1.51		1.51	
Minimum Charge	Job		184		184	
Exterior Outlet w / Wiring						
Demolish	Ea.		20.50		20.50	Cost includes material and labor to install 110V waterproof external receptacle and cover plate with up to 20' of #12/2 wire.
Install	Ea.	28.50	30.50		59	
Demolish and Install	Ea.	28.50	51		79.50	
Reinstall	Ea.		24.48		24.48	
Clean	Ea.		1.51		1.51	
Minimum Charge	Job		184		184	
240V Outlet w / Wiring						
Demolish	Ea.		20.50		20.50	Cost includes material and labor to install 240V receptacle and cover plate with up to 20' of #12/2 wire.
Install	Ea.	63.50	57.50		121	
Demolish and Install	Ea.	63.50	78		141.50	
Reinstall	Ea.		45.83		45.83	
Clean	Ea.		1.51		1.51	
Minimum Charge	Job		184		184	

Finish Electrical

Circuits		Unit	Material	Labor	Equip.	Total	Specification
TV / Cable Jack w / Wiring							
	Demolish	Ea.		20.50		20.50	Includes material and labor to install
	Install	Ea.	14.50	23		37.50	TV/cable outlet box and cover plate
	Demolish and Install	Ea.	14.50	43.50		58	including 20' of wiring.
	Reinstall	Ea.		18.36		18.36	
	Clean	Ea.		1.51		1.51	
	Minimum Charge	Job		184		184	
Phone Jack with Wiring							
	Demolish	Ea.		20.50		20.50	Includes material and labor to install
	Install	Ea.	8.75	14.10		22.85	phone jack and cover plate including
	Demolish and Install	Ea.	8.75	34.60		43.35	up to 20' of telephone cable.
	Reinstall	Ea.		11.30		11.30	
	Clean	Ea.		.94		.94	
	Minimum Charge	Job		184		184	

Light Switch		Unit	Material	Labor	Equip.	Total	Specification
Single							
Switch							
	Demolish	Ea.		4.08		4.08	Includes material and labor to install
	Install	Ea.	5.25	17.50		22.75	single switch and wall plate. Wiring is
	Demolish and Install	Ea.	5.25	21.58		26.83	not included.
	Reinstall	Ea.		13.99		13.99	
	Clean	Ea.		1.51		1.51	
	Minimum Charge	Job		184		184	
Plate							
	Demolish	Ea.		2.04		2.04	Includes material and labor to install a
	Install	Ea.	.34	4.59		4.93	one gang brown wall switch plate.
	Demolish and Install	Ea.	.34	6.63		6.97	
	Reinstall	Ea.		3.67		3.67	
	Clean	Ea.	.07	.30		.37	
	Minimum Charge	Job		184		184	
Double							
Switch							
	Demolish	Ea.		8.15		8.15	Includes material and labor to install
	Install	Ea.	7.70	26		33.70	double switch and wall plate. Wiring
	Demolish and Install	Ea.	7.70	34.15		41.85	is not included.
	Reinstall	Ea.		20.98		20.98	
	Clean	Ea.	.07	4.83		4.90	
	Minimum Charge	Job		184		184	
Plate							
	Demolish	Ea.		2.04		2.04	Includes material and labor to install a
	Install	Ea.	.85	6.95		7.80	two gang brown wall switch plate.
	Demolish and Install	Ea.	.85	8.99		9.84	
	Reinstall	Ea.		5.54		5.54	
	Clean	Ea.	.07	.30		.37	
	Minimum Charge	Job		184		184	
Triple							
Switch							
	Demolish	Ea.		12.35		12.35	Includes material and labor to install
	Install	Ea.	23	30.50		53.50	triple switch and wall plate. Wiring is
	Demolish and Install	Ea.	23	42.85		65.85	not included.
	Reinstall	Ea.		24.48		24.48	
	Clean	Ea.	.07	7.25		7.32	
	Minimum Charge	Job		184		184	

Finish Electrical

Light Switch		Unit	Material	Labor	Equip.	Total	Specification
Plate							
	Demolish	Ea.		2.04		2.04	Includes material and labor to install a
	Install	Ea.	1.25	11.50		12.75	three gang brown wall switch plate.
	Demolish and Install	Ea.	1.25	13.54		14.79	
	Reinstall	Ea.		9.18		9.18	
	Clean	Ea.	.07	.30		.37	
	Minimum Charge	Job		184		184	
Dimmer							
Switch							
	Demolish	Ea.		4.08		4.08	Includes material and labor to install
	Install	Ea.	18.60	21.50		40.10	dimmer switch. Wiring is not included.
	Demolish and Install	Ea.	18.60	25.58		44.18	
	Reinstall	Ea.		17.28		17.28	
	Clean	Ea.		1.51		1.51	
	Minimum Charge	Job		184		184	
Plate							
	Demolish	Ea.		2.04		2.04	Includes material and labor to install a
	Install	Ea.	.34	4.59		4.93	one gang brown wall switch plate.
	Demolish and Install	Ea.	.34	6.63		6.97	
	Reinstall	Ea.		3.67		3.67	
	Clean	Ea.	.07	.30		.37	
	Minimum Charge	Job		184		184	
With Wiring							
Single							
	Demolish	Ea.		20.50		20.50	Includes material and labor to install
	Install	Ea.	9.30	21.50		30.80	single switch and plate including up to
	Demolish and Install	Ea.	9.30	42		51.30	20' of romex cable.
	Reinstall	Ea.		17.18		17.18	
	Clean	Ea.		1.51		1.51	
	Minimum Charge	Job		184		184	
Double							
	Demolish	Ea.		20.50		20.50	Includes material and labor to install
	Install	Ea.	22.50	36.50		59	double switch and plate including up
	Demolish and Install	Ea.	22.50	57		79.50	to 20' of romex cable.
	Reinstall	Ea.		29.38		29.38	
	Clean	Ea.	.07	4.83		4.90	
	Minimum Charge	Job		184		184	
Triple							
	Demolish	Ea.		30.50		30.50	Includes material and labor to install
	Install	Ea.	29.50	41.50		71	triple switch and plate including up to
	Demolish and Install	Ea.	29.50	72		101.50	20' of romex cable.
	Reinstall	Ea.		33.04		33.04	
	Clean	Ea.	.07	7.25		7.32	
	Minimum Charge	Job		184		184	
Dimmer							
	Demolish	Ea.		20.50		20.50	Includes material and labor to install
	Install	Ea.	22.50	25		47.50	dimmer switch and cover plate
	Demolish and Install	Ea.	22.50	45.50		68	including up to 20' of romex cable.
	Reinstall	Ea.		20.19		20.19	
	Clean	Ea.		1.51		1.51	
	Minimum Charge	Job		184		184	

Finish Electrical

Cover for Outlet / Switch

	Unit	Material	Labor	Equip.	Total	Specification
High Grade						
Demolish	Ea.		2.04		2.04	Includes material and labor to install
Install	Ea.	1.98	4.59		6.57	wall plate, stainless steel, 1 gang.
Demolish and Install	Ea.	1.98	6.63		8.61	
Reinstall	Ea.		3.67		3.67	
Clean	Ea.	.07	.30		.37	
Minimum Charge	Job		184		184	
Deluxe Grade						
Demolish	Ea.		2.04		2.04	Includes material and labor to install
Install	Ea.	4.24	4.59		8.83	wall plate, brushed brass, 1 gang.
Demolish and Install	Ea.	4.24	6.63		10.87	
Reinstall	Ea.		3.67		3.67	
Clean	Ea.	.07	.30		.37	
Minimum Charge	Job		184		184	

Residential Light Fixture

	Unit	Material	Labor	Equip.	Total	Specification
Ceiling						
Good Quality						
Demolish	Ea.		8.50		8.50	Includes material and labor to install
Install	Ea.	28.50	9.20		37.70	standard quality incandescent ceiling
Demolish and Install	Ea.	28.50	17.70		46.20	light fixture. Wiring is not included.
Reinstall	Ea.		7.34		7.34	
Clean	Ea.	.17	10.05		10.22	
Minimum Charge	Job		184		184	
Better Quality						
Demolish	Ea.		8.50		8.50	Includes material and labor to install
Install	Ea.	86	19.35		105.35	custom quality incandescent ceiling
Demolish and Install	Ea.	86	27.85		113.85	light fixture. Wiring is not included.
Reinstall	Ea.		15.46		15.46	
Clean	Ea.	.17	10.05		10.22	
Minimum Charge	Job		184		184	
Premium Quality						
Demolish	Ea.		8.50		8.50	Includes material and labor to install
Install	Ea.	253	19.35		272.35	deluxe quality incandescent ceiling
Demolish and Install	Ea.	253	27.85		280.85	light fixture. Wiring is not included.
Reinstall	Ea.		15.46		15.46	
Clean	Ea.	.17	10.05		10.22	
Minimum Charge	Job		184		184	
Recessed Ceiling						
Demolish	Ea.		17		17	Includes material and labor to install
Install	Ea.	39.50	12.25		51.75	standard quality recessed eyeball
Demolish and Install	Ea.	39.50	29.25		68.75	spotlight with housing. Wiring is not
Reinstall	Ea.		9.79		9.79	included.
Clean	Ea.	.03	2.52		2.55	
Minimum Charge	Job		184		184	
Recessed Eyeball						
Demolish	Ea.		17		17	Includes material and labor to install
Install	Ea.	59.50	13.10		72.60	custom quality recessed eyeball
Demolish and Install	Ea.	59.50	30.10		89.60	spotlight with housing. Wiring is not
Reinstall	Ea.		10.49		10.49	included.
Clean	Ea.	.03	2.52		2.55	
Minimum Charge	Job		184		184	

Finish Electrical

Residential Light Fixture	Unit	Material	Labor	Equip.	Total	Specification
Recessed Shower Type						
Demolish	Ea.		17		17	Includes material and labor to install
Install	Ea.	49.50	12.25		61.75	deluxe quality recessed eyeball
Demolish and Install	Ea.	49.50	29.25		78.75	spotlight with housing. Wiring is not
Reinstall	Ea.		9.79		9.79	included.
Clean	Ea.	.03	2.52		2.55	
Minimum Charge	Job		184		184	
Run Wiring						
Install	Ea.	11.50	14.70		26.20	Includes labor and material to install a
Minimum Charge	Job		184		184	electric fixture utility box with
						non-metallic cable.
Fluorescent						
Good Quality						
Demolish	Ea.		18.55		18.55	Includes material and labor to install
Install	Ea.	48.50	52.50		101	standard quality 12" wide by 48" long
Demolish and Install	Ea.	48.50	71.05		119.55	decorative surface mounted fluorescent
Reinstall	Ea.		41.97		41.97	fixture w/acrylic diffuser.
Clean	Ea.	.03	2.52		2.55	
Minimum Charge	Job		184		184	
Better Quality						
Demolish	Ea.		18.55		18.55	Includes material and labor to install
Install	Ea.	99.50	59		158.50	an interior fluorescent lighting fixture.
Demolish and Install	Ea.	99.50	77.55		177.05	
Reinstall	Ea.		47.38		47.38	
Clean	Ea.	.03	2.52		2.55	
Minimum Charge	Job		184		184	
Premium Quality						
Demolish	Ea.		18.55		18.55	Includes material and labor to install
Install	Ea.	103	69.50		172.50	deluxe quality 24" wide by 48" long
Demolish and Install	Ea.	103	88.05		191.05	decorative surface mounted fluorescent
Reinstall	Ea.		55.43		55.43	fixture w/acrylic diffuser.
Clean	Ea.	.03	2.52		2.55	
Minimum Charge	Job		184		184	
Run Wiring						
Install	Ea.	11.50	14.70		26.20	Includes labor and material to install a
Minimum Charge	Job		184		184	electric fixture utility box with
						non-metallic cable.
Exterior						
Good Quality						
Demolish	Ea.		16.30		16.30	Includes material and labor to install
Install	Ea.	34.50	23		57.50	standard quality incandescent outdoor
Demolish and Install	Ea.	34.50	39.30		73.80	light fixture. Wiring is not included.
Reinstall	Ea.		18.36		18.36	
Clean	Ea.	.17	10.05		10.22	
Minimum Charge	Job		184		184	
Better Quality						
Demolish	Ea.		16.30		16.30	Includes material and labor to install
Install	Ea.	99	23		122	custom quality incandescent outdoor
Demolish and Install	Ea.	99	39.30		138.30	light fixture. Wiring is not included.
Reinstall	Ea.		18.36		18.36	
Clean	Ea.	.17	10.05		10.22	
Minimum Charge	Job		184		184	
Premium Quality						
Demolish	Ea.		16.30		16.30	Includes material and labor to install
Install	Ea.	165	92		257	deluxe quality incandescent outdoor
Demolish and Install	Ea.	165	108.30		273.30	light fixture. Wiring is not included.
Reinstall	Ea.		73.44		73.44	
Clean	Ea.	.17	10.05		10.22	
Minimum Charge	Job		184		184	

Finish Electrical

Residential Light Fixture		Unit	Material	Labor	Equip.	Total	Specification
Outdoor w / Pole							
	Demolish	Ea.		16.30		16.30	Includes material and labor to install
	Install	Ea.	121	92		213	post lantern type incandescent outdoor
	Demolish and Install	Ea.	121	108.30		229.30	light fixture. Wiring is not included.
	Reinstall	Ea.		73.44		73.44	
	Clean	Ea.	.17	10.05		10.22	
	Minimum Charge	Job		184		184	
Run Wiring							
	Install	Ea.	11.50	14.70		26.20	Includes labor and material to install a
	Minimum Charge	Job		184		184	electric fixture utility box with
							non-metallic cable.
Track Lighting							
	Demolish	L.F.		5.65		5.65	Includes material and labor to install
	Install	L.F.	13.95	15.30		29.25	track lighting.
	Demolish and Install	L.F.	13.95	20.95		34.90	
	Reinstall	L.F.		12.24		12.24	
	Clean	L.F.	.17	4.83		5	
	Minimum Charge	Job		184		184	

Commercial Light Fixture		Unit	Material	Labor	Equip.	Total	Specification
Fluorescent							
2 Bulb, 2'							
	Demolish	Ea.		16.30		16.30	Includes material and labor to install a
	Install	Ea.	33	46		79	2' long surface mounted fluorescent
	Demolish and Install	Ea.	33	62.30		95.30	light fixture with two 40 watt bulbs.
	Reinstall	Ea.		36.72		36.72	
	Clean	Ea.	.83	17.25		18.08	
	Minimum Charge	Job		184		184	
2 Bulb, 4'							
	Demolish	Ea.		16.30		16.30	Includes material and labor to install a
	Install	Ea.	62	52.50		114.50	4' long surface mounted fluorescent
	Demolish and Install	Ea.	62	68.80		130.80	light fixture with two 40 watt bulbs.
	Reinstall	Ea.		41.97		41.97	
	Clean	Ea.	.83	17.25		18.08	
	Minimum Charge	Job		184		184	
3 Bulb, 4'							
	Demolish	Ea.		16.30		16.30	Includes material and labor to install a
	Install	Ea.	103	52.50		155.50	4' long pendant mounted fluorescent
	Demolish and Install	Ea.	103	68.80		171.80	light fixture with three 34 watt bulbs.
	Reinstall	Ea.		41.97		41.97	
	Clean	Ea.	.83	17.25		18.08	
	Minimum Charge	Job		184		184	
4 Bulb, 4'							
	Demolish	Ea.		16.30		16.30	Includes material and labor to install a
	Install	Ea.	93.50	56.50		150	4' long pendant mounted fluorescent
	Demolish and Install	Ea.	93.50	72.80		166.30	light fixture with four 34 watt bulbs.
	Reinstall	Ea.		45.19		45.19	
	Clean	Ea.	.83	17.25		18.08	
	Minimum Charge	Job		184		184	
2 Bulb, 8'							
	Demolish	Ea.		20.50		20.50	Includes material and labor to install a
	Install	Ea.	96.50	69.50		166	8' long surface mounted fluorescent
	Demolish and Install	Ea.	96.50	90		186.50	light fixture with two 110 watt bulbs.
	Reinstall	Ea.		55.43		55.43	
	Clean	Ea.	.83	17.25		18.08	
	Minimum Charge	Job		184		184	

Finish Electrical

Commercial Light Fixture		Unit	Material	Labor	Equip.	Total	Specification
Run Wiring							
	Install	Ea.	11.50	14.70		26.20	Includes labor and material to install a
	Minimum Charge	Job		184		184	electric fixture utility box with
							non-metallic cable.
Ceiling, Recessed							
Eyeball Type							
	Demolish	Ea.		17		17	Includes material and labor to install
	Install	Ea.	59.50	13.10		72.60	custom quality recessed eyeball
	Demolish and Install	Ea.	59.50	30.10		89.60	spotlight with housing. Wiring is not
	Reinstall	Ea.		10.49		10.49	included.
	Clean	Ea.	.03	2.52		2.55	
	Minimum Charge	Job		184		184	
Shower Type							
	Demolish	Ea.		17		17	Includes material and labor to install
	Install	Ea.	49.50	12.25		61.75	deluxe quality recessed eyeball
	Demolish and Install	Ea.	49.50	29.25		78.75	spotlight with housing. Wiring is not
	Reinstall	Ea.		9.79		9.79	included.
	Clean	Ea.	.03	2.52		2.55	
	Minimum Charge	Job		184		184	
Run Wiring							
	Install	Ea.	11.50	14.70		26.20	Includes labor and material to install a
	Minimum Charge	Job		184		184	electric fixture utility box with
							non-metallic cable.
Ceiling Grid Type							
2 Tube							
	Demolish	Ea.		24		24	Cost includes material and labor to
	Install	Ea.	57	69.50		126.50	install 2 lamp fluorescent light fixture
	Demolish and Install	Ea.	57	93.50		150.50	in 2' x 4' ceiling grid system including
	Reinstall	Ea.		55.43		55.43	lens and tubes.
	Clean	Ea.	.83	17.25		18.08	
	Minimum Charge	Job		184		184	
4 Tube							
	Demolish	Ea.		27		27	Cost includes material and labor to
	Install	Ea.	65	78		143	install 4 lamp fluorescent light fixture
	Demolish and Install	Ea.	65	105		170	in 2' x 4' ceiling grid system including
	Reinstall	Ea.		62.50		62.50	lens and tubes.
	Clean	Ea.	.83	17.25		18.08	
	Minimum Charge	Job		184		184	
Acrylic Diffuser							
	Demolish	Ea.		4.25		4.25	Includes material and labor to install
	Install	Ea.	38.50	11.15		49.65	fluorescent light fixture w/acrylic
	Demolish and Install	Ea.	38.50	15.40		53.90	diffuser.
	Reinstall	Ea.		8.90		8.90	
	Clean	Ea.	.53	3.02		3.55	
	Minimum Charge	Job		184		184	
Louver Diffuser							
	Demolish	Ea.		5.10		5.10	Includes material and labor to install
	Install	Ea.	42	11.15		53.15	fluorescent light fixture w/louver
	Demolish and Install	Ea.	42	16.25		58.25	diffuser.
	Reinstall	Ea.		8.90		8.90	
	Clean	Ea.	.53	3.02		3.55	
	Minimum Charge	Job		184		184	
Run Wiring							
	Install	Ea.	11.50	14.70		26.20	Includes labor and material to install a
	Minimum Charge	Job		184		184	electric fixture utility box with
							non-metallic cable.

Finish Electrical

Commercial Light Fixture	Unit	Material	Labor	Equip.	Total	Specification
Track Lighting						
Demolish	L.F.		5.65		5.65	Includes material and labor to install
Install	L.F.	13.95	15.30		29.25	track lighting.
Demolish and Install	L.F.	13.95	20.95		34.90	
Reinstall	L.F.		12.24		12.24	
Clean	L.F.	.17	4.83		5	
Minimum Charge	Job		184		184	
Emergency						
Demolish	Ea.		29		29	Includes material and labor to install
Install	Ea.	134	92		226	battery-powered emergency lighting
Demolish and Install	Ea.	134	121		255	unit with two (2) lights, including
Reinstall	Ea.		73.44		73.44	wiring.
Clean	Ea.	.41	17.25		17.66	
Minimum Charge	Job		184		184	
Run Wiring						
Install	Ea.	11.50	14.70		26.20	Includes labor and material to install a
Minimum Charge	Job		184		184	electric fixture utility box with
						non-metallic cable.

Fan	Unit	Material	Labor	Equip.	Total	Specification
Ceiling Paddle						
Good Quality						
Demolish	Ea.		22.50		22.50	Includes material and labor to install
Install	Ea.	86.50	36.50		123	standard ceiling fan.
Demolish and Install	Ea.	86.50	59		145.50	
Reinstall	Ea.		29.38		29.38	
Clean	Ea.	.33	15.10		15.43	
Minimum Charge	Job		184		184	
Better Quality						
Demolish	Ea.		22.50		22.50	Includes material and labor to install
Install	Ea.	144	36.50		180.50	better ceiling fan.
Demolish and Install	Ea.	144	59		203	
Reinstall	Ea.		29.38		29.38	
Clean	Ea.	.33	15.10		15.43	
Minimum Charge	Job		184		184	
Premium Quality						
Demolish	Ea.		22.50		22.50	Includes material and labor to install
Install	Ea.	340	46		386	deluxe ceiling fan.
Demolish and Install	Ea.	340	68.50		408.50	
Reinstall	Ea.		36.72		36.72	
Clean	Ea.	.33	15.10		15.43	
Minimum Charge	Job		184		184	
Light Kit						
Install	Ea.	30.50	46		76.50	Includes labor and material to install a
Minimum Charge	Job		184		184	ceiling fan light kit.
Bathroom Exhaust						
Ceiling Mounted						
Demolish	Ea.		18.80		18.80	Cost includes material and labor to
Install	Ea.	43.50	36.50		80	install 60 CFM ceiling mounted
Demolish and Install	Ea.	43.50	55.30		98.80	bathroom exhaust fan, up to 10' of
Reinstall	Ea.		29.29		29.29	wiring with one wall switch.
Clean	Ea.	.17	6.05		6.22	
Minimum Charge	Job		184		184	

Finish Electrical

Fan	Unit	Material	Labor	Equip.	Total	Specification
Wall Mounted						
Demolish	Ea.		18.80		18.80	Cost includes material and labor to
Install	Ea.	58.50	36.50		95	install 110 CFM wall mounted
Demolish and Install	Ea.	58.50	55.30		113.80	bathroom exhaust fan, up to 10' of
Reinstall	Ea.		29.29		29.29	wiring with one wall switch.
Clean	Ea.	.17	6.05		6.22	
Minimum Charge	Job		184		184	
Lighted						
Demolish	Ea.		18.80		18.80	Cost includes material and labor to
Install	Ea.	42	36.50		78.50	install 50 CFM ceiling mounted lighted
Demolish and Install	Ea.	42	55.30		97.30	exhaust fan, up to 10' of wiring with
Reinstall	Ea.		29.29		29.29	one wall switch..
Clean	Ea.	.17	6.05		6.22	
Minimum Charge	Job		184		184	
Blower Heater w / Light						
Demolish	Ea.		18.80		18.80	Includes material and labor to install
Install	Ea.	91	61		152	ceiling blower type ventilator with
Demolish and Install	Ea.	91	79.80		170.80	switch, light and snap-on grill. Includes
Reinstall	Ea.		48.81		48.81	1300 watt/120 volt heater.
Clean	Ea.	.17	6.05		6.22	
Minimum Charge	Job		184		184	
Fan Heater w / Light						
Demolish	Ea.		18.80		18.80	Includes material and labor to install
Install	Ea.	117	61		178	circular ceiling fan heater unit, 1500
Demolish and Install	Ea.	117	79.80		196.80	watt, surface mounted.
Reinstall	Ea.		48.81		48.81	
Clean	Ea.	.17	6.05		6.22	
Minimum Charge	Job		184		184	
Whole House Exhaust						
36" Ceiling Mounted						
Install	Ea.	565	91.50		656.50	Includes material and labor to install a
Clean	Ea.	.04	4.34		4.38	whole house exhaust fan.
Minimum Charge	Job		184		184	
Kitchen Exhaust						
Standard Model						
Install	Ea.	66.50	24.50		91	Includes labor and materials to install
Clean	Ea.	.04	4.34		4.38	a bathroom or kitchen vent fan.
Minimum Charge	Job		184		184	
Low Noise Model						
Install	Ea.	87	24.50		111.50	Includes labor and materials to install
Clean	Ea.	.04	4.34		4.38	a bathroom or kitchen vent fan, low
Minimum Charge	Job		184		184	noise model.

Low Voltage Systems	Unit	Material	Labor	Equip.	Total	Specification
Smoke Detector						
Pre-wired						
Demolish	Ea.		8.50		8.50	Includes material and labor to install
Install	Ea.	19.65	69		88.65	AC type smoke detector including
Demolish and Install	Ea.	19.65	77.50		97.15	wiring.
Reinstall	Ea.		55.11		55.11	
Clean	Ea.	.07	7.55		7.62	
Minimum Charge	Job		184		184	

Finish Electrical

Low Voltage Systems	Unit	Material	Labor	Equip.	Total	Specification
Battery-powered						
Demolish	Ea.		8.50		8.50	Includes material and labor to install
Install	Ea.	24	23		47	battery-powered type smoke detector.
Demolish and Install	Ea.	24	31.50		55.50	
Reinstall	Ea.		18.36		18.36	
Clean	Ea.	.07	7.55		7.62	
Minimum Charge	Job		184		184	
Run Wiring						
Install	Ea.	11.50	14.70		26.20	Includes labor and material to install a
Minimum Charge	Job		184		184	electric fixture utility box with
						non-metallic cable.
Door Bell / Chimes						
Good Quality						
Demolish	Ea.		25.50		25.50	Includes material and labor to install
Install	Ea.	47	92		139	surface mounted 2 note door bell.
Demolish and Install	Ea.	47	117.50		164.50	Wiring is not included.
Reinstall	Ea.		73.44		73.44	
Clean	Ea.	.17	7.55		7.72	
Minimum Charge	Job		184		184	
Better Quality						
Demolish	Ea.		25.50		25.50	Includes material and labor to install
Install	Ea.	61	92		153	standard quality surface mounted 2
Demolish and Install	Ea.	61	117.50		178.50	note door bell. Wiring is not included.
Reinstall	Ea.		73.44		73.44	
Clean	Ea.	.17	7.55		7.72	
Minimum Charge	Job		184		184	
Premium Quality						
Demolish	Ea.		25.50		25.50	Includes material and labor to install
Install	Ea.	69.50	92		161.50	deluxe quality surface mounted 2 note
Demolish and Install	Ea.	69.50	117.50		187	door bell. Wiring is not included.
Reinstall	Ea.		73.44		73.44	
Clean	Ea.	.17	7.55		7.72	
Minimum Charge	Job		184		184	
Run Wiring						
Install	Ea.	11.50	14.70		26.20	Includes labor and material to install a
Minimum Charge	Job		184		184	electric fixture utility box with
						non-metallic cable.
Intercom System						
Intercom (master)						
Demolish	Ea.		68		68	Includes material and labor to install
Install	Ea.	510	184		694	master intercom station with AM/FM
Demolish and Install	Ea.	510	252		762	music and room monitoring. Wiring is
Reinstall	Ea.		146.88		146.88	not included.
Clean	Ea.	.17	20		20.17	
Minimum Charge	Job		184		184	
Intercom (remote)						
Demolish	Ea.		41		41	Includes material and labor to install
Install	Ea.	46	46		92	remote station for master intercom
Demolish and Install	Ea.	46	87		133	system. Wiring is not included.
Reinstall	Ea.		36.72		36.72	
Clean	Ea.	.17	7.55		7.72	
Minimum Charge	Job		184		184	

Finish Electrical

Low Voltage Systems

	Unit	Material	Labor	Equip.	Total	Specification
Security System						
Alarm Panel						
Demolish	Ea.		81.50		81.50	Includes material and labor to install a
Install	Ea.	168	184		352	6 zone burglar alarm panel.
Demolish and Install	Ea.	168	265.50		433.50	
Reinstall	Ea.		146.88		146.88	
Clean	Ea.	.17	7.55		7.72	
Minimum Charge	Job		184		184	
Motion Detector						
Demolish	Ea.		58.50		58.50	Includes material and labor to install a
Install	Ea.	231	160		391	ultrasonic motion detector for a
Demolish and Install	Ea.	231	218.50		449.50	security system.
Reinstall	Ea.		127.72		127.72	
Clean	Ea.	.17	7.55		7.72	
Minimum Charge	Job		184		184	
Door Switch						
Demolish	Ea.		25.50		25.50	Includes material and labor to install a
Install	Ea.	35	69.50		104.50	door / window contact for a security
Demolish and Install	Ea.	35	95		130	system.
Reinstall	Ea.		55.43		55.43	
Clean	Ea.	.17	2.42		2.59	
Minimum Charge	Job		184		184	
Window Switch						
Demolish	Ea.		25.50		25.50	Includes material and labor to install a
Install	Ea.	35	69.50		104.50	door / window contact for a security
Demolish and Install	Ea.	35	95		130	system.
Reinstall	Ea.		55.43		55.43	
Clean	Ea.	.17	2.42		2.59	
Minimum Charge	Job		184		184	
Satellite System						
Dish						
Demolish	Ea.		113		113	Includes material, labor and
Install	Ea.	690	305		995	equipment to install 10' mesh dish.
Demolish and Install	Ea.	690	418		1108	
Reinstall	Ea.		244.80		244.80	
Clean	Ea.	3.28	30		33.28	
Minimum Charge	Job		184		184	
Motor						
Demolish	Ea.		56.50		56.50	Includes material and labor to install a
Install	Ea.	340	153		493	television satellite dish motor unit.
Demolish and Install	Ea.	340	209.50		549.50	
Reinstall	Ea.		122.40		122.40	
Clean	Ea.	.17	7.55		7.72	
Minimum Charge	Job		184		184	
Television Antenna						
40' Pole						
Demolish	Ea.		64		64	Includes material and labor to install
Install	Ea.	23	115		138	stand-alone 1"-2" metal pole, 10' long.
Demolish and Install	Ea.	23	179		202	Cost does not include antenna.
Reinstall	Ea.		91.80		91.80	
Clean	Ea.	.66	30		30.66	
Paint	Ea.	6.80	37		43.80	
Minimum Charge	Job		184		184	
Rotor Unit						
Demolish	Ea.		51		51	Includes material and labor to install a
Install	Ea.	107	46		153	television antenna rotor.
Demolish and Install	Ea.	107	97		204	
Reinstall	Ea.		36.72		36.72	
Clean	Ea.	.66	30		30.66	
Minimum Charge	Job		184		184	

Finish Electrical

Low Voltage Systems	Unit	Material	Labor	Equip.	Total	Specification
Single Booster						Includes material and labor to install
Demolish	Ea.		51		51	standard grade VHF television signal
Install	Ea.	36.50	46		82.50	booster unit and all hardware required
Demolish and Install	Ea.	36.50	97		133.50	for connection to TV cable.
Reinstall	Ea.		36.72		36.72	
Clean	Ea.	.03	3.78		3.81	
Minimum Charge	Job		184		184	

Improvements / Appliances / Treatments

Gazebo

Gutter

Shutter

Window Treatment	Unit	Material	Labor	Equip.	Total	Specification
Drapery						
Good Grade						
Install	L.F.	35	4.83		39.83	Includes material and labor to install
Reinstall	L.F.		3.87		3.87	good quality drapery.
Clean	L.F.	2.53	.12		2.65	
Minimum Charge	Job		121		121	
Premium Grade						
Install	L.F.	62	4.83		66.83	Includes material and labor to install
Reinstall	L.F.		3.87		3.87	premium quality drapery.
Clean	L.F.	2.53	.13		2.66	
Minimum Charge	Job		121		121	
Lining						
Install	L.F.	15.40	4.83		20.23	Includes material and labor to install
Reinstall	L.F.		3.87		3.87	drapery lining.
Clean	L.F.	1.27	.07		1.34	
Minimum Charge	Job		121		121	
Valance Board						
Demolish	L.F.		1.34		1.34	Includes material and labor to install a
Install	L.F.	4.43	9.65		14.08	drapery valance board.
Demolish and Install	L.F.	4.43	10.99		15.42	
Reinstall	L.F.		7.73		7.73	
Clean	L.F.	1.27	.24		1.51	
Minimum Charge	Job		121		121	
Rod						
Demolish	L.F.		3.36		3.36	Includes material and labor to install
Install	L.F.	2.79	5.75		8.54	standard traverse drapery hardware.
Demolish and Install	L.F.	2.79	9.11		11.90	
Reinstall	L.F.		4.58		4.58	
Clean	L.F.	.07	.22		.29	
Paint	L.F.	.06	.57		.63	
Minimum Charge	Job		121		121	
Blinds						
Mini-type						
Demolish	S.F.		.73		.73	Includes material and labor to install
Install	S.F.	2.84	.57		3.41	horizontal 1″ slat blinds.
Demolish and Install	S.F.	2.84	1.30		4.14	
Reinstall	S.F.		.46		.46	
Clean	S.F.	.48	.02		.50	
Minimum Charge	Job		121		121	

Window Treatment	Unit	Material	Labor	Equip.	Total	Specification
Vertical Type						
Demolish	S.F.		.73		.73	Includes material and labor to install
Install	S.F.	8.65	.73		9.38	vertical 3" - 5" cloth or PVC blinds.
Demolish and Install	S.F.	8.65	1.46		10.11	
Reinstall	S.F.		.59		.59	
Clean	S.F.	.25	.02		.27	
Minimum Charge	Job		121		121	
Vertical Type per L.F.						
Demolish	L.F.		1.12		1.12	Includes material and labor to install
Install	L.F.	51.50	3.22		54.72	vertical 3" - 5" cloth or PVC blinds x
Demolish and Install	L.F.	51.50	4.34		55.84	72" long.
Reinstall	L.F.		2.58		2.58	
Clean	L.F.	1.79	.17		1.96	
Minimum Charge	Job		121		121	
Venetian Type						
Demolish	S.F.		.73		.73	Includes material and labor to install
Install	S.F.	4.18	.57		4.75	horizontal 2" slat blinds.
Demolish and Install	S.F.	4.18	1.30		5.48	
Reinstall	S.F.		.46		.46	
Clean	S.F.	.25	.02		.27	
Minimum Charge	Job		121		121	
Wood Shutter						
27" x 36", 4 Panel						
Demolish	Ea.		23.50		23.50	Includes material and labor to install
Install	Pr.	80	19.90		99.90	moveable louver, up to 33" x 36", 4
Demolish and Install	Pr.	80	43.40		123.40	panel shutter.
Reinstall	Pr.		15.91		15.91	
Clean	Ea.	.07	7.25		7.32	
Paint	Ea.	1.09	21		22.09	
Minimum Charge	Job		169		169	
33" x 36", 4 Panel						
Demolish	Ea.		23.50		23.50	Includes material and labor to install
Install	Pr.	80	19.90		99.90	moveable louver, up to 33" x 36", 4
Demolish and Install	Pr.	80	43.40		123.40	panel shutter.
Reinstall	Pr.		15.91		15.91	
Clean	Ea.	.07	7.25		7.32	
Paint	Ea.	1.13	23		24.13	
Minimum Charge	Job		169		169	
39" x 36", 4 Panel						
Demolish	Ea.		23.50		23.50	Includes material and labor to install
Install	Pr.	90	19.90		109.90	moveable louver, up to 47" x 36", 4
Demolish and Install	Pr.	90	43.40		133.40	panel shutter.
Reinstall	Pr.		15.91		15.91	
Clean	Ea.	.07	7.25		7.32	
Paint	Ea.	1.13	23		24.13	
Minimum Charge	Job		169		169	
47" x 36", 4 Panel						
Demolish	Ea.		23.50		23.50	Includes material and labor to install
Install	Pr.	90	19.90		109.90	moveable louver, up to 47" x 36", 4
Demolish and Install	Pr.	90	43.40		133.40	panel shutter.
Reinstall	Pr.		15.91		15.91	
Clean	Ea.	.07	7.25		7.32	
Paint	Ea.	1.18	24.50		25.68	
Minimum Charge	Job		169		169	

Improvements / Appliances / Treatments

Stairs

Stairs	Unit	Material	Labor	Equip.	Total	Specification
Concrete Steps						
Precast						
Demolish	Ea.		46.50	29.50	76	Includes material and labor to install
Install	Flight	298	116	41	455	precast concrete front entrance stairs,
Demolish and Install	Flight	298	162.50	70.50	531	4' wide, with 48" platform and 2
Clean	Ea.		3.36		3.36	risers.
Paint	Ea.	2.33	37		39.33	
Minimum Charge	Job		300		300	
Metal						
Concrete Filled						
Demolish	Ea.		13.40		13.40	Includes material and labor to install
Install	Riser	405	56.50	3.27	464.77	steel, cement filled metal pan stairs
Demolish and Install	Riser	405	69.90	3.27	478.17	with picket rail.
Clean	Ea.	.10	2.13		2.23	
Paint	Ea.	.34	2.20		2.54	
Minimum Charge	Job		207		207	
Landing						
Demolish	S.F.		3.36		3.36	Includes material and labor to install
Install	S.F.	84	6.65	.39	91.04	pre-erected conventional steel pan
Demolish and Install	S.F.	84	10.01	.39	94.40	landing.
Clean	S.F.	.10	.43		.53	
Paint	S.F.	.12	.63		.75	
Minimum Charge	Job		207		207	
Railing						
Demolish	L.F.		1.85		1.85	Includes material and labor to install
Install	L.F.	11.10	9.60	.56	21.26	primed steel pipe.
Demolish and Install	L.F.	11.10	11.45	.56	23.11	
Clean	L.F.	.66	1.51		2.17	
Paint	L.F.	.65	4.94		5.59	
Minimum Charge	Job		207		207	

Elevator

Elevator	Unit	Material	Labor	Equip.	Total	Specification
2 Stop Residential						
Install	Ea.	9350	4375		13725	Includes labor, material and
Clean	Ea.	.75	15.10		15.85	equipment to install 2 stop elevator for
Minimum Charge	Job		435		435	residential use.
2 Stop Commercial						
Install	Ea.	34800	8725		43525	Includes labor, material and
Clean	Ea.	.75	15.10		15.85	equipment to install 2 stop elevator for
Minimum Charge	Job		435		435	apartment or commercial building with
						2,000 lb/13 passenger capacity.
3 Stop Commercial						
Install	Ea.	42800	13400		56200	Includes labor, material and
Clean	Ea.	.75	15.10		15.85	equipment to install 3 stop elevator for
Minimum Charge	Job		435		435	apartment or commercial building with
						2,000 lb/13 passenger capacity.
4 Stop Commercial						
Install	Ea.	51000	18200		69200	Includes labor, material and
Clean	Ea.	.75	15.10		15.85	equipment to install 4 stop elevator for
Minimum Charge	Job		435		435	apartment or commercial building with
						2,000 lb/13 passenger capacity.

Elevator	Unit	Material	Labor	Equip.	Total	Specification
5 Stop Commercial						
Install	Ea.	59000	23000		82000	Includes labor, material and
Clean	Ea.	.75	15.10		15.85	equipment to install 5 stop elevator for
Minimum Charge	Job		435		435	apartment or commercial building with
						2,000 lb/13 passenger capacity.

Rangetop	Unit	Material	Labor	Equip.	Total	Specification
4-burner						
Demolish	Ea.		24.50		24.50	Cost includes material and labor to
Install	Ea.	211	61		272	install 4-burner electric surface unit.
Demolish and Install	Ea.	211	85.50		296.50	
Reinstall	Ea.		48.96		48.96	
Clean	Ea.	.04	21.50		21.54	
Minimum Charge	Job		60.50		60.50	
6-burner						
Demolish	Ea.		24.50		24.50	Includes material and labor to install a
Install	Ea.	495	73.50		568.50	6-burner cooktop unit.
Demolish and Install	Ea.	495	98		593	
Reinstall	Ea.		58.75		58.75	
Clean	Ea.	.04	21.50		21.54	
Minimum Charge	Job		60.50		60.50	
Premium Brand w / Grill						
Demolish	Ea.		24.50		24.50	Includes material and labor to install
Install	Ea.	920	73.50		993.50	electric downdraft cooktop with grill.
Demolish and Install	Ea.	920	98		1018	
Reinstall	Ea.		58.75		58.75	
Clean	Ea.	.04	21.50		21.54	
Minimum Charge	Job		60.50		60.50	

Rangehood	Unit	Material	Labor	Equip.	Total	Specification
Vented						
30"						
Demolish	Ea.		22.50		22.50	Includes material and labor to install a
Install	Ea.	42.50	73.50		116	vented range hood.
Demolish and Install	Ea.	42.50	96		138.50	
Reinstall	Ea.		58.75		58.75	
Clean	Ea.	.04	6.20		6.24	
Minimum Charge	Job		60.50		60.50	
36"						
Demolish	Ea.		22.50		22.50	Includes material and labor to install a
Install	Ea.	91.50	73.50		165	vented range hood.
Demolish and Install	Ea.	91.50	96		187.50	
Reinstall	Ea.		58.75		58.75	
Clean	Ea.	.04	6.20		6.24	
Minimum Charge	Job		60.50		60.50	
42"						
Demolish	Ea.		22.50		22.50	Includes material and labor to install a
Install	Ea.	255	86		341	vented range hood.
Demolish and Install	Ea.	255	108.50		363.50	
Reinstall	Ea.		68.77		68.77	
Clean	Ea.	.04	6.20		6.24	
Minimum Charge	Job		60.50		60.50	

Improvements / Appliances / Treatments

Rangehood	Unit	Material	Labor	Equip.	Total	Specification
Ventless						
30″						
Demolish	Ea.		22.50		22.50	Includes material and labor to install a
Install	Ea.	43	46		89	ventless range hood.
Demolish and Install	Ea.	43	68.50		111.50	
Reinstall	Ea.		36.72		36.72	
Clean	Ea.	.04	6.20		6.24	
Minimum Charge	Job		60.50		60.50	
36″						
Demolish	Ea.		22.50		22.50	Includes material and labor to install a
Install	Ea.	45	73.50		118.50	ventless range hood.
Demolish and Install	Ea.	45	96		141	
Reinstall	Ea.		58.75		58.75	
Clean	Ea.	.04	6.20		6.24	
Minimum Charge	Job		60.50		60.50	
42″						
Demolish	Ea.		22.50		22.50	Includes material and labor to install a
Install	Ea.	229	61		290	ventless range hood.
Demolish and Install	Ea.	229	83.50		312.50	
Reinstall	Ea.		48.96		48.96	
Clean	Ea.	.04	6.20		6.24	
Minimum Charge	Job		60.50		60.50	

Oven	Unit	Material	Labor	Equip.	Total	Specification
Free Standing						
Demolish	Ea.		24.50		24.50	Cost includes material and labor to
Install	Ea.	445	48.50		493.50	install 30″ free standing electric range
Demolish and Install	Ea.	445	73		518	with self-cleaning oven.
Reinstall	Ea.		38.66		38.66	
Clean	Ea.	.04	21.50		21.54	
Minimum Charge	Job		60.50		60.50	
Single Wall Oven						
Demolish	Ea.		24.50		24.50	Includes material and labor to install a
Install	Ea.	435	92		527	built-in oven.
Demolish and Install	Ea.	435	116.50		551.50	
Reinstall	Ea.		73.44		73.44	
Clean	Ea.	.04	21.50		21.54	
Minimum Charge	Job		60.50		60.50	
Double Wall Oven						
Demolish	Ea.		24.50		24.50	Includes material and labor to install a
Install	Ea.	1050	169		1219	conventional double oven.
Demolish and Install	Ea.	1050	193.50		1243.50	
Reinstall	Ea.		135.20		135.20	
Clean	Ea.	.04	21.50		21.54	
Minimum Charge	Job		60.50		60.50	
Drop-in Type w / Range						
Demolish	Ea.		24.50		24.50	Includes material and labor to install a
Install	Ea.	505	61		566	built-in cooking range with oven.
Demolish and Install	Ea.	505	85.50		590.50	
Reinstall	Ea.		48.96		48.96	
Clean	Ea.	.04	21.50		21.54	
Minimum Charge	Job		60.50		60.50	

Improvements / Appliances / Treatments

Oven	Unit	Material	Labor	Equip.	Total	Specification
Hi/Lo w / Range, Microwave						
Demolish	Ea.		24.50		24.50	Includes material and labor to install a
Install	Ea.	1675	340		2015	double oven, one conventional, one
Demolish and Install	Ea.	1675	364.50		2039.50	microwave.
Reinstall	Ea.		270.40		270.40	
Clean	Ea.	.04	21.50		21.54	
Minimum Charge	Job		60.50		60.50	
Hi/Lo w/ Range / Double Oven						
Demolish	Ea.		24.50		24.50	Includes material and labor to install a
Install	Ea.	1050	169		1219	conventional double oven.
Demolish and Install	Ea.	1050	193.50		1243.50	
Reinstall	Ea.		135.20		135.20	
Clean	Ea.	.04	21.50		21.54	
Minimum Charge	Job		60.50		60.50	
Countertop Microwave						
Demolish	Ea.		8.40		8.40	Includes material and labor to install a
Install	Ea.	119	30		149	1.0 cubic foot microwave oven.
Demolish and Install	Ea.	119	38.40		157.40	
Reinstall	Ea.		24.16		24.16	
Clean	Ea.	.04	10.85		10.89	
Minimum Charge	Job		60.50		60.50	
Cabinet / Wall Mounted Microwave						
Demolish	Ea.		16.80		16.80	Includes material and labor to install a
Install	Ea.	565	92		657	space saver microwave oven.
Demolish and Install	Ea.	565	108.80		673.80	
Reinstall	Ea.		73.44		73.44	
Clean	Ea.	.04	10.85		10.89	
Minimum Charge	Job		60.50		60.50	

Dryer	Unit	Material	Labor	Equip.	Total	Specification
Basic Grade						
Demolish	Ea.		11.20		11.20	Includes material and labor to install
Install	Ea.	315	195		510	electric or gas dryer. Vent kit not
Demolish and Install	Ea.	315	206.20		521.20	included.
Reinstall	Ea.		155.86		155.86	
Clean	Ea.	.04	21.50		21.54	
Minimum Charge	Job		60.50		60.50	
Good Grade						
Demolish	Ea.		11.20		11.20	Includes material and labor to install
Install	Ea.	395	195		590	electric or gas dryer. Vent kit not
Demolish and Install	Ea.	395	206.20		601.20	included.
Reinstall	Ea.		155.86		155.86	
Clean	Ea.	.04	21.50		21.54	
Minimum Charge	Job		60.50		60.50	
Premium Grade						
Demolish	Ea.		11.20		11.20	Includes material and labor to install
Install	Ea.	890	292		1182	electric or gas dryer. Vent kit not
Demolish and Install	Ea.	890	303.20		1193.20	included.
Reinstall	Ea.		233.79		233.79	
Clean	Ea.	.04	21.50		21.54	
Minimum Charge	Job		60.50		60.50	

Improvements / Appliances / Treatments

Dryer	Unit	Material	Labor	Equip.	Total	Specification
Washer / Dryer Combination Unit						
Demolish	Ea.		11.20		11.20	Includes material and labor to install
Install	Ea.	885	48.50		933.50	electric washer/dryer combination
Demolish and Install	Ea.	885	59.70		944.70	unit, 27" wide, stackable or unitized.
Reinstall	Ea.		38.66		38.66	
Clean	Ea.	.04	21.50		21.54	
Minimum Charge	Job		60.50		60.50	

Washer	Unit	Material	Labor	Equip.	Total	Specification
Basic Grade						
Demolish	Ea.		11.20		11.20	Includes material and labor to install
Install	Ea.	315	124		439	washing machine. New hoses not
Demolish and Install	Ea.	315	135.20		450.20	included.
Reinstall	Ea.		98.99		98.99	
Clean	Ea.	.04	21.50		21.54	
Minimum Charge	Job		60.50		60.50	
Good Grade						
Demolish	Ea.		11.20		11.20	Includes material and labor to install
Install	Ea.	525	186		711	washing machine. New hoses not
Demolish and Install	Ea.	525	197.20		722.20	included.
Reinstall	Ea.		148.48		148.48	
Clean	Ea.	.04	21.50		21.54	
Minimum Charge	Job		60.50		60.50	
Premium Grade						
Demolish	Ea.		11.20		11.20	Includes material and labor to install
Install	Ea.	1150	370		1520	washing machine with digital
Demolish and Install	Ea.	1150	381.20		1531.20	computer readouts and large capacity.
Reinstall	Ea.		296.96		296.96	New hoses not included.
Clean	Ea.	.04	21.50		21.54	
Minimum Charge	Job		60.50		60.50	
Washer / Dryer Combination Unit						
Demolish	Ea.		11.20		11.20	Includes material and labor to install
Install	Ea.	885	48.50		933.50	electric washer/dryer combination
Demolish and Install	Ea.	885	59.70		944.70	unit, 27" wide, stackable or unitized.
Reinstall	Ea.		38.66		38.66	
Clean	Ea.	.04	21.50		21.54	
Minimum Charge	Job		60.50		60.50	

Refrigerator	Unit	Material	Labor	Equip.	Total	Specification
12 Cubic Foot						
Demolish	Ea.		33.50		33.50	Includes material and labor to install a
Install	Ea.	500	48.50		548.50	refrigerator.
Demolish and Install	Ea.	500	82		582	
Reinstall	Ea.		38.66		38.66	
Clean	Ea.	.04	21.50		21.54	
Minimum Charge	Job		60.50		60.50	
16 Cubic Foot						
Demolish	Ea.		33.50		33.50	Includes material and labor to install a
Install	Ea.	505	53.50		558.50	refrigerator.
Demolish and Install	Ea.	505	87		592	
Reinstall	Ea.		42.95		42.95	
Clean	Ea.	.04	21.50		21.54	
Minimum Charge	Job		60.50		60.50	

Improvements / Appliances / Treatments

Refrigerator	Unit	Material	Labor	Equip.	Total	Specification
18 Cubic Foot						
Demolish	Ea.		33.50		33.50	Includes material and labor to install a
Install	Ea.	605	60.50		665.50	refrigerator.
Demolish and Install	Ea.	605	94		699	
Reinstall	Ea.		48.32		48.32	
Clean	Ea.	.04	21.50		21.54	
Minimum Charge	Job		60.50		60.50	
21 Cubic Foot						
Demolish	Ea.		33.50		33.50	Includes material and labor to install a
Install	Ea.	735	69		804	refrigerator.
Demolish and Install	Ea.	735	102.50		837.50	
Reinstall	Ea.		55.22		55.22	
Clean	Ea.	.04	21.50		21.54	
Minimum Charge	Job		60.50		60.50	
24 Cubic Foot						
Demolish	Ea.		33.50		33.50	Includes material and labor to install a
Install	Ea.	980	69		1049	refrigerator.
Demolish and Install	Ea.	980	102.50		1082.50	
Reinstall	Ea.		55.22		55.22	
Clean	Ea.	.04	21.50		21.54	
Minimum Charge	Job		60.50		60.50	
27 Cubic Foot						
Demolish	Ea.		33.50		33.50	Includes material and labor to install a
Install	Ea.	1725	48.50		1773.50	refrigerator.
Demolish and Install	Ea.	1725	82		1807	
Reinstall	Ea.		38.66		38.66	
Clean	Ea.	.04	21.50		21.54	
Minimum Charge	Job		60.50		60.50	
Ice Maker						
Install	Ea.	68	46.50		114.50	Includes labor and material to install
Minimum Charge	Job		60.50		60.50	an automatic ice maker in a
						refrigerator.

Freezer	Unit	Material	Labor	Equip.	Total	Specification
15 Cubic Foot						
Demolish	Ea.		33.50		33.50	Cost includes material and labor to
Install	Ea.	445	48.50		493.50	install 15 cubic foot freezer.
Demolish and Install	Ea.	445	82		527	
Reinstall	Ea.		38.66		38.66	
Clean	Ea.	.04	21.50		21.54	
Minimum Charge	Job		60.50		60.50	
18 Cubic Foot						
Demolish	Ea.		33.50		33.50	Cost includes material and labor to
Install	Ea.	590	48.50		638.50	install 18 cubic foot freezer.
Demolish and Install	Ea.	590	82		672	
Reinstall	Ea.		38.66		38.66	
Clean	Ea.	.04	21.50		21.54	
Minimum Charge	Job		60.50		60.50	
21 Cubic Foot						
Demolish	Ea.		33.50		33.50	Includes material and labor to install a
Install	Ea.	530	48.50		578.50	15 to 23 cubic foot deep freezer.
Demolish and Install	Ea.	530	82		612	
Reinstall	Ea.		38.66		38.66	
Clean	Ea.	.04	21.50		21.54	
Minimum Charge	Job		60.50		60.50	

Improvements / Appliances / Treatments

Freezer	Unit	Material	Labor	Equip.	Total	Specification
24 Cubic Foot						
Demolish	Ea.		33.50		33.50	Includes material and labor to install a
Install	Ea.	590	96.50		686.50	24 cubic foot deep freezer.
Demolish and Install	Ea.	590	130		720	
Reinstall	Ea.		77.31		77.31	
Clean	Ea.	.04	21.50		21.54	
Minimum Charge	Job		60.50		60.50	
Chest-type						
Demolish	Ea.		33.50		33.50	Cost includes material and labor to
Install	Ea.	445	48.50		493.50	install 15 cubic foot freezer.
Demolish and Install	Ea.	445	82		527	
Clean	Ea.	.04	21.50		21.54	
Minimum Charge	Job		60.50		60.50	

Dishwasher	Unit	Material	Labor	Equip.	Total	Specification
4 Cycle						
Demolish	Ea.		27		27	Includes material and labor to install
Install	Ea.	298	116		414	good grade automatic dishwasher.
Demolish and Install	Ea.	298	143		441	
Reinstall	Ea.		92.60		92.60	
Clean	Ea.	.04	21.50		21.54	
Minimum Charge	Job		60.50		60.50	

Garbage Disposal	Unit	Material	Labor	Equip.	Total	Specification
In Sink Unit						
Demolish	Ea.		26		26	Cost includes material and labor to
Install	Ea.	91.50	46.50		138	install 1/2 HP custom disposal unit.
Demolish and Install	Ea.	91.50	72.50		164	
Reinstall	Ea.		37.04		37.04	
Clean	Ea.	.02	6.20		6.22	
Minimum Charge	Job		60.50		60.50	

Trash Compactor	Unit	Material	Labor	Equip.	Total	Specification
4 to 1 Compaction						
Demolish	Ea.		25.50		25.50	Includes material and labor to install a
Install	Ea.	625	67.50		692.50	trash compactor.
Demolish and Install	Ea.	625	93		718	
Reinstall	Ea.		54.08		54.08	
Clean	Ea.	.04	15.10		15.14	
Minimum Charge	Job		60.50		60.50	

Kitchenette Unit	Unit	Material	Labor	Equip.	Total	Specification
Range, Refrigerator and Sink						
Demolish	Ea.		102		102	Includes material and labor to install
Install	Ea.	3100	385		3485	range/refrigerator/sink unit, 72"
Demolish and Install	Ea.	3100	487		3587	wide, base unit only.
Reinstall	Ea.		308.67		308.67	
Clean	Ea.	.83	10.85		11.68	
Minimum Charge	Job		60.50		60.50	

Bath Accessories		Unit	Material	Labor	Equip.	Total	Specification
Towel Bar							
	Demolish	Ea.		5.60		5.60	Includes material and labor to install
	Install	Ea.	34	14.10		48.10	towel bar up to 24" long.
	Demolish and Install	Ea.	34	19.70		53.70	
	Reinstall	Ea.		11.27		11.27	
	Clean	Ea.		3.02		3.02	
	Minimum Charge	Job		169		169	
Toothbrush Holder							
	Demolish	Ea.		5.60		5.60	Includes material and labor to install a
	Install	Ea.	20.50	16.90		37.40	surface mounted tumbler and
	Demolish and Install	Ea.	20.50	22.50		43	toothbrush holder.
	Reinstall	Ea.		13.52		13.52	
	Clean	Ea.		3.02		3.02	
	Minimum Charge	Job		169		169	
Grab Rail							
	Demolish	Ea.		8.40		8.40	Includes material and labor to install a
	Install	Ea.	57	14.70		71.70	24" long grab bar. Blocking is not
	Demolish and Install	Ea.	57	23.10		80.10	included.
	Reinstall	Ea.		11.76		11.76	
	Clean	Ea.	.33	5.05		5.38	
	Minimum Charge	Job		169		169	
Soap Dish							
	Demolish	Ea.		5.60		5.60	Includes material and labor to install a
	Install	Ea.	22	16.90		38.90	surface mounted soap dish.
	Demolish and Install	Ea.	22	22.50		44.50	
	Reinstall	Ea.		13.52		13.52	
	Clean	Ea.		3.02		3.02	
	Minimum Charge	Job		169		169	
Toilet Paper Roller							
	Demolish	Ea.		5.60		5.60	Includes material and labor to install a
	Install	Ea.	12.85	11.25		24.10	surface mounted toilet tissue dispenser.
	Demolish and Install	Ea.	12.85	16.85		29.70	
	Reinstall	Ea.		9.01		9.01	
	Clean	Ea.	.33	7.55		7.88	
	Minimum Charge	Job		169		169	
Dispenser							
Soap							
	Demolish	Ea.		7.45		7.45	Includes material and labor to install a
	Install	Ea.	51.50	16.90		68.40	surface mounted soap dispenser.
	Demolish and Install	Ea.	51.50	24.35		75.85	
	Reinstall	Ea.		13.52		13.52	
	Clean	Ea.	.33	7.55		7.88	
	Minimum Charge	Job		169		169	
Towel							
	Demolish	Ea.		8.40		8.40	Includes material and labor to install a
	Install	Ea.	41.50	21		62.50	surface mounted towel dispenser.
	Demolish and Install	Ea.	41.50	29.40		70.90	
	Reinstall	Ea.		16.90		16.90	
	Clean	Ea.	.33	10.05		10.38	
	Minimum Charge	Job		169		169	
Seat Cover							
	Demolish	Ea.		5.60		5.60	Includes material and labor to install a
	Install	Ea.	30.50	22.50		53	surface mounted toilet seat cover
	Demolish and Install	Ea.	30.50	28.10		58.60	dispenser.
	Reinstall	Ea.		18.03		18.03	
	Clean	Ea.	.33	7.55		7.88	
	Minimum Charge	Job		169		169	

Bath Accessories

Bath Accessories	Unit	Material	Labor	Equip.	Total	Specification
Sanitary Napkin						
Demolish	Ea.		14.15		14.15	Includes material and labor to install
Install	Ea.	94	35.50		129.50	surface mounted sanitary napkin
Demolish and Install	Ea.	94	49.65		143.65	dispenser.
Reinstall	Ea.		28.46		28.46	
Clean	Ea.	.17	10.05		10.22	
Minimum Charge	Job		169		169	
Partition						
Urinal						
Demolish	Ea.		16.80		16.80	Includes material and labor to install a
Install	Ea.	218	84.50		302.50	urinal screen.
Demolish and Install	Ea.	218	101.30		319.30	
Reinstall	Ea.		67.60		67.60	
Clean	Ea.	.50	6.70		7.20	
Minimum Charge	Job		169		169	
Toilet						
Demolish	Ea.		49.50		49.50	Includes material and labor to install
Install	Ea.	515	135		650	baked enamel standard size partition,
Demolish and Install	Ea.	515	184.50		699.50	floor or ceiling mounted with one wall
Reinstall	Ea.		108.16		108.16	and one door.
Clean	Ea.	3.28	15.10		18.38	
Minimum Charge	Job		169		169	
Handicap						
Demolish	Ea.		49.50		49.50	Includes material and labor to install
Install	Ea.	830	135		965	baked enamel handicap partition,
Demolish and Install	Ea.	830	184.50		1014.50	floor or ceiling mounted with one wall
Reinstall	Ea.		108.16		108.16	and one door.
Clean	Ea.	3.28	15.10		18.38	
Minimum Charge	Job		169		169	

Fire Prevention / Protection

Fire Prevention / Protection	Unit	Material	Labor	Equip.	Total	Specification
Fire Extinguisher						
5# Carbon Dioxide						
Install	Ea.	143			143	Includes labor and material to install one wall mounted 5 lb. carbon dioxide factory charged unit, complete with hose, horn and wall mounting bracket.
10# Carbon Dioxide						
Install	Ea.	176			176	Includes labor and material to install one wall mounted 10 lb. carbon dioxide factory charged unit, complete with hose, horn and wall mounting bracket.
15# Carbon Dioxide						
Install	Ea.	209			209	Includes labor and material to install one wall mounted 15 lb. carbon dioxide factory charged unit, complete with hose, horn and wall mounting bracket.

Improvements / Appliances / Treatments

Fire Prevention / Protection		Unit	Material	Labor	Equip.	Total	Specification
5# Dry Chemical	Install	Ea.	66			66	Includes labor and material to install one wall mounted 5 lb. dry chemical factory charged unit, complete with hose, horn and wall mounting bracket.
20# Dry Chemical	Install	Ea.	121			121	Includes labor and material to install one wall mounted 20 lb. dry chemical factory charged unit, complete with hose, horn and wall mounting bracket.

Gutters		Unit	Material	Labor	Equip.	Total	Specification
Galvanized							
	Demolish	L.F.		1.12		1.12	Cost includes material and labor to install 5" galvanized steel gutter with enamel finish, fittings, hangers, and corners.
	Install	L.F.	1.23	3.12		4.35	
	Demolish and Install	L.F.	1.23	4.24		5.47	
	Reinstall	L.F.		2.49		2.49	
	Clean	L.F.		.32		.32	
	Paint	L.F.	.18	.91		1.09	
	Minimum Charge	Job		187		187	
Aluminum							
	Demolish	L.F.		1.12		1.12	Cost includes material and labor to install 5" aluminum gutter with enamel finish, fittings, hangers, and corners.
	Install	L.F.	1.46	3.12		4.58	
	Demolish and Install	L.F.	1.46	4.24		5.70	
	Reinstall	L.F.		2.49		2.49	
	Clean	L.F.		.32		.32	
	Paint	L.F.	.18	.91		1.09	
	Minimum Charge	Job		187		187	
Copper							
	Demolish	L.F.		1.12		1.12	Cost includes material and labor to install 5" half-round copper gutter with fittings, hangers, and corners.
	Install	L.F.	5.80	3.12		8.92	
	Demolish and Install	L.F.	5.80	4.24		10.04	
	Reinstall	L.F.		2.49		2.49	
	Clean	L.F.		.32		.32	
	Minimum Charge	Job		187		187	
Plastic							
	Demolish	L.F.		1.12		1.12	Cost includes material and labor to install 4" white vinyl gutter fittings, hangers, and corners.
	Install	L.F.	.94	3.07		4.01	
	Demolish and Install	L.F.	.94	4.19		5.13	
	Reinstall	L.F.		2.46		2.46	
	Clean	L.F.		.32		.32	
	Paint	L.F.	.18	.91		1.09	
	Minimum Charge	Job		187		187	

Downspouts		Unit	Material	Labor	Equip.	Total	Specification
Galvanized							
	Demolish	L.F.		.77		.77	Cost includes material and labor to install 4" diameter downspout with enamel finish.
	Install	L.F.	1.54	2.58		4.12	
	Demolish and Install	L.F.	1.54	3.35		4.89	
	Reinstall	L.F.		2.06		2.06	
	Clean	L.F.		.32		.32	
	Paint	L.F.	.18	.91		1.09	
	Minimum Charge	Job		187		187	

Improvements / Appliances / Treatments

Downspouts

	Unit	Material	Labor	Equip.	Total	Specification
Aluminum						
Demolish	L.F.		.77		.77	Includes material and labor to install
Install	L.F.	2.75	2.67		5.42	aluminum enameled downspout.
Demolish and Install	L.F.	2.75	3.44		6.19	
Reinstall	L.F.		2.14		2.14	
Clean	L.F.		.32		.32	
Paint	L.F.	.18	.91		1.09	
Minimum Charge	Job		187		187	
Copper						
Demolish	L.F.		.77		.77	Includes material and labor to install
Install	L.F.	4.84	1.97		6.81	round copper downspout.
Demolish and Install	L.F.	4.84	2.74		7.58	
Reinstall	L.F.		1.57		1.57	
Clean	L.F.		.32		.32	
Paint	L.F.	.18	.91		1.09	
Minimum Charge	Job		187		187	
Plastic						
Demolish	L.F.		.77		.77	Cost includes material and labor to
Install	L.F.	.79	1.78		2.57	install 2" x 3" vinyl rectangular
Demolish and Install	L.F.	.79	2.55		3.34	downspouts.
Reinstall	L.F.		1.42		1.42	
Clean	L.F.		.32		.32	
Paint	L.F.	.18	.91		1.09	
Minimum Charge	Job		187		187	

Shutters

	Unit	Material	Labor	Equip.	Total	Specification
Exterior						
25" High						
Demolish	Ea.		8.40		8.40	Includes material and labor to install
Install	Pr.	97.50	34		131.50	pair of exterior shutters 25" x 16",
Demolish and Install	Pr.	97.50	42.40		139.90	unpainted, installed as fixed shutters
Reinstall	Pr.		33.80		33.80	on concrete or wood frame.
Clean	Ea.	.07	7.25		7.32	
Paint	Ea.	1.44	15.95		17.39	
Minimum Charge	Job		169		169	
39" High						
Demolish	Ea.		8.40		8.40	Includes material and labor to install
Install	Pr.	105	34		139	pair of shutters 39" x 16", unpainted,
Demolish and Install	Pr.	105	42.40		147.40	installed as fixed shutters on concrete
Reinstall	Pr.		33.80		33.80	or wood frame.
Clean	Ea.	.07	7.25		7.32	
Paint	Ea.	2.23	21.50		23.73	
Minimum Charge	Job		169		169	
51" High						
Demolish	Ea.		8.40		8.40	Includes material and labor to install
Install	Pr.	117	34		151	pair of shutters 51" x 16", unpainted,
Demolish and Install	Pr.	117	42.40		159.40	installed as fixed shutters on concrete
Reinstall	Pr.		33.80		33.80	or wood frame.
Clean	Ea.	.07	7.25		7.32	
Paint	Ea.	2.92	22.50		25.42	
Minimum Charge	Job		169		169	

Shutters	Unit	Material	Labor	Equip.	Total	Specification
59" High						
Demolish	Ea.		8.40		8.40	Includes material and labor to install
Install	Pr.	164	34		198	pair of shutters 59" x 16", unpainted,
Demolish and Install	Pr.	164	42.40		206.40	installed as fixed shutters on concrete
Reinstall	Pr.		33.80		33.80	or wood frame.
Clean	Ea.	.07	7.25		7.32	
Paint	Ea.	3.36	24.50		27.86	
Minimum Charge	Job		169		169	
67" High						
Demolish	Ea.		8.40		8.40	Includes material and labor to install
Install	Pr.	165	37.50		202.50	pair of shutters 67" x 16", unpainted,
Demolish and Install	Pr.	165	45.90		210.90	installed as fixed shutters on concrete
Reinstall	Pr.		37.56		37.56	or wood frame.
Clean	Ea.	.07	7.25		7.32	
Paint	Ea.	3.84	27.50		31.34	
Minimum Charge	Job		169		169	
Interior Wood						
23" x 24", 2 Panel						
Demolish	Ea.		8.40		8.40	Cost includes material and labor to
Install	Ea.	81	19.90		100.90	install 23" x 24", 4 panel interior
Demolish and Install	Ea.	81	28.30		109.30	moveable pine shutter, 1-1/4" louver,
Reinstall	Ea.		15.91		15.91	knobs and hooks and hinges.
Clean	Ea.	.07	7.25		7.32	
Paint	Ea.	1.09	21		22.09	
Minimum Charge	Job		169		169	
27" x 24", 4 Panel						
Demolish	Ea.		8.40		8.40	Cost includes material and labor to
Install	Ea.	94	19.90		113.90	install 27" x 24", 4 panel interior
Demolish and Install	Ea.	94	28.30		122.30	moveable pine shutter, 1-1/4" louver,
Reinstall	Ea.		15.91		15.91	knobs and hooks and hinges.
Clean	Ea.	.07	7.25		7.32	
Paint	Ea.	1.09	21		22.09	
Minimum Charge	Job		169		169	
31" x 24", 4 Panel						
Demolish	Ea.		8.40		8.40	Cost includes material and labor to
Install	Ea.	108	19.90		127.90	install 31" x 24", 4 panel interior
Demolish and Install	Ea.	108	28.30		136.30	moveable pine shutter, 1-1/4" louver,
Reinstall	Ea.		15.91		15.91	knobs and hooks and hinges.
Clean	Ea.	.07	7.25		7.32	
Paint	Ea.	1.09	21		22.09	
Minimum Charge	Job		169		169	
35" x 24", 4 Panel						
Demolish	Ea.		8.40		8.40	Cost includes material and labor to
Install	Ea.	108	19.90		127.90	install 35" x 24", 4 panel interior
Demolish and Install	Ea.	108	28.30		136.30	moveable pine shutter, 1-1/4" louver,
Reinstall	Ea.		15.91		15.91	knobs and hooks and hinges.
Clean	Ea.	.07	7.25		7.32	
Paint	Ea.	1.09	21		22.09	
Minimum Charge	Job		169		169	
39" x 24", 4 Panel						
Demolish	Ea.		8.40		8.40	Cost includes material and labor to
Install	Ea.	124	19.90		143.90	install 39" x 24", 4 panel interior
Demolish and Install	Ea.	124	28.30		152.30	moveable pine shutter, 1-1/4" louver,
Reinstall	Ea.		15.91		15.91	knobs and hooks and hinges.
Clean	Ea.	.07	7.25		7.32	
Paint	Ea.	1.09	21		22.09	
Minimum Charge	Job		169		169	

Improvements / Appliances / Treatments

Shutters	Unit	Material	Labor	Equip.	Total	Specification
47" x 24", 4 Panel						
Demolish	Ea.		8.40		8.40	Cost includes material and labor to
Install	Ea.	138	19.90		157.90	install 47" x 24", 4 panel interior
Demolish and Install	Ea.	138	28.30		166.30	moveable pine shutter, 1-1/4" louver,
Reinstall	Ea.		15.91		15.91	knobs and hooks and hinges.
Clean	Ea.	.07	7.25		7.32	
Paint	Ea.	1.13	23		24.13	
Minimum Charge	Job		169		169	

Awning	Unit	Material	Labor	Equip.	Total	Specification
Standard Aluminum						
36" Wide x 30" Deep						
Demolish	Ea.		8.40		8.40	Includes material and labor to install
Install	Ea.	169	28		197	one aluminum awning with
Demolish and Install	Ea.	169	36.40		205.40	weather-resistant finish.
Reinstall	Ea.		22.53		22.53	
Clean	Ea.		10.05		10.05	
Minimum Charge	Job		169		169	
48" Wide x 30" Deep						
Demolish	Ea.		8.40		8.40	Includes material and labor to install
Install	Ea.	195	34		229	one aluminum awning with
Demolish and Install	Ea.	195	42.40		237.40	weather-resistant finish.
Reinstall	Ea.		27.04		27.04	
Clean	Ea.		10.05		10.05	
Minimum Charge	Job		169		169	
60" Wide x 30" Deep						
Demolish	Ea.		16.80		16.80	Includes material and labor to install
Install	Ea.	219	37.50		256.50	one aluminum awning with
Demolish and Install	Ea.	219	54.30		273.30	weather-resistant finish.
Reinstall	Ea.		30.04		30.04	
Clean	Ea.		10.05		10.05	
Minimum Charge	Job		169		169	
72" Wide x 30" Deep						
Demolish	Ea.		16.80		16.80	Includes material and labor to install
Install	Ea.	243	42.50		285.50	one aluminum awning with
Demolish and Install	Ea.	243	59.30		302.30	weather-resistant finish.
Reinstall	Ea.		33.80		33.80	
Clean	Ea.		10.05		10.05	
Minimum Charge	Job		169		169	
Roll-up Type						
Demolish	S.F.		.60		.60	Includes material and labor to install
Install	S.F.	7.05	3.38		10.43	an aluminum roll-up awning.
Demolish and Install	S.F.	7.05	3.98		11.03	
Reinstall	S.F.		2.70		2.70	
Clean	S.F.	.03	.50		.53	
Minimum Charge	Job		169		169	
Canvas						
30" Wide						
Demolish	Ea.		5.05		5.05	Includes material and labor to install a
Install	Ea.	183	28		211	canvas window awning.
Demolish and Install	Ea.	183	33.05		216.05	
Reinstall	Ea.		22.53		22.53	
Clean	Ea.		3.02		3.02	
Minimum Charge	Job		169		169	

Improvements / Appliances / Treatments

Awning		Unit	Material	Labor	Equip.	Total	Specification
36" Wide							
	Demolish	Ea.		5.05		5.05	Includes material and labor to install a
	Install	Ea.	195	34		229	canvas window awning.
	Demolish and Install	Ea.	195	39.05		234.05	
	Reinstall	Ea.		27.04		27.04	
	Clean	Ea.		3.31		3.31	
	Minimum Charge	Job		169		169	
42" Wide							
	Demolish	Ea.		5.05		5.05	Includes material and labor to install a
	Install	Ea.	207	42.50		249.50	canvas window awning.
	Demolish and Install	Ea.	207	47.55		254.55	
	Reinstall	Ea.		33.80		33.80	
	Clean	Ea.		3.93		3.93	
	Minimum Charge	Job		169		169	
48" Wide							
	Demolish	Ea.		5.05		5.05	Includes material and labor to install a
	Install	Ea.	219	48.50		267.50	canvas window awning.
	Demolish and Install	Ea.	219	53.55		272.55	
	Reinstall	Ea.		38.63		38.63	
	Clean	Ea.		4.56		4.56	
	Minimum Charge	Job		169		169	
Cloth Canopy Cover							
8' x 10'							
	Demolish	Ea.		283		283	Includes material and labor to install
	Install	Ea.	375	169		544	one 8' x 10' cloth canopy patio cover
	Demolish and Install	Ea.	375	452		827	with front bar and tension support
	Reinstall	Ea.		135.20		135.20	rafters and 9" valance.
	Clean	Ea.	2.62	40.50		43.12	
	Minimum Charge	Job		340		340	
8' x 15'							
	Demolish	Ea.		283		283	Includes material and labor to install
	Install	Ea.	505	169		674	one 8' x 15' cloth canopy patio cover
	Demolish and Install	Ea.	505	452		957	with front bar and tension support
	Reinstall	Ea.		135.20		135.20	rafters and 9" valance.
	Clean	Ea.	2.62	60.50		63.12	
	Minimum Charge	Job		340		340	

Vent		Unit	Material	Labor	Equip.	Total	Specification
Small Roof Turbine							
	Demolish	Ea.		8.40		8.40	Includes material and labor to install
	Install	Ea.	57	33.50		90.50	spinner ventilator, wind driven,
	Demolish and Install	Ea.	57	41.90		98.90	galvanized, 4" neck diam, 180 CFM.
	Reinstall	Ea.		26.93		26.93	
	Clean	Ea.		5.35		5.35	
	Minimum Charge	Job		186		186	
Large Roof Turbine							
	Demolish	Ea.		6.70		6.70	Includes material and labor to install
	Install	Ea.	66.50	48		114.50	spinner ventilator, wind driven,
	Demolish and Install	Ea.	66.50	54.70		121.20	galvanized, 8" neck diam, 360 CFM.
	Reinstall	Ea.		38.47		38.47	
	Clean	Ea.		6.05		6.05	
	Minimum Charge	Job		186		186	

Improvements / Appliances / Treatments

Vent		Unit	Material	Labor	Equip.	Total	Specification
Roof Ventilator							
	Demolish	Ea.		8.40		8.40	Includes material and labor to install a
	Install	Ea.	163	42		205	roof ventilator.
	Demolish and Install	Ea.	163	50.40		213.40	
	Reinstall	Ea.		33.66		33.66	
	Clean	Ea.	.66	6.05		6.71	
	Paint	Ea.	1.36	7.40		8.76	
	Minimum Charge	Job		186		186	
Gable Louvered							
	Demolish	Ea.		16.80		16.80	Includes material and labor to install a
	Install	Ea.	37.50	11.25		48.75	vinyl gable end vent.
	Demolish and Install	Ea.	37.50	28.05		65.55	
	Reinstall	Ea.		9.01		9.01	
	Clean	Ea.		3.22		3.22	
	Paint	Ea.	.90	4.56		5.46	
	Minimum Charge	Job		169		169	
Ridge Vent							
	Demolish	L.F.		.87		.87	Includes material and labor to install a
	Install	L.F.	2.93	2.41		5.34	mill finish aluminum ridge vent strip.
	Demolish and Install	L.F.	2.93	3.28		6.21	
	Minimum Charge	Job		169		169	
Foundation							
	Demolish	Ea.		6		6	Includes material and labor to install
	Install	Ea.	21	11.35		32.35	galvanized foundation block vent.
	Demolish and Install	Ea.	21	17.35		38.35	
	Reinstall	Ea.		9.08		9.08	
	Clean	Ea.	.17	6.05		6.22	
	Minimum Charge	Job		169		169	
Frieze							
	Demolish	L.F.		.60		.60	Includes material and labor to install
	Install	L.F.	.85	.77		1.62	pine frieze board.
	Demolish and Install	L.F.	.85	1.37		2.22	
	Reinstall	L.F.		.61		.61	
	Clean	L.F.	.03	.20		.23	
	Minimum Charge	Job		169		169	
Water Heater Cap							
	Demolish	Ea.		12.90		12.90	Includes material and labor to install a
	Install	Ea.	25.50	47		72.50	galvanized steel water heater cap.
	Demolish and Install	Ea.	25.50	59.90		85.40	
	Reinstall	Ea.		37.40		37.40	
	Clean	Ea.	1.14	15.10		16.24	

Cupolas		Unit	Material	Labor	Equip.	Total	Specification
22" x 22"							
	Demolish	Ea.		162		162	Includes material and labor to install a
	Install	Ea.	360	84.50		444.50	cedar cupola with aluminum roof
	Demolish and Install	Ea.	360	246.50		606.50	covering.
	Reinstall	Ea.		67.60		67.60	
	Minimum Charge	Job		169		169	
29" x 29"							
	Demolish	Ea.		162		162	Includes material and labor to install a
	Install	Ea.	490	84.50		574.50	cedar cupola with aluminum roof
	Demolish and Install	Ea.	490	246.50		736.50	covering.
	Reinstall	Ea.		67.60		67.60	
	Minimum Charge	Job		169		169	

Improvements / Appliances / Treatments

Cupolas

	Unit	Material	Labor	Equip.	Total	Specification
35" x 35"						
Demolish	Ea.		162		162	Includes material and labor to install a
Install	Ea.	670	113		783	cedar cupola with aluminum roof
Demolish and Install	Ea.	670	275		945	covering.
Reinstall	Ea.		90.13		90.13	
Minimum Charge	Job		169		169	
47" x 47"						
Demolish	Ea.		162		162	Includes material and labor to install a
Install	Ea.	1175	169		1344	cedar cupola with aluminum roof
Demolish and Install	Ea.	1175	331		1506	covering.
Reinstall	Ea.		135.20		135.20	
Minimum Charge	Job		169		169	
Weather Vane						
Install	Ea.	143	42.50		185.50	Includes minimum labor and
Minimum Charge	Job		169		169	equipment to install residential type weathervane.

Aluminum Carport

	Unit	Material	Labor	Equip.	Total	Specification
Natural Finish						
Demolish	S.F.		1.68		1.68	Includes material, labor and
Install	S.F.	23.50	5.60	.97	30.07	equipment to install an aluminum
Demolish and Install	S.F.	23.50	7.28	.97	31.75	carport. Foundations are not included.
Reinstall	S.F.		4.49	.78	5.27	
Clean	S.F.	.03	.18		.21	
Minimum Charge	Job		340		340	
Enamel Finish						
Demolish	S.F.		1.68		1.68	Includes material, labor and
Install	S.F.	23.50	5.60	.97	30.07	equipment to install an aluminum
Demolish and Install	S.F.	23.50	7.28	.97	31.75	carport. Foundations are not included.
Reinstall	S.F.		4.49	.78	5.27	
Clean	S.F.	.03	.18		.21	
Minimum Charge	Job		340		340	
20' x 20' Complete						
Demolish	Ea.		168		168	Includes material, labor and
Install	Ea.	7475	560	97	8132	equipment to install an aluminum
Demolish and Install	Ea.	7475	728	97	8300	carport. Foundations are not included.
Reinstall	Ea.		448.80	77.57	526.37	
Clean	Ea.	.18	60.50		60.68	
Minimum Charge	Job		340		340	
24' x 24' Complete						
Demolish	Ea.		168		168	Includes material, labor and
Install	Ea.	10200	750	129	11079	equipment to install an aluminum
Demolish and Install	Ea.	10200	918	129	11247	carport. Foundations are not included.
Reinstall	Ea.		598.40	103.42	701.82	
Clean	Ea.	.18	121		121.18	
Minimum Charge	Job		340		340	
Metal Support Posts						
Demolish	Ea.		2.24		2.24	Includes material and labor to install
Install	Ea.	16.60	42.50		59.10	metal support posts for a carport.
Demolish and Install	Ea.	16.60	44.74		61.34	
Reinstall	Ea.		33.80		33.80	
Clean	Ea.	.18	2.52		2.70	
Paint	Ea.	.12	5.95		6.07	
Minimum Charge	Job		340		340	

Improvements / Appliances / Treatments

Storage Shed		Unit	Material	Labor	Equip.	Total	Specification
Aluminum							
Pre-fab (8' x 10')							
	Demolish	Ea.		49		49	Includes material and labor to install
	Install	Ea.	273	485		758	an aluminum storage shed.
	Demolish and Install	Ea.	273	534		807	
	Clean	Ea.	1.09	30		31.09	
	Minimum Charge	Ea.		415		415	
Pre-fab (10' x 12')							
	Demolish	Ea.		94.50		94.50	Includes material and labor to install
	Install	Ea.	460	485		945	an aluminum storage shed.
	Demolish and Install	Ea.	460	579.50		1039.50	
	Clean	Ea.	1.09	33.50		34.59	
	Minimum Charge	Ea.		415		415	
Wood							
Pre-fab (8' x 10')							
	Demolish	Ea.		49		49	Includes material and labor to install a
	Install	Ea.	760	485		1245	pre-fabricated wood storage shed.
	Demolish and Install	Ea.	760	534		1294	
	Clean	Ea.	1.09	30		31.09	
	Minimum Charge	Ea.		415		415	
Pre-fab (10' x 12')							
	Demolish	Ea.		94.50		94.50	Includes material and labor to install a
	Install	Ea.	990	485		1475	pre-fabricated wood storage shed.
	Demolish and Install	Ea.	990	579.50		1569.50	
	Clean	Ea.	1.09	33.50		34.59	
	Minimum Charge	Ea.		415		415	

Gazebo		Unit	Material	Labor	Equip.	Total	Specification
Good Grade							
	Demolish	S.F.		2.68		2.68	Includes material and labor to install a
	Install	S.F.	18.90	25.50		44.40	gazebo.
	Demolish and Install	S.F.	18.90	28.18		47.08	
	Reinstall	S.F.		20.28		20.28	
	Clean	S.F.	.13	1.61		1.74	
	Minimum Charge	Job		169		169	
Custom Grade							
	Demolish	S.F.		2.68		2.68	Includes material and labor to install a
	Install	S.F.	30.50	25.50		56	gazebo.
	Demolish and Install	S.F.	30.50	28.18		58.68	
	Reinstall	S.F.		20.28		20.28	
	Clean	S.F.	.13	1.61		1.74	
	Minimum Charge	Job		169		169	
Bench Seating							
	Demolish	L.F.		4.47		4.47	Includes material and labor to install
	Install	L.F.	2.18	16.90		19.08	pressure treated bench seating.
	Demolish and Install	L.F.	2.18	21.37		23.55	
	Reinstall	L.F.		13.52		13.52	
	Clean	L.F.	.13	1.21		1.34	
	Paint	L.F.	.51	2.37		2.88	
	Minimum Charge	Job		169		169	
Re-screen							
	Install	S.F.	.59	1.35		1.94	Includes labor and material to
	Minimum Charge	Job		169		169	re-screen wood frame.

Improvements / Appliances / Treatments

Pool	Unit	Material	Labor	Equip.	Total	Specification
Screen Enclosure						
Re-screen						
Install	SF Wall		4.15		4.15	Includes material and labor to install a
Minimum Charge	Job		169		169	screen enclosure for a pool.
Screen Door						
Demolish	Ea.		6.70		6.70	Includes material and labor to install
Install	Ea.	138	56.50		194.50	wood screen door with aluminum cloth
Demolish and Install	Ea.	138	63.20		201.20	screen. Frame and hardware not
Reinstall	Ea.		45.07		45.07	included.
Clean	Ea.	.12	8.70		8.82	
Paint	Ea.	3.03	11.85		14.88	
Minimum Charge	Job		169		169	
Heater / Motor						
Demolish	Ea.		32.50		32.50	Includes material and labor to install
Install	Ea.	2275	415		2690	swimming pool heater, not incl. base
Demolish and Install	Ea.	2275	447.50		2722.50	or pad, elec., 12 kW, 4,800 gal
Reinstall	Ea.		332.70		332.70	pool, incl. pump.
Clean	Ea.	3.28	30		33.28	
Minimum Charge	Job		187		187	
Filter System						
Demolish	Ea.		53.50		53.50	Includes material, labor and
Install	Total	595	297		892	equipment to install sand filter system
Demolish and Install	Total	595	350.50		945.50	tank including hook-up to existing
Reinstall	Total		237.57		237.57	equipment. Pump and motor not
Minimum Charge	Job		186		186	included.
Pump / Motor						
Demolish	Ea.		32.50		32.50	Includes material and labor to install a
Install	Ea.	510	74		584	22 GPM pump.
Demolish and Install	Ea.	510	106.50		616.50	
Reinstall	Ea.		59.39		59.39	
Clean	Ea.	3.28	30		33.28	
Minimum Charge	Job		187		187	
Pump Water						
Install	Day		335	71	406	Includes labor and equipment to pump
Minimum Charge	Job		156		156	water from a swimming pool.
Replaster						
Install	SF Surf	1.07	8		9.07	Includes labor and material to
Minimum Charge	Job		156		156	replaster gunite pool with a surface
						area of 500 to 600 S.F.
Solar Panels						
Demolish	S.F.		1.37		1.37	Includes material and labor to install
Install	S.F.	21.50	2.44		23.94	solar panel.
Demolish and Install	S.F.	21.50	3.81		25.31	
Reinstall	S.F.		1.95		1.95	
Minimum Charge	Ea.		415		415	

Construction Clean-up	Unit	Material	Labor	Equip.	Total	Specification
Final						
Clean	S.F.		.21		.21	Includes labor for final site/job
Minimum Charge	Job		121		121	clean-up including detail cleaning.

Location Factors

Costs shown in *Means cost data publications* are based on National Averages for materials and installation. To adjust these costs to a specific location, simply multiply the base cost by the factor for that city. The data is arranged alphabetically by state and postal zip code numbers. For a city not listed, use the factor for a nearby city with similar economic characteristics.

STATE	CITY	Residential
ALABAMA		
350-352	Birmingham	.86
354	Tuscaloosa	.73
355	Jasper	.71
356	Decatur	.76
357-358	Huntsville	.84
359	Gadsden	.73
360-361	Montgomery	.75
362	Anniston	.68
363	Dothan	.74
364	Evergreen	.70
365-366	Mobile	.79
367	Selma	.72
368	Phenix City	.72
369	Butler	.71
ALASKA		
995-996	Anchorage	1.27
997	Fairbanks	1.29
998	Juneau	1.27
999	Ketchikan	1.28
ARIZONA		
850,853	Phoenix	.86
852	Mesa/Tempe	.84
855	Globe	.80
856-857	Tucson	.84
859	Show Low	.82
860	Flagstaff	.87
863	Prescott	.82
864	Kingman	.83
865	Chambers	.81
ARKANSAS		
716	Pine Bluff	.75
717	Camden	.65
718	Texarkana	.69
719	Hot Springs	.63
720-722	Little Rock	.80
723	West Memphis	.73
724	Jonesboro	.72
725	Batesville	.70
726	Harrison	.71
727	Fayetteville	.67
728	Russellville	.69
729	Fort Smith	.76
CALIFORNIA		
900-902	Los Angeles	1.06
903-905	Inglewood	1.05
906-908	Long Beach	1.03
910-912	Pasadena	1.05
913-916	Van Nuys	1.08
917-918	Alhambra	1.09
919-921	San Diego	1.04
922	Palm Springs	1.03
923-924	San Bernardino	1.05
925	Riverside	1.07
926-927	Santa Ana	1.06
928	Anaheim	1.07
930	Oxnard	1.07
931	Santa Barbara	1.07
932-933	Bakersfield	1.04
934	San Luis Obispo	1.09
935	Mojave	1.05
936-938	Fresno	1.09
939	Salinas	1.12
940-941	San Francisco	1.22
942,956-958	Sacramento	1.11
943	Palo Alto	1.17
944	San Mateo	1.21
945	Vallejo	1.14
946	Oakland	1.20
947	Berkeley	1.23
948	Richmond	1.24
949	San Rafael	1.21
950	Santa Cruz	1.14
951	San Jose	1.19

STATE	CITY	Residential
952	Stockton	1.09
953	Modesto	1.08
954	Santa Rosa	1.14
955	Eureka	1.08
959	Marysville	1.10
960	Redding	1.09
961	Susanville	1.10
COLORADO		
800-802	Denver	.95
803	Boulder	.94
804	Golden	.92
805	Fort Collins	.91
806	Greeley	.79
807	Fort Morgan	.94
808-809	Colorado Springs	.91
810	Pueblo	.92
811	Alamosa	.89
812	Salida	.91
813	Durango	.93
814	Montrose	.88
815	Grand Junction	.93
816	Glenwood Springs	.91
CONNECTICUT		
060	New Britain	1.10
061	Hartford	1.10
062	Willimantic	1.11
063	New London	1.10
064	Meriden	1.10
065	New Haven	1.10
066	Bridgeport	1.11
067	Waterbury	1.11
068	Norwalk	1.11
069	Stamford	1.12
D.C.		
200-205	Washington	.93
DELAWARE		
197	Newark	1.00
198	Wilmington	1.00
199	Dover	1.00
FLORIDA		
320,322	Jacksonville	.78
321	Daytona Beach	.85
323	Tallahassee	.72
324	Panama City	.67
325	Pensacola	.75
326,344	Gainesville	.77
327-328,347	Orlando	.85
329	Melbourne	.87
330-332,340	Miami	.83
333	Fort Lauderdale	.83
334,349	West Palm Beach	.83
335-336,346	Tampa	.86
337	St. Petersburg	.76
338	Lakeland	.82
339,341	Fort Myers	.80
342	Sarasota	.84
GEORGIA		
300-303,399	Atlanta	.88
304	Statesboro	.66
305	Gainesville	.74
306	Athens	.74
307	Dalton	.70
308-309	Augusta	.76
310-312	Macon	.77
313-314	Savannah	.79
315	Waycross	.71
316	Valdosta	.70
317,398	Albany	.75
318-319	Columbus	.79
HAWAII		
967	Hilo	1.22

LOCATION FACTORS

Location Factors

STATE	CITY	Residential
968	Honolulu	1.23
STATES & POSS.		
969	Guam	1.73
IDAHO		
832	Pocatello	.87
833	Twin Falls	.74
834	Idaho Falls	.72
835	Lewiston	.98
836-837	Boise	.89
838	Coeur d'Alene	.94
ILLINOIS		
600-603	North Suburban	1.13
604	Joliet	1.13
605	South Suburban	1.13
606-608	Chicago	1.14
609	Kankakee	.99
610-611	Rockford	1.04
612	Rock Island	.97
613	La Salle	1.03
614	Galesburg	.98
615-616	Peoria	1.00
617	Bloomington	.96
618-619	Champaign	.98
620-622	East St. Louis	1.00
623	Quincy	.97
624	Effingham	.98
625	Decatur	.99
626-627	Springfield	.99
628	Centralia	1.01
629	Carbondale	.96
INDIANA		
460	Anderson	.93
461-462	Indianapolis	.95
463-464	Gary	1.03
465-466	South Bend	.94
467-468	Fort Wayne	.92
469	Kokomo	.94
470	Lawrenceburg	.88
471	New Albany	.87
472	Columbus	.93
473	Muncie	.92
474	Bloomington	.96
475	Washington	.92
476-477	Evansville	.92
478	Terre Haute	.92
479	Lafayette	.93
IOWA		
500-503,509	Des Moines	.93
504	Mason City	.77
505	Fort Dodge	.75
506-507	Waterloo	.80
508	Creston	.82
510-511	Sioux City	.88
512	Sibley	.73
513	Spencer	.75
514	Carroll	.74
515	Council Bluffs	.81
516	Shenandoah	.75
520	Dubuque	.84
521	Decorah	.76
522-524	Cedar Rapids	.94
525	Ottumwa	.84
526	Burlington	.88
527-528	Davenport	.98
KANSAS		
660-662	Kansas City	.97
664-666	Topeka	.79
667	Fort Scott	.86
668	Emporia	.72
669	Belleville	.78
670-672	Wichita	.81
673	Independence	.86
674	Salina	.76
675	Hutchinson	.68
676	Hays	.83
677	Colby	.76
678	Dodge City	.82
679	Liberal	.68

STATE	CITY	Residential
KENTUCKY		
400-402	Louisville	.92
403-405	Lexington	.84
406	Frankfort	.81
407-409	Corbin	.66
410	Covington	.92
411-412	Ashland	.93
413-414	Campton	.68
415-416	Pikeville	.76
417-418	Hazard	.66
420	Paducah	.89
421-422	Bowling Green	.89
423	Owensboro	.81
424	Henderson	.92
425-426	Somerset	.66
427	Elizabethtown	.87
LOUISIANA		
700-701	New Orleans	.85
703	Thibodaux	.83
704	Hammond	.78
705	Lafayette	.81
706	Lake Charles	.82
707-708	Baton Rouge	.83
710-711	Shreveport	.78
712	Monroe	.73
713-714	Alexandria	.74
MAINE		
039	Kittery	.80
040-041	Portland	.91
042	Lewiston	.91
043	Augusta	.83
044	Bangor	.89
045	Bath	.81
046	Machias	.82
047	Houlton	.86
048	Rockland	.82
049	Waterville	.81
MARYLAND		
206	Waldorf	.84
207-208	College Park	.86
209	Silver Spring	.85
210-212	Baltimore	.90
214	Annapolis	.86
215	Cumberland	.87
216	Easton	.68
217	Hagerstown	.86
218	Salisbury	.75
219	Elkton	.82
MASSACHUSETTS		
010-011	Springfield	1.05
012	Pittsfield	1.03
013	Greenfield	1.02
014	Fitchburg	1.11
015-016	Worcester	1.12
017	Framingham	1.13
018	Lowell	1.14
019	Lawrence	1.14
020-022, 024	Boston	1.20
023	Brockton	1.13
025	Buzzards Bay	1.11
026	Hyannis	1.10
027	New Bedford	1.13
MICHIGAN		
480,483	Royal Oak	1.03
481	Ann Arbor	1.04
482	Detroit	1.10
484-485	Flint	.97
486	Saginaw	.94
487	Bay City	.95
488-489	Lansing	.97
490	Battle Creek	.92
491	Kalamazoo	.91
492	Jackson	.92
493,495	Grand Rapids	.82
494	Muskegon	.88
496	Traverse City	.80
497	Gaylord	.83
498-499	Iron Mountain	.90

Location Factors

STATE	CITY	Residential
MINNESOTA		
550-551	Saint Paul	1.14
553-555	Minneapolis	1.17
556-558	Duluth	1.11
559	Rochester	1.06
560	Mankato	1.02
561	Windom	.83
562	Willmar	.85
563	St. Cloud	1.09
564	Brainerd	.99
565	Detroit Lakes	.97
566	Bemidji	.97
567	Thief River Falls	.95
MISSISSIPPI		
386	Clarksdale	.61
387	Greenville	.68
388	Tupelo	.63
389	Greenwood	.65
390-392	Jackson	.73
393	Meridian	.66
394	Laurel	.62
395	Biloxi	.75
396	McComb	.73
397	Columbus	.64
MISSOURI		
630-631	St. Louis	1.01
633	Bowling Green	.95
634	Hannibal	.87
635	Kirksville	.80
636	Flat River	.95
637	Cape Girardeau	.87
638	Sikeston	.82
639	Poplar Bluff	.83
640-641	Kansas City	1.04
644-645	St. Joseph	.95
646	Chillicothe	.86
647	Harrisonville	.96
648	Joplin	.85
650-651	Jefferson City	.89
652	Columbia	.88
653	Sedalia	.85
654-655	Rolla	.89
656-658	Springfield	.87
MONTANA		
590-591	Billings	.87
592	Wolf Point	.83
593	Miles City	.86
594	Great Falls	.88
595	Havre	.80
596	Helena	.87
597	Butte	.82
598	Missoula	.83
599	Kalispell	.82
NEBRASKA		
680-681	Omaha	.90
683-685	Lincoln	.78
686	Columbus	.69
687	Norfolk	.77
688	Grand Island	.77
689	Hastings	.76
690	Mccook	.70
691	North Platte	.75
692	Valentine	.66
693	Alliance	.65
NEVADA		
889-891	Las Vegas	1.01
893	Ely	.89
894-895	Reno	.96
897	Carson City	.96
898	Elko	.95
NEW HAMPSHIRE		
030	Nashua	.90
031	Manchester	.90
032-033	Concord	.86
034	Keene	.73
035	Littleton	.81
036	Charleston	.70

STATE	CITY	Residential
037	Claremont	.72
038	Portsmouth	.84
NEW JERSEY		
070-071	Newark	1.15
072	Elizabeth	1.18
073	Jersey City	1.14
074-075	Paterson	1.15
076	Hackensack	1.14
077	Long Branch	1.15
078	Dover	1.14
079	Summit	1.15
080,083	Vineland	1.09
081	Camden	1.10
082,084	Atlantic City	1.13
085-086	Trenton	1.13
087	Point Pleasant	1.11
088-089	New Brunswick	1.15
NEW MEXICO		
870-872	Albuquerque	.85
873	Gallup	.85
874	Farmington	.85
875	Santa Fe	.85
877	Las Vegas	.85
878	Socorro	.85
879	Truth/Consequences	.84
880	Las Cruces	.82
881	Clovis	.85
882	Roswell	.85
883	Carrizozo	.85
884	Tucumcari	.86
NEW YORK		
100-102	New York	1.36
103	Staten Island	1.28
104	Bronx	1.30
105	Mount Vernon	1.17
106	White Plains	1.20
107	Yonkers	1.22
108	New Rochelle	1.21
109	Suffern	1.14
110	Queens	1.28
111	Long Island City	1.31
112	Brooklyn	1.33
113	Flushing	1.30
114	Jamaica	1.30
115,117,118	Hicksville	1.21
116	Far Rockaway	1.29
119	Riverhead	1.22
120-122	Albany	.96
123	Schenectady	.96
124	Kingston	1.04
125-126	Poughkeepsie	1.06
127	Monticello	1.06
128	Glens Falls	.89
129	Plattsburgh	.94
130-132	Syracuse	.97
133-135	Utica	.94
136	Watertown	.90
137-139	Binghamton	.94
140-142	Buffalo	1.06
143	Niagara Falls	1.02
144-146	Rochester	1.00
147	Jamestown	.89
148-149	Elmira	.86
NORTH CAROLINA		
270,272-274	Greensboro	.74
271	Winston-Salem	.74
275-276	Raleigh	.75
277	Durham	.74
278	Rocky Mount	.64
279	Elizabeth City	.62
280	Gastonia	.75
281-282	Charlotte	.75
283	Fayetteville	.72
284	Wilmington	.73
285	Kinston	.62
286	Hickory	.62
287-288	Asheville	.73
289	Murphy	.66

Location Factors

STATE	CITY	Residential
NORTH DAKOTA		
580-581	Fargo	.80
582	Grand Forks	.76
583	Devils Lake	.79
584	Jamestown	.74
585	Bismarck	.80
586	Dickinson	.76
587	Minot	.80
588	Williston	.77
OHIO		
430-432	Columbus	.95
433	Marion	.93
434-436	Toledo	1.00
437-438	Zanesville	.91
439	Steubenville	.96
440	Lorain	1.00
441	Cleveland	1.01
442-443	Akron	.99
444-445	Youngstown	.95
446-447	Canton	.94
448-449	Mansfield	.95
450	Hamilton	.95
451-452	Cincinnati	.95
453-454	Dayton	.92
455	Springfield	.94
456	Chillicothe	.97
457	Athens	.89
458	Lima	.91
OKLAHOMA		
730-731	Oklahoma City	.80
734	Ardmore	.78
735	Lawton	.81
736	Clinton	.77
737	Enid	.77
738	Woodward	.76
739	Guymon	.66
740-741	Tulsa	.78
743	Miami	.82
744	Muskogee	.72
745	Mcalester	.74
746	Ponca City	.77
747	Durant	.76
748	Shawnee	.75
749	Poteau	.78
OREGON		
970-972	Portland	1.02
973	Salem	1.02
974	Eugene	1.01
975	Medford	1.00
976	Klamath Falls	1.00
977	Bend	1.03
978	Pendleton	1.00
979	Vale	.99
PENNSYLVANIA		
150-152	Pittsburgh	.99
153	Washington	.94
154	Uniontown	.91
155	Bedford	.88
156	Greensburg	.96
157	Indiana	.92
158	Dubois	.90
159	Johnstown	.91
160	Butler	.93
161	New Castle	.93
162	Kittanning	.94
163	Oil City	.90
164-165	Erie	.96
166	Altoona	.89
167	Bradford	.91
168	State College	.92
169	Wellsboro	.90
170-171	Harrisburg	.95
172	Chambersburg	.89
173-174	York	.92
175-176	Lancaster	.92
177	Williamsport	.85
178	Sunbury	.92
179	Pottsville	.91
180	Lehigh Valley	1.02
181	Allentown	1.04

STATE	CITY	Residential
182	Hazleton	.91
183	Stroudsburg	.92
184-185	Scranton	.96
186-187	Wilkes-Barre	.93
188	Montrose	.90
189	Doylestown	1.06
190-191	Philadelphia	1.15
193	Westchester	1.10
194	Norristown	1.08
195-196	Reading	.97
PUERTO RICO		
009	San Juan	.84
RHODE ISLAND		
028	Newport	1.09
029	Providence	1.09
SOUTH CAROLINA		
290-292	Columbia	.73
293	Spartanburg	.71
294	Charleston	.72
295	Florence	.66
296	Greenville	.70
297	Rock Hill	.65
298	Aiken	.84
299	Beaufort	.67
SOUTH DAKOTA		
570-571	Sioux Falls	.77
572	Watertown	.72
573	Mitchell	.75
574	Aberdeen	.76
575	Pierre	.75
576	Mobridge	.72
577	Rapid City	.76
TENNESSEE		
370-372	Nashville	.84
373-374	Chattanooga	.76
375,380-381	Memphis	.85
376	Johnson City	.71
377-379	Knoxville	.74
382	Mckenzie	.73
383	Jackson	.70
384	Columbia	.71
385	Cookeville	.68
TEXAS		
750	Mckinney	.74
751	Waxahachie	.75
752-753	Dallas	.83
754	Greenville	.68
755	Texarkana	.73
756	Longview	.67
757	Tyler	.74
758	Palestine	.66
759	Lufkin	.71
760-761	Fort Worth	.82
762	Denton	.76
763	Wichita Falls	.79
764	Eastland	.72
765	Temple	.74
766-767	Waco	.77
768	Brownwood	.68
769	San Angelo	.71
770-772	Houston	.85
773	Huntsville	.68
774	Wharton	.70
775	Galveston	.84
776-777	Beaumont	.82
778	Bryan	.73
779	Victoria	.73
780	Laredo	.72
781-782	San Antonio	.80
783-784	Corpus Christi	.76
785	McAllen	.75
786-787	Austin	.79
788	Del Rio	.65
789	Giddings	.69
790-791	Amarillo	.77
792	Childress	.75
793-794	Lubbock	.75
795-796	Abilene	.74

STATE	CITY	Residential
797	Midland	.76
798-799,885	El Paso	.75
UTAH		
840-841	Salt Lake City	.82
842,844	Ogden	.81
843	Logan	.81
845	Price	.72
846-847	Provo	.82
VERMONT		
050	White River Jct.	.73
051	Bellows Falls	.75
052	Bennington	.74
053	Brattleboro	.74
054	Burlington	.79
056	Montpelier	.81
057	Rutland	.81
058	St. Johnsbury	.75
059	Guildhall	.74
VIRGINIA		
220-221	Fairfax	.85
222	Arlington	.87
223	Alexandria	.90
224-225	Fredericksburg	.75
226	Winchester	.70
227	Culpeper	.77
228	Harrisonburg	.67
229	Charlottesville	.73
230-232	Richmond	.81
233-235	Norfolk	.82
236	Newport News	.81
237	Portsmouth	.78
238	Petersburg	.79
239	Farmville	.69
240-241	Roanoke	.72
242	Bristol	.68
243	Pulaski	.66
244	Staunton	.69
245	Lynchburg	.70
246	Grundy	.67
WASHINGTON		
980-981,987	Seattle	1.01
982	Everett	1.04
983-984	Tacoma	1.00
985	Olympia	1.00
986	Vancouver	.97
988	Wenatchee	.92
989	Yakima	.96
990-992	Spokane	1.00
993	Richland	.98
994	Clarkston	.97
WEST VIRGINIA		
247-248	Bluefield	.88
249	Lewisburg	.89
250-253	Charleston	.97
254	Martinsburg	.86
255-257	Huntington	.97
258-259	Beckley	.90
260	Wheeling	.94
261	Parkersburg	.92
262	Buckhannon	.93
263-264	Clarksburg	.92
265	Morgantown	.93
266	Gassaway	.92
267	Romney	.88
268	Petersburg	.90
WISCONSIN		
530,532	Milwaukee	1.07
531	Kenosha	1.06
534	Racine	1.04
535	Beloit	1.01
537	Madison	1.01
538	Lancaster	.99
539	Portage	.98
540	New Richmond	1.01
541-543	Green Bay	1.03
544	Wausau	.96
545	Rhinelander	.96
546	La Crosse	.95

STATE	CITY	Residential
547	Eau Claire	1.00
548	Superior	1.01
549	Oshkosh	.97
WYOMING		
820	Cheyenne	.76
821	Yellowstone Nat. Pk.	.72
822	Wheatland	.73
823	Rawlins	.71
824	Worland	.70
825	Riverton	.71
826	Casper	.75
827	Newcastle	.70
828	Sheridan	.74
829-831	Rock Springs	.75
CANADIAN FACTORS (reflect Canadian Currency)		
ALBERTA		
	Calgary	1.05
	Edmonton	1.04
	Fort McMurray	1.02
	Lethbridge	1.03
	Lloydminster	1.02
	Medicine Hat	1.03
	Red Deer	1.03
BRITISH COLUMBIA		
	Kamloops	1.00
	Prince George	1.00
	Vancouver	1.07
	Victoria	1.00
MANITOBA		
	Brandon	.99
	Portage la Prairie	.99
	Winnipeg	1.00
NEW BRUNSWICK		
	Bathurst	.90
	Dalhousie	.90
	Fredericton	.98
	Moncton	.90
	Newcastle	.90
	Saint John	.99
NEWFOUNDLAND		
	Corner Brook	.92
	St. John's	.92
NORTHWEST TERRITORIES		
	Yellowknife	1.01
NOVA SCOTIA		
	Dartmouth	.93
	Halifax	.93
	New Glasgow	.92
	Sydney	.91
	Yarmouth	.92
ONTARIO		
	Barrie	1.09
	Brantford	1.11
	Cornwall	1.10
	Hamilton	1.11
	Kingston	1.11
	Kitchener	1.05
	London	1.09
	North Bay	1.08
	Oshawa	1.09
	Ottawa	1.11
	Owen Sound	1.08
	Peterborough	1.08
	Sarnia	1.11
	St. Catharines	1.04
	Sudbury	1.02
	Thunder Bay	1.07
	Toronto	1.14
	Windsor	1.08
PRINCE EDWARD ISLAND		
	Charlottetown	.87
	Summerside	.87

Location Factors

STATE	CITY	Residential
QUEBEC		
	Cap-de-la-Madeleine	1.10
	Charlesbourg	1.10
	Chicoutimi	1.10
	Gatineau	1.09
	Laval	1.09
	Montreal	1.10
	Quebec	1.12
	Sherbrooke	1.09
	Trois Rivieres	1.10
SASKATCHEWAN		
	Moose Jaw	.90
	Prince Albert	.90
	Regina	.92
	Saskatoon	.90
YUKON		
	Whitehorse	.89

Abbreviations

A	Area
ASTM	American Society for Testing and Materials
B.F.	Board feet
Carp.	Carpenter
C.F.	Cubic feet
CWJ	Composite wood joist
C.Y.	Cubic yard
Ea.	Each
Equip.	Equipment
Exp.	Exposure
Ext.	Exterior
F	Fahrenheit
Ft.	Foot, feet
Gal.	Gallon
Hr.	Hour
in.	Inch, inches
Inst.	Installation
Int.	Interior
Lb.	Pound
L.F.	Linear feet
LVL	Laminated veneer lumber
Mat.	Material
Max.	Maximum
MBF	Thousand board feet
MBM	Thousand feet board measure
MSF	Thousand square feet
Min.	Minimum
O.C.	On center
O&P	Overhead and profit
OWJ	Open web wood joist
Oz.	Ounce
Pr.	Pair
Quan.	Quantity
S.F.	Square foot
Sq.	Square, 100 square feet
S.Y.	Square yard
V.L.F.	Vertical linear feet
'	Foot, feet
"	Inch, inches
°	Degrees

Reed Construction Data, Inc.

Reed Construction Data, Inc., a leading worldwide provider of total construction information solutions, is comprised of three main product groups designed specifically to help construction professionals advance their businesses with timely, accurate and actionable project, product and cost data. Reed Construction Data is a division of Reed Business Information, a member of the Reed Elsevier plc group of companies.

The Project, Product and Cost & Estimating divisions offer a variety of innovative products and services designed for the full spectrum of design, construction and manufacturing professionals. Through its International companies, Reed Construction Data's reputation for quality construction market data is growing worldwide.

Cost Information
RSMeans, the undisputed market leader and authority on construction costs, publishes current cost and estimating information in annual cost books and on the *Means CostWorks* CD-ROM. RSMeans furnishes the construction industry with a rich library of corresponding reference books and a series of professional seminars that are designed to sharpen professional skills and maximize the effective use of cost estimating and management tools. RSMeans also provides construction cost consulting for owners, manufacturers, designers, and contractors.

Project Data
Reed Construction Data provides complete, accurate and relevant project information through all stages of construction. Customers are supplied industry data through leads, project reports, contact lists, plans and specifications surveys, market penetration analyses and sales evaluation reports. Any of these products can pinpoint a county, look at a state, or cover the country. Data is delivered via paper, e-mail, CD-ROM or the Internet.

Building Product Information
The First Source suite of products is the only integrated building product information system offered to the commercial construction industry for comparing and specifying building products. These print and online resources include *First Source,* CSI's SPEC-DATA™, CSI's MANU-SPEC™, First Source CAD, and Manufacturer Catalogs. Written by industry professionals and organized using CSI's MasterFormat™, construction professionals use this information to make better design decisions. FirstSourceONL.com combines Reed Construction Data's project, product and cost data with news and information from Reed Business Information's *Building Design & Construction* and *Consulting-Specifying Engineer magazines.* This industry-focused site offers easy and unlimited access to vital information for all construction professionals.

International
Reed Construction Data Canada serves the Canadian construction market with reliable and comprehensive project and product information services that cover all facets of construction. Core services include: *BuildSource, BuildSpec, BuildSelect,* product selection and specification tools available in print and on the Internet; Building Reports, a national construction project lead service; CanaData, statistical and forecasting information; *Daily Commercial News,* a construction newspaper reporting on news and projects in Ontario; and *Journal of Commerce,* reporting news in British Columbia and Alberta.

BIMSA/Mexico provides construction project news, product information, cost data, seminars and consulting services to construction professionals in Mexico. Its subsidiary, PRISMA, provides job costing software.

Byggfakta Scandinavia AB, founded in 1936, is the parent company for the leaders of customized construction market data for Denmark, Estonia, Finland, Norway and Sweden. Each company fully covers the local construction market and provides information across several platforms including via subscription, on an ad-hoc basis, electronically and on paper.

Cordell Building Information Services, with its complete range of project and cost and estimating services, is Australia's specialist in the construction information industry. Cordell provides in-depth and historical information on all aspects of construction projects and estimation, including several customized reports, construction and sales leads and detailed cost information.

For more information, please visit our Web site at www.reedconstructiondata.com.

Reed Construction Data, Inc., Corporate Office
30 Technology Parkway South
Norcross, GA 30092-2912
(800) 322-6996
(800) 895-8661 (fax)
info@reedbusiness.com
www.reedconstructiondata.com

 Reed Construction Data®

Contractor's Pricing Guides

For more information
visit Means Web site
at www.rsmeans.com

ontractor's Pricing Guide:
esidential Detailed Costs 2006

ery aspect of residential construction, from
erhead costs to residential lighting and wiring, is in
re. All the detail you need to accurately estimate
e costs of your work with or without markups–
oor-hours, typical crews and equipment are
cluded as well. When you need a detailed
timate, this publication has all the costs to help
u come up with a complete, on the money, price
u can rely on to win profitable work.

9.95 per copy
ver 350 pages, with charts and tables, 8-1/2 x 11
talog No. 60336   ISBN 0-87629-808-0

Contractor's Pricing Guide:
Residential Repair &
Remodeling Costs 2006

This book provides total unit price costs for every
aspect of the most common repair & remodeling
projects. Organized in the order of construction by
component and activity, it includes demolition and
installation, cleaning, painting, and more.

With simplified estimating methods; clear, concise
descriptions; and technical specifications for each
component, the book is a valuable tool for
contractors who want to speed up their estimating
time, while making sure their costs are on target.

$39.95 per copy
Over 300 pages, illustrated, 8-1/2 x 11
Catalog No. 60346   ISBN 0-87629-807-2

ontractor's Pricing Guide:
esidential Square Foot
osts 2006

ow available in one concise volume, all you need
 know to plan and budget the cost of new homes.
you are looking for a quick reference, the model
me section contains costs for over 250 different
es and types of residences, with hundreds of
sily applied modifications. If you need even
ore detail, the Assemblies Section lets you build
ur own costs or modify the model costs further.
ndreds of graphics are provided, along with
ms and procedures to help you get it right.

9.95 per copy
ver 300 pages, illustrated, 8-1/2 x 11
talog No. 60326   ISBN 0-87629-809-9

Plumbing Estimating Methods
New 3rd Edition
By Joseph Galeno and Sheldon Greene

This newly updated, practical guide walks you
through a complete plumbing estimate, from basic
materials and installation methods through to
change order analysis. It contains everything
needed for plumbing estimating and covers:

- Residential, commercial, industrial, and medical
  systems

- Updated, expanded information using *Means
  Plumbing Cost Data*

- Sample takeoff and estimate forms

- Detailed illustrations of systems & components

$29.98 per copy
Over 330 pages
Catalog No. 67283B   ISBN 0-87629-704-1

reen Building: Project Planning &
ost Estimating
y RSMeans and Contributing Authors

itten by a team of leading experts in sustainable design,
s is a complete guide to planning and estimating green
ilding projects, a growing trend in building design and
nstruction. It explains:

- How to select and specify green products

- How the project team works differently on a
  green versus a traditional building project

- What criteria your building needs to meet to get
  a LEED, Energy Star, or other recognized rating for green
  buildings

atures an extensive Green Building Cost Data section,
ich details the available products, how they are specified,
d how much they cost.

9.95 per copy
0 pages, illustrated, hardcover
talog No. 67338   ISBN 0-87629-659-2

Residential & Light
Commercial Construction
Standards, 2nd Edition
By RSMeans and Contributing Authors

New, updated second edition of this unique
collection of industry standards that define quality
construction. For contractors, subcontractors,
owners, developers, architects, engineers,
attorneys, and insurance personnel, this book
provides authoritative requirements and
recommendations compiled from the nation's
leading professional associations, industry
publications, and building code organizations.

$59.95 per copy
600 pages, illustrated, softcover
Catalog No. 67322A   ISBN 0-87629-658-4

**For more information
visit Means Web site
at www.rsmeans.com**

Annual Cost Guides

Means Building Construction Cost Data 2006

Offers you unchallenged unit price reliability in an easy-to-use arrangement. Whether used for complete, finished estimates or for periodic checks, it supplies more cost facts better and faster than any comparable source. Over 23,000 unit prices for 2006. The City Cost Indexes now cover over 930 areas, for indexing to any project location in North America.

**$126.95 per copy
Over 700 pages, softcover
Catalog No. 60016   ISBN 0-87629-786-6**

Means Open Shop Building Construction Cost Data 2006

The open-shop version of the *Means Building Construction Cost Data*. More than 22,000 reliable unit cost entries based on open shop trade labor rates. Eliminates time-consuming searches for these prices. The first book with open shop labor rates and crews. Labor information is itemized by labor-hours, crew, hourly/daily output, equipment, overhead and profit. For contractors, owners and facility managers.

**$126.95 per copy
Over 700 pages, softcover
Catalog No. 60156   ISBN 0-87629-797-1**

Means Plumbing Cost Data 2006

Comprehensive unit prices and assemblies for plumbing, irrigation systems, commercial and residential fire protection, point-of-use water heaters, and the latest approved materials. This publication and its companion, *Means Mechanical Cost Data*, provide full-range cost estimating coverage for all the mechanical trades.

**$126.95 per copy
Over 550 pages, softcover
Catalog No. 60216   ISBN 0-87629-796-3**

Means Residential Cost Data 2006

Speeds you through residential construction pricing with more than 100 illustrated complete house square-foot costs. Alternate assemblies cost selections are located on adjoining pages, so that you can develop tailor-made estimates in minutes. Complete data for detailed unit cost estimates is also provided.

**$107.95 per copy
Over 600 pages, softcover
Catalog No. 60176   ISBN 0-87629-803-x**

Means Electrical Cost Data 2006

Pricing information for every part of electrical cost planning: unit and systems costs with design tables; engineering guides and illustrated estimating procedures; complete labor-hour, materials, and equipment costs for better scheduling and procurement. With the latest products and construction methods used in electrical work. More than 15,000 unit and systems costs, clear specifications and drawings.

**$126.95 per copy
Over 450 pages, softcover
Catalog No. 60036   ISBN 0-87629-790-4**

Means Repair & Remodeling Cost Data 2006

Commercial/Residential

You can use this valuable tool to estimate commercial and residential renovation and remodeling. By using the specialized costs in this manual, you'll find it's not necessary to force fit prices for new construction into remodeling cost planning. Provides comprehensive unit costs, building systems costs, extensive labor data and estimating assistance for every kind of building improvement.

**$107.95 per copy
Over 650 pages, softcover
Catalog No. 60046   ISBN 0-87629-795-5**

Means Site Work & Landscape Cost Data 2006

Hard-to-find costs are presented in an easy-to-use format for every type of site work and landscape construction. Costs are organized, described, and laid out for earthwork, utilities, roads and bridges, as well as grading, planting, lawns, trees, irrigation systems, and site improvements.

**$126.95 per copy
Over 600 pages, softcover
Catalog No. 60286   ISBN 0-87629-792-0**

Means Light Commercial Cost Data 2006

Specifically addresses the light commercial market which is an increasingly specialized niche in the industry. Aids you, the owner/designer/contractor, in preparing all types of estimates, from budgets to detailed bids. Includes new advances in methods and materials. Assemblies section allows you to evaluate alternatives in early stages of design/planning.

**$107.95 per copy
Over 650 pages, softcover
Catalog No. 60186   ISBN 0-87629-804-8**

Books for Builders

For more information
visit Means Web site
at www.rsmeans.com

Builder's Essentials:
Plan Reading & Material Takeoff
By Wayne J. DelPico

A complete course in reading and interpreting building plans—and performing quantity takeoffs to professional standards.

This book shows and explains, in clear language and with over 160 illustrations, typical working drawings encountered by contractors in residential and light commercial construction. The author describes not only how all common features are presented, but how to translate that information into a material list. Organized by CSI division, each chapter uses plans, details and tables, and a summary checklist.

$35.95 per copy
Over 420 pages, illustrated, softcover
Catalog No. 67307   ISBN 0-87629-348-8

Builder's Essentials:
Best Business Practices for Builders & Remodelers:
An Easy-to-Use Checklist System
By Thomas N. Frisby

A comprehensive guide covering all aspects of running a construction business, with more than 40 user–friendly checklists. This book provides expert guidance on: increasing your revenue and keeping more of your profit; planning for long-term growth; keeping good employees and managing subcontractors.

$29.95 per copy
Over 220 pages, softcover
Catalog No. 67329   ISBN 0-87629-619-3

Means Estimating Handbook, 2nd Edition

Updated new Second Edition answers virtually any estimating technical question – all organized by CSI MasterFormat. This comprehensive reference covers the full spectrum of technical data required to estimate construction costs. The book includes information on sizing, productivity, equipment requirements, code-mandated specifications, design standards, and engineering factors.

This reference—widely used in the industry for tasks ranging from routine estimates to special cost analysis projects—has been completely updated with new and expanded technical information and an enhanced format for ease-of-use.

$99.95 per copy
Over 900 pages, hardcover
Catalog No. 67276A   ISBN 0-87629-699-1

Builder's Essentials:
Framing & Rough Carpentry, 2nd Edition

A complete, illustrated do-it-yourself course on framing and rough carpentry. The book covers walls, floors, stairs, windows, doors, and roofs, as well as nailing patterns and procedures. Additional sections are devoted to equipment and material handling, standards, codes, and safety requirements.

The "framer-friendly" approach includes easy-to-follow, step-by-step instructions. This practical guide will benefit both the carpenter's apprentice and the experienced carpenter, and sets a uniform standard for framing crews.

Also available in Spanish!

$24.95 per copy
Over 125 pages, illustrated, softcover
Catalog No. 67298A   ISBN 0-87629-617-7
Spanish 67298AS ISBN 0-87629-654-1

Interior Home Improvement Costs, New 9th Edition

Estimates for the most popular remodeling and repair projects-from small, do-it-yourself jobs-to major renovations and new construction.

Includes:
- Kitchens & Baths
- New Living Space from Your Attic, Basement or Garage
- New Floors, Paint & Wallpaper
- Tearing Out or Building New Walls
- Closets, Stairs & Fireplaces
- New Energy-Saving Improvements, Home Theatres, and Much More!

$24.95 per copy
Over 250 pages, illustrated, softcover
Catalog No. 67308E   ISBN 0-87629-743-2

Exterior Home Improvement Costs, New 9th Edition

Estimates for the most popular remodeling and repair projects-from small, do-it-yourself jobs, to major renovations and new construction.

Includes:
- Curb Appeal Projects-Landscaping, Patios, Porches, Driveways and Walkways
- New Windows and Doors
- Decks, Greenhouses, and Sunrooms
- Room Additions and Garages
- Roofing, Siding, and Painting
- "Green" Improvements to Save Energy & Water

$24.95 per copy
Over 250 pages, illustrated, softcover
Catalog No. 67309E   ISBN 0-87629-742-4

For more information
visit Means Web site
at www.rsmeans.com

Books for Builders

Designing & Building with the IBC, Second Edition

By Rolf Jensen & Associates, Inc.

This new second edition (formerly titled, *From Model Codes to the IBC: A Transitional Guide*) is an essential reference on the latest building codes for architects and engineers, building officials and AHJs, contractors and homebuilders, and manufacturers and building owners. It provides complete coverage of the new *International Building Code*®(IBC) 2003, with expert advice and user-friendly, side-by-side comparison to the IBC 2000 and model building codes.

$99.95 per copy
Over 1,000 pages, softcover
Catalog No. 67328A   ISBN 0-87629-616-9

Historic Preservation
Project Planning & Estimating

by Swanke Hayden Connell Architects

Managing Historic Restoration, Rehabilitation, and Preservation Building Projects and Determining and Controlling Their Costs

The authors explain:
- How to determine whether a structure qualifies as historic
- Where to obtain funding and other assistance
- How to evaluate and repair more than 75 historic building materials
- How to properly research, document, and manage the project to meet code, agency, and other special requirements
- How to approach the upgrade of major building systems

$99.95 per copy
Over 675 pages, hardcover
Catalog No. 67323   ISBN 0-87629-573-1

Builder's Essentials:
Advanced Framing Methods

By Scot Simpson

A highly illustrated, "framer-friendly" approach to advanced framing elements. Provides expert, but easy-to-interpret, instruction for laying out and framing complex walls, roofs, and stairs, and special requirements for earthquake and hurricane protection. Also helps bring framers up to date on the latest building code changes, and provides tips on the lead framer's role and responsibilities, how to prepare for a job, and how to get the crew started.

$24.95 per copy
250 pages, illustrated, softcover
Catalog No. 67330   ISBN 0-87629-618-5

Unit Price Estimating Methods
New 3rd Edition

This new edition includes up-to-date cost data and estimating examples, updated to reflect changes to the CSI numbering system and new features of Means cost data. It describes the most productive, universally accepted ways to estimate, and uses checklists and forms to illustrate shortcuts and timesavers. A model estimate demonstrates procedures. A new chapter explores computer estimating alternatives.

$59.95 per copy
Over 350 pages, illustrated, hardcover
Catalog No. 67303A   ISBN 0-87629-662-2

Means Spanish/English Construction Dictionary

By RSMeans, The International Conference of Building Officials (ICBO), and Rolf Jensen & Associates (RJA)

Designed to facilitate communication among Spanish- and English-speaking construction personnel—improving performance and job-site safety. Features the most common words and phrases used in the construction industry, with easy-to-follow pronunciations. Includes extensive building systems and tools illustrations.

$22.95 per copy
250 pages, illustrated, softcover
Catalog No. 67327   ISBN 0-87629-578-2

Means Illustrated Construction Dictionary, 3rd Edition

Long regarded as the Industry's finest, the *Means Illustrated Construction Dictionary* is now even better. With the addition of over 1,000 new terms and hundreds of new illustrations, it is the clear choice for the most comprehensive and current information.

The companion CD-ROM that comes with this new edition adds many extra features: larger graphics, expanded definitions, and links to both CSI MasterFormat numbers and product information.

Contains 19,000 construction words, terms, phrases, symbols, weights, measures, and equivalents. 1,000 new entries; 1,200 helpful illustrations, easy-to-use format with thumbtabs.

$99.95 per copy
Over 790 pages, illustrated, hardcover
Catalog No. 67292A   ISBN 0-87629-538-3

MeansData™

CONSTRUCTION COSTS FOR SOFTWARE APPLICATIONS
Your construction estimating software is only as good as your cost data.

A proven construction cost database is a mandatory part of any estimating package. The following list of software providers can offer you MeansData™ as an added feature for their estimating systems. See the table below for what types of products and services they offer (match their numbers). Visit online at **www.rsmeans.com/demosource/** for more information and free demos. Or call their numbers listed below.

1. **3D International**
   713-871-7000
   venegas@3di.com

2. **4Clicks-Solutions, LLC**
   719-574-7721
   mbrown@4clicks-solutions.com

3. **Aepco, Inc.**
   301-670-4642
   blueworks@aepco.com

4. **Applied Flow Technology**
   800-589-4943
   info@aft.com

5. **ArenaSoft Estimating**
   888-370-8806
   info@arenasoft.com

6. **BSD - Building Systems Design, Inc.**
   888-273-7638
   bsd@bsdsoftlink.com

7. **CMS - Computerized Micro Solutions**
   800-255-7407
   cms@proest.com

8. **Corecon Technologies, Inc.**
   714-895-7222
   sales@corecon.com

9. **CorVet Systems**
   301-622-9069
   sales@corvetsys.com

10. **Estimating Systems, Inc.**
    800-967-8572
    esipulsar@adelphia.net

11. **MC² - Management Computer**
    800-225-5622
    vkeys@mc2-ice.com

12. **Maximus Asset Solutions**
    800-659-9001
    assetsolutions@maximus.com

13. **Shaw Beneco Enterprises, Inc.**
    877-719-4748
    inquire@beneco.com

14. **Timberline Software Corp.**
    800-628-6583
    product.info@timberline.com

15. **US Cost, Inc.**
    800-372-4003
    sales@uscost.com

16. **Vanderweil Facility Advisors**
    617-451-5100
    info@VFA.com

17. **WinEstimator, Inc.**
    800-950-2374
    sales@winest.com

TYPE	1	2	3	4	5	6	7	8	9	10	11	12	13	14	15	16	17
BID					●	●	●		●		●			●			●
Estimating		●			●	●	●	●	●	●	●		●	●	●		●
DOC/JOC/SABER		●			●	●			●	●		●	●	●			●
ID/IQ		●							●	●		●	●				●
Asset Mgmt.												●				●	●
Facility Mgmt.	●		●									●				●	
Project Mgmt.	●	●						●	●			●	●				
TAKE-OFF					●		●	●	●		●			●	●		
EARTHWORK							●				●						
Pipe Flow				●													
HVAC/Plumbing					●		●		●								
Roofing					●		●		●								
Design	●				●									●	●		
Other Offers Links:																	
Accounting/HR		●			●							●	●				
Scheduling					●									●	●		●
CAD														●			●
PDA														●	●		●
Lt. Versions		●					●							●	●		
Consulting	●	●			●		●		●		●			●	●	●	●
Training		●			●	●	●	●	●	●		●	●	●	●	●	●

Qualified re-seller applications now being accepted. Call Carol Polio, Ext. 5107.

FOR MORE INFORMATION
CALL 1-800-448-8182, EXT. 5107 OR FAX 1-800-632-6732

For more information
visit Means Web site
at www.rsmeans.com

New Titles

Preventive Maintenance for Multi-Family Housing
by John C. Maciha

With 24/7 occupancy, multi-family buildings can be some of the toughest to maintain. Prepared by one of the nation's leading experts on multi-family housing, Preventive Maintenance for Multi-Family Housing puts easy-to-use guidelines right at your fingertips for the what, when, why, and how much of multi-family preventive maintenance.

This complete PM system for apartment and condominium communities features expert guidance, checklists for buildings and grounds maintenance tasks and their frequencies, a reusable wall chart to track maintenance, and a dedicated Web site featuring customizable electronic forms. A must-have for anyone involved with multi-family housing maintenance and upkeep.

$89.95 per copy
225 pages
Catalog No. 67346

Job Order Contracting
Expediting Construction Project Delivery
by Allen Henderson

Expert guidance to help you implement JOC—fast becoming the preferred project delivery method for repair and renovation, minor new construction, and maintenance projects in the public sector and in many states and municipalities. The author, a leading JOC expert and practitioner, shows how to:

- Establish a JOC program
- Evaluate proposals and award contracts
- Handle general requirements and estimating
- Partner for maximum benefits

This book will quickly bring you up to speed on JOC (also known as Delivery Order Contracting and SABER) so you can achieve faster project starts with fewer delays, cost overruns, and quality disputes.

$89.95 per copy
192 pages, illustrated, Hardcover
Catalog No. 67348

Kitchen & Bath Project Costs:
Planning & Estimating Successful Projects

Project estimates for 35 of the most popular kitchen and bath renovations... from replacing a single fixture to whole-room remodels. Each estimate includes:

- All materials needed for the project
- Labor-hours to install (and demolish/remove) each item
- Subcontractor costs for certain trades and services
- An allocation for overhead and profit

PLUS! Takeoff and pricing worksheets—forms you can photocopy or access electronically from the book's Web site; alternate materials—unit costs for different finishes and fixtures, location factors—easy multipliers to adjust the costs to your location; and expert guidance on estimating methods, project design, contracts, marketing, working with homeowners, and tips for each of the estimated projects.

$29.95 per copy
Over 175 pages
Catalog No. 67347

Home Addition & Renovation Project Costs

New! This essential home remodeling reference gives you 35 project estimates and guidance for some of the most popular home renovation and addition projects... from opening up a simple interior wall to adding an entire second story. Each estimate includes a floor plan, color photos, and detailed costs. Use the project estimates as backup for pricing, to check your own estimates, or as a cost reference for preliminary discussion with homeowners.
Includes:

- Case studies—with creative solutions and design ideas.
- Alternate materials costs—so you can match the estimates to the particulars of your projects.
- Location factors—easy multipliers to adjust the book's costs to your own location.

$29.95 per copy
Over 200 pages, illustrated,
Softcover
Catalog No. 67349

The Practice of Cost Segregation Analysis
by Bruce A. Desrosiers and Wayne J. DelPico

This expert guide walks you through the practice of cost segregation analysis, which enables property owners to defer taxes and benefit from "accelerated cost recovery" through depreciation deductions on assets that are properly identified and classified. Cost segregation practice requires knowledge of both tax law and the construction process. In this book, the authors share their expertise in these areas with tax and accounting professionals, cost segregation consultants, facility owners, architects, and general contractors—providing guidance on major aspects of a professional, defensible cost segregation study.

With a glossary of terms, sample cost segregation estimates for various building types, key information resources, and updates via a dedicated Web site, this book is a critical resource for anyone involved in cost segregation analysis.

$99.95 per copy
Over 225 pages
Catalog No. 67345